THE PRIMARY AFFERENT NEURON

A Survey of Recent
Morpho-Functional Aspects

THE PRIMARY AFFERENT NEURON

A Survey of Recent Morpho-Functional Aspects

Edited by

Wolfgang Zenker and Winfried L. Neuhuber

Institute of Anatomy
University of Zurich–Irchel
Zurich, Switzerland

PLENUM PRESS • NEW YORK AND LONDON

Library of Congress Cataloging in Publication Data

The primary afferent neuron: a survey of recent morpho-functional aspects / edited by Wolfgang Zenker and Winfried L. Neuhuber.

    p.   cm.

    "Based on the proceedings of a symposium on the primary afferent neuron: a survey of recent morpho-functional aspects, held in conjunction with the Eighty-Third Congregation of the Anatomical Society, March 24, 1988, in Zurich, Switzerland" —T.p. verso.

    Includes bibliographical references.

    ISBN 0-306-43480-6

    1. Spinal ganglia—Physiology—Congresses. 2. Spinal ganglia—Ultrastructure—Congresses. 3. Afferent pathways—Congresses. 4. Sensorimotor integration—Congresses. I. Zenker, W. (Wolfgang). 1925– . II. Neuhuber, Winfried L.

QP374.P75   1990

599′.0188—dc20

89-70967

CIP

Based on the proceedings of a symposium on The Primary Afferent Neuron: A Survey of Recent Morpho-Functional Aspects, held in conjunction with the Eighty-Third Congregation of the Anatomical Society, March 24, 1988, in Zurich, Switzerland

© 1990 Plenum Press, New York
A Division of Plenum Publishing Corporation
233 Spring Street, New York, N.Y. 10013

Printed in the United States of America

PREFACE

This book is based on contributions presented at the symposion *"The Primary Afferent Neuron : A survey of recent morpho-functional aspects,"* held in Zürich on March 24th, 1988 in connection with the 83rd Congregation of the German Anatomical Society. Members of the Anatomical Society as well as non-member researchers were invited to join a circle of specialists to discuss the topic of primary afferents. In addition, some aspects which had not been dealt with at the Symposion because of shortage of time are represented by invited reviews included in this volume.

As scientific research on the primary afferent neuron is so extensive, it is impossible to take inventory of all the present activities on this subject. This book attempts to provide an overview of various aspects of high actuality and, in particular, shows how morphological research contributes to our present-day concepts of the primary afferent neuron.

Although fundamental knowledge on morphology and physiology of the spinal ganglion cell seems to be well established, many questions of the past and the present still await conclusive answers. Thus, many question-marks determine the conceptual layout of this book (Fig. 1):

First, the peripheral sensory field, and, in particular, the *peripheral sensory receptors* are discussed. Which morphological traits of the corpuscular receptors can be correlated with specific sensory functions? Which structure-to-function relationships have been revealed by the recent morphological and physiological investigations of the "free" nerve endings of small myelinated and unmyelinated afferent fibers? In view of recent reports on the localization and the role of specifically active *peptides* such as substance P and calcitonin gene-related peptide, the "efferent", or local effectory function of the peripheral processes of primary afferent neurons is reassessed. This efferent component of certain small spinal ganglion cells fits well into pathophysiological concepts of inflammation. The question is addressed whether these functional aspects can be supported by immunohistochemical findings.

This leads us, **second**, to the spinal ganglion cell body and its initial segment, the *trophic center* of the primary afferent neuron. The concept of the functional individuality of single spinal ganglion cells and its morphological equivalents is reassessed. What are the properties of the various cytoplasmic proteins and peptides which have been visualized by immunocytochemistry? What is known about the regulatory mechanisms of these substances? Furthermore, current

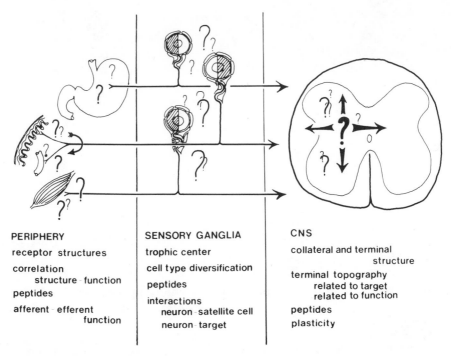

PERIPHERY	SENSORY GANGLIA	CNS
receptor structures	trophic center	collateral and terminal structure
correlation structure - function	cell type diversification	terminal topography related to target related to function
peptides	peptides	
afferent - efferent function	interactions neuron - satellite cell neuron - target	peptides plasticity

Fig. 1. The primary afferent neuron: The catalogue of questions addressed at this symposion.

research on the importance of peripheral and central targets for development and determination of spinal ganglion neurons calls for re-evaluation of the present-day concepts on this subject. What do we know about the development of the primary afferent neuron at all? Which mechanisms determine the expression of specific substances during embryologic development? What is to be learned from *in vitro* experiments?

Third, after reviewing the characteristics of soma and peripheral process, we take a look at the *central neurite* of the spinal ganglion cell. Where does it terminate? Are there specific patterns of collateral and terminal distribution within the spinal cord for afferents from skin, muscles and viscera? Which are the ultrastructural features of afferent terminals? What is our current understanding of the synaptic connectivity of these afferent terminals? Finally, the effects of experimental interventions on the spinal ganglion neuron are described, and current concepts of spinal cord *plasticity* and implications for *regeneration* are discussed.

Some of the questions have been answered conclusively, whereas for many others only speculations can be offered. However, we hope that the explanations and concepts provided will enable the reader to assess the state of the art in research on the primary afferent neuron.

Publication of this volume would have been impossible without the continuous effort and expert help of Dr. Hanspeter Lipp whose skill and experience in word processing and page-setting techniques was invaluable. Likewise, we would like to express our gratitude to Dr. Christine Zenker for typing and processing many of the manuscripts, to Mrs Verena Liebhart for secretarial help, and to Dr. Sergei Bankoul for numerous advices in the use of the word processing system. Thanks are due to Plenum Publishing Corporation for kindly providing the opportunity to publish this book, and to the Swiss National Foundation for financial support of the Symposion (Grant 3.658-0.87).

W. Zenker

W. L. Neuhuber

CONTENTS

COMPARATIVE AND FUNCTIONAL ASPECTS OF THE HISTOLOGICAL ORGANIZATION OF CUTANEOUS RECEPTORS IN VERTEBRATES

K. H. Andres and M. v Düring

Ruhr-Universität Bochum, Institut für Anatomie/
Abteilung Neuroanatomie

Bochum, F.R.G

INTRODUCTION

The skin of vertebrates represents and functions as a highly developed sensory organ enabling the organism to explore the near environment. Skin receptors are also used as sensors for body protection, reflexes and cutaneous neuro-vegetative reactions. The same receptor may be used for different functions but there are also receptors with one specific sensory function.

In the vertebrate class of *fish* the lateral line system composed of secondary sensory hair cells and its synaptically linked afferent nerve fiber plays the important role in the transduction of mechanosensory stimuli and information from the skin to the central nervous system. Cutaneous electroreceptors and taste buds may additionally support the explorative function of the skin in several species of fish. *Aquatic amphibia* exhibit lateral line organs and most *aquatic urodeles* differentiate electroreceptors too (Fritzsch and Wahnschaffe, 1983). Additionally, to the above mentioned receptor systems which use secondary sensory cells for stimulus transduction, fish and amphibia have simple afferent terminals of primary sensory neurons. The terminals, however, often lack conspicuous morphological differentiation.

Terrestrial vertebrates lacking the highly sensitive lateral line system develop and differentiate several new types of cutaneous receptors with characteristic morphological and functional properties; for example: epidermal discoid vesicle receptors, lamellated corpuscular receptors, hair follicle receptors and thermo-receptors (Andres, 1966, 1973; Andres and v. Düring, 1973; v. Düring, 1974 a, b; Iggo and Andres, 1982). Also fish mostly living on the sea ground with benthic feeding behavior, as for example *triglidae* differentiate conspicuous cutaneous afferent nerve endings in their free movable breast fin rays.

Describing and analyzing morphological criteria and structural specifica-tions, we propose a morphological classification. Combined morphological and physiological experiments show that there is a good correlation between structure and function (Chambers et al., 1972; Hensel et al., 1974). Immuno-cytochemistry using either fluorescence- or peroxidase-labeled antibodies to label individual cellular components within single afferent nerve fibers, such as catecholamines or neuropeptides (Hökfelt et al., 1980) support the diversity of peripheral tissue receptors. Tract tracing studies of individual nerve fibers with HRP allow to recognize the different spinal and supraspinal termination sites (Brown, 1981; Craig et al., 1981, 1982; Mense and Craig, 1988). Finally, this progress indicates that the tissue receptors function as selective peripheral encoding devices, able to extract information about various parameters of cutaneous stimuli thus substantiating a reaffirmation of the old idea of "specific nerve energies" (Müller, 1844).

The following article deals with the electron microscopical analysis of cutaneous receptors of primary sensory neurons, receptor organs in specialized skin regions, comparative aspects and functional implications.

To characterize cutaneous nerve terminals of primary afferent neurons of all classes of vertebrates we propose to use the following parameters and *morphological criteria:*

1. The receptor terminal is formed by the afferent axon which has its cell body in or near the dorsal root ganglion or cranial nerve ganglion.

2. The afferent axons enter the spinal or supraspinal levels via the dorsal roots or the cranial nerves. Only a minority of sensory fibers enter the spinal cord via the ventral roots (Clifton et al.,1976).

3. The receptor terminal exhibits, beside cytoskeletal structures such as microtubules and neurofilaments, and profiles of the endoplasmic reticulum, mitochondria, clear and dense core vesicles and a fine filamentous ground substance (the receptor matrix) beneath the axonal membrane (Andres and v. Düring, 1973).

4. The receptor terminal is devoid of a myelin sheath.

In order to differentiate cutaneous receptors of vertebrates and to correlate their probable sensory function we need further morphological parameters:

1. Classification of the afferent axon according to diameter, myelination and conduction velocity (A ß, A ∂, C; Gasser and Grundfest, 1939).

2. Size of the receptive field and arborization of the afferent stem axon (Hagbarth and Vallbo, 1967).

3. Topography and location of the receptor terminal, e.g., ubiquitous, restricted areas, association with hairs or glands.

4. Specializations of the Schwann cell covering.

5. Differentiation of the perineural and endoneural sheath exhibits some highlights of terminal character.

6. Composition and differentiation of the surrounding tissue involved in the receptor complex (connective tissue, smooth muscle cells, ground substance,

cells of the connective tissue as mast cells, histiocyts, vascularization).

7. Composition of different receptor types in circumscribed skin areas, e.g., papula, papilla, dome, hair.

C-FIBER TERMINALS

Afferent and efferent C-fibers often share one Schwann cell per segment on the way to their target organs (Andres et al., 1985). In addition to this fact the long unmyelinated segments of thin myelinated afferent nerve fibers may also share the Schwann cells of C-fiber bundles (Andres et al., 1987). Without experimental elimination of the efferent C-fiber component it seems up to now almost impossible to distinguish afferent and efferent C-fibers morphologically. By tracing C-fibers in serial semi- and ultrathin sections and mapping cytoskeletal elements of the axon, afferent and efferent axons, however, appear to be different in their receptive or, in the case of efferent C-fibers, their secretory sites. Receptive zones of the afferent C-fiber occur at several sites in its terminal course and exhibit the filamentous receptor matrix, vesicles and one or two mitochondria per cross section. At these specialized locations of the afferent nerve fiber, the axonal membrane often lacks its Schwann cell covering; sometimes the basement lamella is also absent. The axon is slightly enlarged. The inconspicuous morphological appearance of the afferent C-fiber and its terminals explains the poor morphological description. The topography and the location of afferent C-fibers within the skin is distributed in the epidermis, in the connective tissue of the papillary and reticular layer of the dermis and in the endoneural connective tissue of nerve fiber bundles. The C-fiber terminals of the dermis are often located in close relation to the vascular bed. The functional interpretation of morphologically identified afferent C-fiber terminals remains up to now speculative. The specific sensory function as uni- or polymodal cutaneous receptors probably depends on membrane receptors.

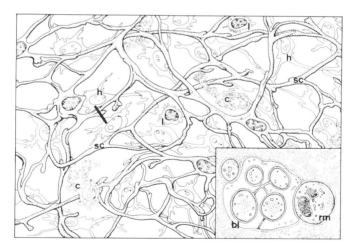

Fig. 1. Schematic representation of the superficial nerve fiber plexus in the skin of birds represented by the Schwann cell tubes (sc) forming a three dimensional net. Inset shows the cross sectioned Schwann cell tube marked by an arrow with C-fibers and terminals. Schwann cell (sc), receptor matrix of the terminal (rm), basement lamella (bl), histiocytes (h), lymphocytes (l), chromatophores (c).

In *birds*, the skin is always thin, tight, poor in capillarization and smooth in its contour except for some papillae. The lamina propria composed of densely packed collagen fibers contains a prominent C-fiber plexus. This plexus consists of Schwann cells forming a net of anastomising tubes, containing numerous C-fibers of different sizes. The unmyelinated nerve fibers either lie invaginated in Schwann cell grooves or are locally exposed with most of their surface area facing the surrounding connective tissue (Fig. 1). These receptive sites of the axonal segment are characterized by the receptor matrix, vesicles and mitochondria in addition to microtubules and neurofilaments. Unmyelinated segments of thin myelinated axons contribute to this nerve fiber plexus too, but they can only be recognized by serial sectional tracing analyses.

ELECTRORECEPTORS

Generally known electroreceptors in the skin of some vertebrates are the ampullae of Lorenzini of chondrychthyes, ampullary organs of sarcopterygian and teleost fishes and the tuberous organs in gymnotids and mormyrids. In primitive aquatic urodela too, electroreceptors of the ampullary type occur. All these electroreceptors are composed of a secondary sensory cell synaptically linked to afferent nerve fiber terminals of the eighth cranial nerve (Bullock and Heiligenberg, 1986).

Contrary to electroreceptors of fish and amphibia, the recently discovered electrosensitivity of *monotremes* (Scheich et al., 1986; Gregory et al., 1988) is mediated by the nerve terminals of primary afferent neurons of the fifth cranial nerve (Andres and v. Düring, 1984, 1988; Iggo et al., 1988; Fig. 2a). They are topographically restricted to the intraepidermal duct segments of specialized eccrine sweat glands of the bill with mucous and serous secretion. The secretion rich in electrolytes within the gland duct and the superficial mucous containing blossom pores may function as specific conductors for the electric current, where vectors of current densities may act as specific stimuli to the nerve terminals. In *platypus* the electroreceptor system is highly organized. More than 70 000 glands with mucous (type ms and nsm) and serous secretion (type ss) are distributed in the skin of the platypus bill and their pores build up a special pattern on the bill's surface. Ms and ss glands forming club-shaped bulbs of their basal intraepidermal duct segment contain the afferent electrosensory nerve terminals. Most of the myelinated axons are finally located at the basal epidermal layer of the club-shaped bulb where they loose their myelin sheath and terminate with exceptionally short and thin nerve endings between the epidermal cells of the center of the bulb. This morphological appearance of afferent nerve fiber termination is unique in all observed cutaneous receptors of vertebrates. While the ms gland duct segment is innervated by up to 30 afferent myelinated axons, the ss gland duct is supplied by about 22 afferent myelinated nerve fibers. In *echidna* we find comparable afferent nerve terminals at specialized intraepidermal gland duct segments on the bill tip (Fig. 2b). In this animal only one type of sensory gland occurs, but as in platypus, innervated and non innervated gland ducts build up a unique pattern. Physiological experiments have recently shown that the bill of echidna is electrosensitive too (Proske et al., 1988), which confirms our functional interpretation deduced form morphological findings.

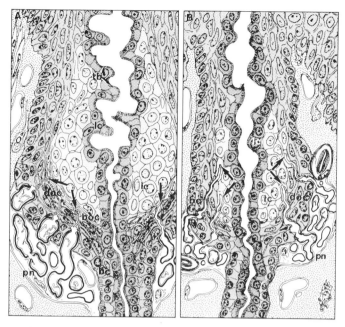

Fig. 2. Semi-schematic representation of the electrosensitive nerve terminals located in the epithelium of the intraepidermal duct segment of the large mucous gland of the platypus bill (**A**) and the muco-serous gland of the echidna bill (**B**). Myelinated axons are finally located within the basal epidermal layer or at the epidermal border. They terminate as short and thin axonal terminals (arrows) within the bright outer core (boc). Basal epidermal cell layer (bc), dark outer core (doc), inner core (ic), lining cells (lc) of the duct, terminal bar (tb), perineural sheath (pn).

THERMORECEPTORS

A combined physiological and morphologcical experiment has revealed that the *cold receptor* in the skin of the cat's nose is a free nerve terminal supplied by a myelinated nerve fiber (group III) terminating in a spray of up to eight expanded nerve terminals in the basal layer of the epidermis (Hensel et al., 1974). The receptor terminals have an intraepidermal location and lack any desmosomes or attachment plaques to the epidermal cell membrane.

Warm receptors in mammals have so far not been identified morphologically, but there is strong evidence from physiological experiments that the main axons are unmyelinated (C-fibers) and terminate mostly in the subcutaneous layer.

Structure and function of the *infrared* sensory receptors and organs in snakes are the best known warm receptors. In *Boa* and *Python* they form intraepidermal and free dermal receptor endings rich in mitochondria.They are located in the skin of the scales or in grooves of the scales of the head and they stem from group III myelinated axons (v. Düring, 1974b; Hensel, 1975; v. Düring and Miller, 1978). In *Crotalus* and *Agkistrodon* the nerve endings expand up to 15 µm within the thin dermal layer of the pit membrane (Terashima et al., 1970). Deep invaginations of the axonal membrane enlarge its surface area. A dense capillary plexus is included in the receptive complex.

The morphological features which characterize physiologically identified temperature receptors are 1) the exposed location of the nerve terminal close to the skin surface, 2) the absence of any mechanical structure linking the expanded axon terminal to its surrounding tissue (e.g., tonofibrils in the epithelium or collagen and elastic fiber systems of the connective tissue) and 3) a distinct capillary plexus restricted to the location of the nerve endings. In the following paragraph we describe afferent nerve terminals and their topography which possibly have thermoreceptive function.

In the subepidermal area of the *pigeon's* breast region we observe a localized capillary plexus associated with numerous free nerve terminals in close relation to the wall of the capillaries. In the tip region of dermal tongue papillae of red fronted parrots we observe a similar arrangement of a capillary plexus/loops and expanded nerve terminals which resembles the warm receptors in *Boidae.* Kürten (1984), in a systematical physiological and morphological investigation, analyzed the skin area of the nose leaf of *Desmodus rotundus* and mapped a distinct distribution of mechano- and thermoreceptive points. In our material of the nose leaf of *Desmodus rotundus* we confirm such thermoreceptive nerve terminals too, which in some morphological aspects are comparable to the warm receptors of reptiles. A further location of such thermoreceptive nerve terminals in *Desmodus rotundus* is the large medial plane of the underlip where the thermoreceptive endings are located between prominent columns of Merkel cell complexes (Fig. 3a). *Rhinolophus rouxii,* not specialized for detecting fresh blood droplets as *Desmodus rotundus,* exhibits a similar horseshoe-like nose leaf and the same broad plane of its underlip with identical columns of Merkel cell complexes, but we do not observe the nerve terminals possibly involved in thermoreceptive function (Fig. 3b). This morphological result may indicate that the great blood-sucking bat, *Desmodus rotundus,* requires a prominent and distinct thermoreceptor necessary for its specific prey detection.

MECHANORECEPTORS

Morphologically, mechanoreceptive terminals of primary afferent neurons are characterized by an intimate contact of supporting structural elements (e.g., tonofibrils of epidermal cells, collagen or elastic fibrils or fibers of the connective tissue compartments involved in the receptive field) with localized areas of the axonal receptor membrane of the nerve terminal.

These connections seem to be used as the intimate transducers to transmit the specific mechanical stimuli towards the axonal membrane. Myelinated mechanoreceptive fibers exhibit in the course of the receptor axons (unmyelinated segments) or at their terminal expansions finger-, spine- or crest-like axonal protrusions penetrating or invading into the surrounding tissue structure. These axonal protrusions mostly lack a Schwann cell covering so that the axonal membrane faces directly the stimulus transmitting tissue elements. These sites and the expanded nerve terminal itself, exhibit the receptor matrix. In this way, changes in stretch or pressure of the surrounding tissue elements, by mediation of the stimulus transducing tissue structures, will act directly on the molecular organization of the receptor proteins of these axonal membrane segments, transforming the stimulus into electrical depolarization. These steps are the primary processes for information mechanisms in the central neuronal control of the skin.

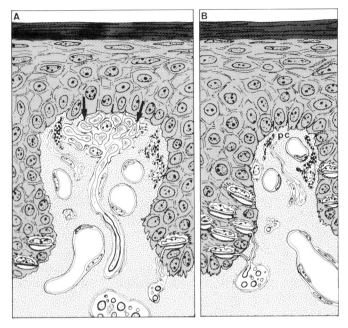

Fig. 3. Semi-schematic representation of cross sections through the surface of the under lip plane of *Desmodus rotundus* (A) and of *Rhinolophus rouxii* (B). **A.** Between the Merkel cell columns of epidermal rods large subepidermal convolutes of receptor terminals are present (arrows). **B.** These terminals do not occur in the papillary tissue (pc) in *Rhinolophus rouxii.*

The differentiation of a capsule separating the receptor complex from tissue compartments not involved in the stimulus transduction, allows the registration of a specific stretch direction and futhermore, the capsule of receptor corpuscles guarantees a specific elasticity and turgor. The turgor and elasticity of the corpuscular receptors support the rapidly adapting character of their discharge pattern.

Comparative morphological studies of visceroreceptors of vertebrates have indicated that their morphology mostly resembles cutaneous receptors if the stimuli are identical as for example stretch or tension (v. Düring et al., 1974; Andres and v. Düring, 1985; Andres et al., 1987; v. Düring and Andres, 1988).

Intraepidermal mechanic terminals

In the epidermis of teleosts and amphibia we observe axons terminating in the intercellular space. The nerve fiber profiles do not exhibit morphological criteria which allow their unequivocal classification as typical mechanoreceptor terminals.

Discoid vesicle receptor

The discoid vesicle receptor is ubiquitously found in the epidermal layer of reptiles, birds and monotremes. The nerve terminals are deeply invaginated into the epidermal cells. The afferent nerve fiber is myelinated (group III) and arborizes into several unmyelinated branches which, after traversing the epithelium, usually are distally expanded to form a discoid enlargement of up to 8 µm in diameter forming the discoid vesicle receptor. Beside a small border of fine filamentous receptor matrix containing dense core and clear vesicles, the nerve terminal may contain different amounts of glycogen granules and some osmiophilic bodies. The nerve terminal is surrounded by condensed bundles of tonofibrils, arranged in a basket-like manner. Between the axonal membrane and the covering epithelial cell membrane, the intercellular space, apart from the tonofibril bundles, is narrowed and richer in contrast, in this way anchoring the epithelial tonofibril system to the axonal membrane of the discoid vesicle terminal.

In *reptiles* the discoid vesicle receptor represents a typical cutaneous intraepidermal nerve terminal mostly localized below the keratinized outer epidermal layer. The nerve terminals are often concentrated in the epidermis of different types of touch papillae or maculae of crocodiles, lizards as *Agamae* and *Geckonidae*, monitors, iguanas and snakes (v. Düring and Miller, 1978).

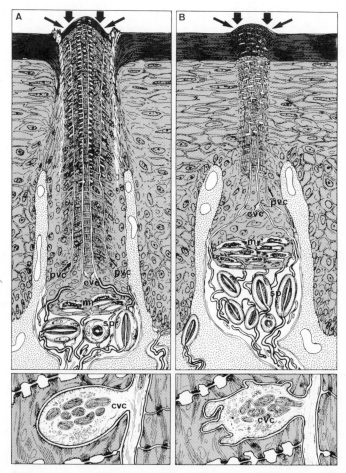

Fig. 4. Schematic representations of longitudinal sections through push rods of platypus (**A**) and of echidna (**B**). Both push rods contain four types of receptors: The central (cvc) and the peripheral (pvc) discoid vesicle chain receptors, Merkel cell receptor (mr) and paciniform corpuscles (sp). Arrows indicate directions of moving or bending of the rod induced by different tactile stimuli.

In *birds* the discoid vesicle receptors have a scattered distribution in the thin epidermal layer and occur in a denser arrangement in the epidermal follicular sheath of small feathers as the filoplumae, the ciliar feathers of the eyelash and the beard feathers of the basis of the bill and in papules surrounding feather follicles.

In *mammals* typical discoid vesicle receptors are only observed in monotremes where they are localized in the epidermal layer of the vestibulum of the nose and in the epidermis of the hairy skin behind the flap of the bill. In this location the discoid vesicle receptor is associated to papules. A specialized form of the discoid vesicle receptor is the "vesicle chain receptor" of the push rods in *platypus* (Fig. 4a; Wilson and Martin, 1893) and *echidna* (Fig. 4b). Along the intraepidermal course of the receptor axon up to the rod's tip, up to 70 discoid vesicle terminals branch off as appendages in a pearlchain-like arrangement. A dense network of tonofibrils surrounds each terminal in a basket-like fashion. In echidna the discoid vesicle receptor terminals differentiate finger-like protrusions exhibiting the receptor matrix. In cross sections of the epidermal rod we observe two types of vesicle chain receptors, one located in the centrum of the rod and the other type surrounds the first in a circular fashion. Both types of discoid vesicle chain receptors are supplied by myelinated afferent nerve fibers different in size (Andres and v. Düring, 1984). The polarized localization of the vesicle chain receptors in the epidermal rod may indicate a specialized direction detection of the mechanical stimuli.

Mechanoreceptors of the connective tissue compartments

Mechanoreceptors in the connective tissue compartment of the vertebrate dermis occur as free or as encapsulated nerve terminals and exhibit a spectrum of structural differentiation and appearance. In this paragraph we will focus on the myelinated afferent nerve fibers and their terminations. Cyclostomes lack myelinated nerve fibers but they differentiate unmyelinated axons of related axon diameters.

The most simple type of the connective tissue receptor found in all classes of vertebrates is represented by free and branched receptor axons exhibiting *lanceolate* segments and terminals. Receptive sites of the receptor axons are in direct or indirect contact to specific fiber compartments of the dermis. The main axon of this receptor type belongs to the range of group III fibers. We designate this type of receptor as *group III / free stretch receptor*. This type of cutaneous stretch receptor was first described in the sinus hair as a branched lanceolate terminal (Andres, 1966). The length of the receptor axons is variable but usually it measures up to 300 µm. During the course of the most peripheral myelinated segments, the axon exhibits large elongated nodal segments measuring up to 50 µm. From these unmyelinated segments receptoraxons arborize and terminate in the surrounding connective tissue.

Free stretch receptors of the dermis and subdermis exhibit the same structural features but the supplying axons mostly belong to the group II nerve fibers. The arborization of the receptor axon is mainly restricted to a distinct bundle of collagen fibers. We classify this type as *group II / free stretch receptor*. The size of the main axon with a conduction velocity of about 60 m/sec implies a functional integration into direct reflexes.

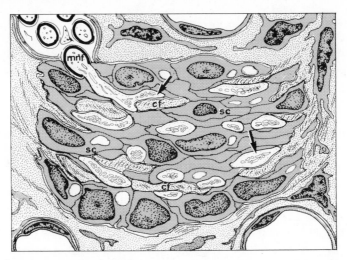

Fig. 5. Semi-schematic representation of the structural elements forming the "sandwich-receptor" of trigla. Schwann cell (sc), nerve terminals (arrows), afferent myelinated nerve fiber (mnf), collagen fibers (cf).

The spindle shaped *Ruffini* ending ranges in length from 0.5-2 mm and resembles in its fine structural organization the free stretch receptors. The receptor complex, however, is surrounded by a fluid-filled endoneural capsule space. The capsule composed of three to five perineural cells, surrounds the receptor complex except for the pole region, where collagen fiber bundles enter the spindle (Chambers et al., 1972). The collagen fibers divide into finer collagen bundles and fibrils within the capsule space. The afferent myelinated nerve fiber is of the group II range and we classify this type as group II / encapsulated stretch receptor.

Mechanoreceptors with Schwann cell differentiation

A further specialization of mechanoreceptors of the connective tissue are modifications of the Schwann cell (Andres, 1966; Andres and v. Düring, 1973).

Mechanoreceptors with Schwann cell cushion: The free movable fin rays of *Trigla lucerna* exhibit the ray tip thickenings as a specialized type of mechano-receptor. The receptor complex are composed of layers of enlarged receptor terminals intercalated with puffed Schwann cells, comparable to a sandwich (sandwich-receptor, Fig. 5). In mammals the lanceolate palisade nerve terminals surrounding the hair root sheath below the sebaceous gland are embedded between cushion-like Schwann cells. A comparable structural arrangement is applied to the straight lanceolate nerve terminals of the sinus hair. The *Meissner* and the *Krause* corpuscles display coiled receptor axons in combination with puffed Schwann cells. Collagen fibrils of the surrounding tissue are involved in the receptive complex. Physiological experiments confirm the rapidly adapting discharge pattern of the Meissner corpuscle (Lindblom, 1965; Vallbo and Hagbarth, 1968). The appearance of the Meissner corpuscle is restricted to the papillary ridges of the hairless skin of primates.

Mechanoreceptors covered by lamellated Schwann cells: Phylogenetically the lamellated mechanoreceptors occur for the first time in *anura* (v. Düring and

Seiler, 1974). In all following classes of vertebrates they occur in different sizes, forms and locations. A simple type of lamellated receptor with only the Schwann cell lamellation and the surrounding basement lamella is frequent in snakes and in *Xenopus*. Most types of the lamellated mechanoreceptors are encapsulated by a perineural sheath. Larger corpuscles exhibit beside the lamellated inner core, a more or less prominent capsule space. *Herbst* and *mini-Herbst* corpuscles represent the lamellated receptor types in birds. The capsule space of the Herbst corpuscle contains sparse endoneural fibrocyts but a distinct and loose network of fine connective tissue fibrillae. Mini-Herbst corpuscles occur in the papillae of the bill organ and in the lamina propria of the tongue's surface of parrots. They are differentiated by a capsule space with some sparse fibrillae. A differentiated proteoglycan composition of the mini-Herbst corpuscles becomes obvious in their varying metachromatic histochemical reactions which may indicate a functional differentiation of Herbst corpuscles.

The capsule space of large and small lamellated receptor corpuscles in monotremes contains a unique cellular outer core. The cytoplasm of the large cells of the outer core is filled with regularly arranged wide spread smooth endoplasmatic reticulum.

Unmyelinated efferent nerve fibers may supply the large lamellated corpuscles. The preterminal fibers penetrate into the capsule and terminate between the lamellae of the inner core (Santini, 1968; Andres and Schwarztkopff, 1973).

Merkel cell receptors

Epidermal Merkel cells differentiate from the stratified squamous epithelial cells (English et al., 1980; Eglmeier, 1987) and are found in the skin of all vertebrate classes (v. Düring, 1974; Breathnach, 1980; Fox et al., 1980) except in *Chondrychthies* which lack this type of receptor. In reptiles Merkel cells have not been found in *squamatae*. The Merkel cell-neurite complex consists of the Merkel cell characterized by finger-like cell processes invading surrounding epithelial cells, dense core vesicles and the afferent nerve terminal forming a disc below the cell. Synaptoid contacts are established between the Merkel cell and the afferent nerve terminal (for review see Iggo and Andres, 1982; Hartschuh et al., 1986). Recent neurophysiological and further experimental studies including a specific elimination of the Merkel cells do not establish a receptor- or mechanosensory transducer function of the Merkel cell (Gottschaldt and Vahle-Hinz, 1981; Diamond et al., 1988). The discovery of multiple messenger candidates within the Merkel cells in the last years (for review see Hartschuh and Weihe, 1988) and the presence of "synaptoid junctions" may indicate modulatory or mediator functions. In this context we observe one or a few Merkel cells associated with the receptor axon of a typical free stretch receptor in the skin of birds. Nevertheless the biological function of the Merkel cell still remains obscure.

CUTANEOUS SENSORY RECEPTOR ORGANS

Cutaneous vertebrate sensory receptors may be organized into complex tactile sense organs or touch papillae. In this organs various structural elements of the skin like epidermal domes, papillary connective tissue arrangements of papulae, papillae and warts but also epidermal differentiations like horns, hairs, vibrissae, bristles, feathers or skin ridges are used in manifold ways for stimulus transduction to the different types of receptor terminals.

Cutaneous sensory organs in fish and amphibia

In addition to the lateral line organ as a highly differentiated cutaneous mechanoreceptive system, fish have Merkel cell receptors but they lack other more complex nerve terminals of primary afferent neurons. A striking exception is found in *Triglidae* which exhibit a benthic habitat. The lower three pectoral fin rays of these fishes are free and movable and are used as feelers for detecting and locating food. Each of these free fin rays are supplied by up to 9000 thin myelinated nerve fibers. One group of axons terminate as free nerve terminals within the intercellular space of the epidermis. A second group of sensory nerve fibers terminates as "sandwich receptor" situated in the apical dermal thickening of the ray. Large stretch receptors terminate within ligaments connecting the skin to the bone. Finally a bundle of about 75 large myelinated axons terminate as free periostal stretch receptors controlling the bending angle of the individual ray.

Aquatic urodela which exhibit lateral line neuromast like fish do not have prominent primary sensory skin receptor organs. Mechanosensory organs occur in *Xenopus laevis* as cone organs, and semiaquatic frogs develop runt warts for mating behavior. Below the keratinized caps of the cone organ, free intraepidermal receptor terminals are distributed within the intercellular space of the spinous cells. In the runt wart of frogs the receptor terminals are located within the connective tissue below the epidermal cone and are in close contact with fibrocytes.

Cutaneous receptor organs of reptiles

Many of the reptilian cutaneous receptors may be organized into tactile sense organs or touch papillae which penetrate the thick and dense corial and the keratinized epidermal layers of the skin. Various degrees of complexity have been shown to exist in lizards by v. Düring and Miller (1979). In general, these sensory organs are covered by a thin keratinized epidermal layer in contrast to the surrounding epidermal surface, and the connective tissue of the papillae is loose and movable contrary to the usually tautly organized dermal layer. On the top of touch papillae of agamids and gecko, bristles of the uppermost keratinized cells are connected to the intraepidermal vesicle receptors by prominent tonofibrils. The adhesive skin on the underside of the toes of geckos exhibits a dense sensory innervation pattern. A prominent erectile cavernous vascular system extends below this specialized skin region. The blood pressure within the cavernous system in the toe skin of geckos may modify the tension of the skin and herewith the efficacy of the adhesive filaments and may also modulate the sensitivity of the receptor terminals. The most prominent touch papillae occur on the facial scales of crocodiles. These papillae regularly contain dermal Merkel cell neurite complexes which is exceptional in the skin of reptiles. The receptor complexes are arranged in tightly packed Merkel cell columns which are in close contact with expanded nerve terminals interleaved between the cells.

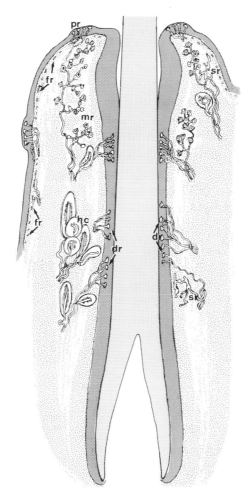

Fig. 6. Topographical distribution of the different types of sensory receptors of a larger filopluma of a pigeon. Subepidermal nerve fiber plexus (fr) with free nerve terminals in the connective tissue. Papular receptor (pr) with intraepidermal discoid receptors. Discoid receptors (dr) within the epidermal follicle sheath. Merkel cell receptor (mr) in a pedicle-like arrangement in the follicular connective tissue sheath. Small Herbst corpuscle (hc) in the deeper part of the follicular connective tissue sheath. Free stretch receptor (sr) with large myelinated axons in follicular connective tissue sheath.

Cutaneous receptor organs of birds

Feathers are important sensory organs with an exceptionally good lever function for stimulus transduction. Therefore the follicular wall of wing-, tail- and contour feathers are supplied with sensory terminals. Fine structural studies reveal that large Herbst corpuscles and free stretch receptors (group II / free stretch receptor) are linked to the elastic muscular tying of the follicle. The stiff bristles of barbets exhibit in their follicular wall a tight connective tissue sheath which contain medium-sized Herbst corpuscles arranged in a dense pattern. The epidermal edge of the follicle opening of these feathers shows intraepidermal discoid receptors. Filoplumes lack smooth muscles attached to their follicles. The topography of filoplumes is always adjoined to contour feather follicles and they are thought to be concerned with controlling the position of the larger feathers. Fine structural studies show that the filoplumes are richly supplied with different types of cutaneous receptors (Fig. 6). Three types of receptors are commonly present: 1) the intraepidermal discoid receptor, 2) the Merkel pannicle receptor, 3) a group III / free branched stretch receptor connected to the glassy membrane of the follicle epithelium. The follicle wall of large filoplumes exhibits in addition small Herbst corpuscles.

Cutaneous receptor organs of monotremes

The push rod organs of the *platypus* which were discovered by Poulton (1885, 1894) and Wilson and Martin (1893) are found in the outer and inner skin of the bill. Electron microscopic studies of the fine structural organization of push rods confirmed the suggestion from the classical light microscopical descriptions that the push rod functions as a complex mechanosensitive organ (Bohringer, 1981; Andres and v. Düring, 1984). The push rod is built up by an epidermal column of flattened keratinocytes. The densely packed tonofibrils and the interconnecting desmosomes may contribute to the rigidity and elasticity of the rod. Each rod is separated by a sheath of light spinous cells. Due to the lack of the tonofibril system within this sheath the rod may be moved up and down or may be bent in different directions by stimuli perpendicular or tangential to the skin surface. There are four types of sensory receptors associated with the rod: 1) the central discoid vesicle chain receptor, 2) the peripheral discoid vesicle chain receptor and 3) the Merkel cell receptor of the basal epidermal foot of the rod, and 4) the lamellated encapsulated receptor. The direction of the intracorpuscular receptoraxon being identical to the long axis of the ellipsoid corpuscle is orientated strictly parallelly or strictly perpendicularly to the skin surface. Reconstructions from serial sections demonstrate that the corpuscles with receptoraxons parallelly to the bill surface are orientated rectangularly to each

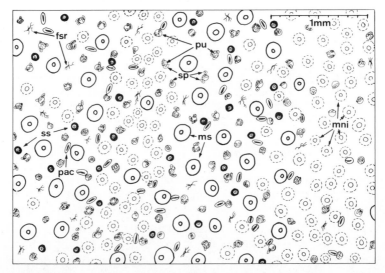

Fig. 7. The graphical reconstruction in a horizontal projection shows the distribution of sensory organs and elementary receptors in a skin segment from the outer face of the under bill of platypus. Push rods (pu) including the small basal paciniform corpuscles proper (sp). Large paciniform corpuscles (pac). Ducts of the sensory mucous glands (ms), ducts of the sensory serous glands (ss), ducts of the non sensory innervated mucous glands (mni). Large subepidermal free stretch receptor (fsr).

other. Including the perpendicularly orientated corpuscles, the paciniform receptors of the push rod seem to build up a three dimensional stimulus detecting system. Recently, fine structural studies of the skin of the echidna bill revealed a similar push rod system as in platypus. In detail there are some differences as the epidermal rod basis contains a larger Merkel cell receptor complex arranged in three layers of Merkel cells. The number of the paccinian corpuscles is up to three times higher. The arrangement of the corpuscles is less geometrical than in platypus. The discoid vesicle chain receptors in echidna is less distinctly separated in a central and a peripheral type.

Pattern and arrangement of cutaneous receptors

As we know, skin sensation is not restricted to a single receptor activity but is mediated by the integration of signals from many receptors of different sensory qualities. In the contiguous stations of this central processing a fixed topographic arrangement of the sensory receptors in the periphery and a corresponding somatotopic organization in the central nervous system is an important condition.Therefore more attention should be paid to the specific pattern array of the different skin receptors, the density of the elementary receptors depending on the skin region and the size of the receptive fields. New results in this context may give a better background for the discussion about resolution of skin regions or for stereognostic and esthesiognostic interpretations of sensory information from the skin. In this way the graphical reconstruction of a skin segment of the platypus bill (Fig. 7) may impart insight into a very highly organized sensory skin region. In this case the terminals of primary sensory ganglion cells of the fifth nerve build up a complex pattern of different sensory nerve fibers which may function as a mechano- and electroreceptive eye for the platypus while hunting prey on the grounds of seas, rivers and creaks even in muddy waters.

Acknowledgements: Supported by the Deutsche Forschungsgemeinschaft (SFB 114, SP Nociception und Schmerz)

REFERENCES

Andres, K. H., 1966, Über die Feinstruktur der Rezeptoren an Sinushaaren, *Z. Zellforsch.*, 75:339

Andres, K. H., 1973, Morphological criteria for the differentiation of mechanoreceptors in vertebrates, *in:* "Symposium Mechanorezeption," J. Schwartzkopff, ed., Abhdlg. Rhein. Westf. Akad. Wiss. Westdeutscher Verlag

Andres, K. H., and Düring, v. M., 1973, Morphology of cutaneous receptors, *in:* "Hd. Sensory Physiology Vol II Somatosensory System," A. Iggo, ed., Springer Verlag, Berlin Heidelberg, New York

Andres, K. H., and Düring, v. M., 1984, The platypus bill. A structural and functional model of a pattern-like arrangement of different cutaneous sensory receptors, *in:* "Sensory Receptor Mechanisms," W. Hamann and A. Iggo, eds., World Scientific Publ. Co., Singapore

Andres, K. H., and Düring, v. M., 1985, Rezeptoren und nervöse Versorgung des bronchopulmonalen Systems, *Bochumer Treff,* 31

Andres, K. H., Düring, v. M., Muszynski, K., and Schmidt, R. F., 1987, Nerve fibres and their terminals of the dura mater encephali of the rat, *Anat. Embryol.,* 175:289

Andres, K. H., and Düring, v. M., 1988, Comparative anatomy of vertebrate electroreceptors, *in:* " Progress in Brain Res. Vol 74 , Transduction and Cellular Mechanisms in Sensory Receptors," W. Hamann and A. Iggo, eds., Elsevier Science Pub., Amsterdam

Bohringer, R. C., 1981, Cutaneous receptors in the bill of the platypus *(Ornithorhynchus anatinus), Austr. Mam.,* 4:93

Breathnach, A. S., 1980, The mammalian and avian Merkel cell, *in:* " The Skin of Vertebrates," R. I. C. Spearman and P. A. Riley, eds., Academic Press, London

Brown, A. G., 1981, The primary afferent input: some basic principles, in: "Spinal cord sensation," A. G. Brown and M. Réthelyi, eds., Scottish Academic Press

Bullock, T. H, and Heiligenberg, W., 1986, "Electroreception," Wiley, New York

Chambers, M. R., Andres, K. H., Düring, v. M., and Iggo, A., 1972, The structure and function of the slowly adapting Type II mechanoreceptor in hairy skin, Q. J. Exp. Physiol., 57:417

Clifton, G.L., Coggeshall, R.E., Vance, W. H., and Willis, W. D., 1976, Receptive fields of unmyelinated ventral root afferent fibres in the cat, J. Physiol. (Lond.), 256:573

Craig, A. D., and Burton, H., 1981, Spinal and medullary lamina I projection to nucleus submedius in medial thalamus: a possible pain center, J. Neurophysiol., 45:443

Craig, A. D., and Kniffki, K. D., 1982, Lumbosacral lamina I cells projecting to medial and/or lateral thalamus in the cat, Soc. Neurosci. Abstr., 8:95

Diamond, J., Mills, L. R. , and Mearow, K. M., 1988, Evidence that the Merkel cell is not the transducer in the mechanosensory Merkel cell-neurite complex, in: "Progress in Brain Res. Vol. 74, Transduction and Cellular Mechanisms in Sensory Receptors," W. Hamann and A. Iggo, eds., Elsevier Science Pub., Amsterdam

Düring, v. M., 1974a, The ultrastructure of cutaneous receptors in the skin of Caiman crocodilus, Abhdlg. Rhein. Westf. Akad. Wissensch., 53:123

Düring, v. M., 1974b, The radiant heat receptor and other tissue receptors in the scales of the upper jaw of Boa constrictor, Z. Anat. Entw. Gesch., 145:299

Düring, v. M., and Andres, K. H., 1988, Structure and functional anatomy of visceroreceptors in the mammalian respiratory system, in: "Progress in Brain Res Vol 74, Transduction and Cellular Mechanisms in Sensory Receptors," W. Hamann and A. Iggo, eds., Elsevier Science Pub., Amsterdam

Düring, v. M., Andres, K. H., and Iravani, J., 1974, The fine structure of the pulmonary stretch receptor in the rat, Z. Anat. Entw. Gesch., 143:215

Düring, v. M., and Miller, M. R., 1978, Sensory nerve endings of the skin and deeper structures, in: "Biology of the reptilia Vol 9, Neurology A," C. Gans, ed., Academic Press London, New York

Düring, v. M., and Seiler, W., 1974, The fine structure of lamellated receptors in the skin of Rana esculenta, Z. Anat . Entw. Gesch., 144:165

Eglmeier, W., 1987, The development of the Merkel cells in the tentacles of Xenopus laevis larvae, Anat. Embryol., 176:493

English, K. B., Burgess, P. R., and Norman, D. K., 1980, Development of rat Merkel cells, J. Comp. Neurol., 194:475

Gasser, H. S., and Grundfest, H., 1939, Axon diameters in relation to the spike dimension and the conduction velocity in mammalian A-fibers, Amer. J. Physiol., 127:393

Fox, H., Lane, E. B., and Whithear, M., 1980, Sensory nerve endings and receptors in fish and amphibians, in: "The Skin of Vertebrates," R. I. C. Spearman and P. A. Riley, eds., Academic Press, London

Fritzsch, B., and Wahnschaffe, U., 1983, The electroreceptive ampullary organs of urodeles, Cell Tissue Res., 229:483

Gottschaldt, K. M., and Vahle-Hinz, C. H., 1981, Merkel cell receptors: structure and transducer function, Science., 214:183

Gregory, J. E., Iggo, A., McIntyre, A. K., and Proske, U., 1988, Receptors in the bill of the platypus, J. Physiol. (Lond.), 400:349

Hagbarth, K. E., and Vallbo, A. B., 1967, Mechanoreceptor activity recorded percutaneously with semi-microelectrodes in human peripheral nerves, Acta Physiol. Scand., 69:121

Hartschuh, W., Weihe, E., and Reinecke, M., 1986, The Merkel cell, in: "Biology of the integument Vol. 2, Vertebrates," J. Bereiter-Hahn, A. G. Matoltsy, and K. S. Richards, eds., Springer Verlag, Berlin, Heidelberg, New York, Tokyo

Hartschuh, W., and Weihe, E., 1988, Multiple messenger candidates and marker substances in the mammalian Merkel cell-axon complex: a light and electron microscopic immunohistochemical study, in: "Progress in Brain Res Vol 74, Transduction and Cellular Mechanisms in Sensory Receptors," W. Hamann and A. Iggo, eds., Elsevier Science Pub., Amsterdam

Hensel, H., Andres, K. H., and Düring, v. M., 1974, Structure and function of cold receptors, Pfluegers Arch., 352:1

Hensel, H., 1975, Static and dynamic activity of warm receptors in Boa constrictor, Pfluegers Arch., 353:191

Hökfelt, T., Johansson, O., Ljungdahl, A. A., Lundberg, J. M., and Schultzberg, M., 1980, Peptidergic neurones, Nature, 284:515

Iggo, A., and Andres, K. H., 1982, Morphology of cutaneous receptors, Ann. Rev. Neurosci., 5:1

Iggo, A., Proske, U., McIntyre, A. K., and Gregory, J. E., 1988, Cutaneous electroreceptors in the platypus: a new mammalian receptor, in: "Progress in Brain Res Vol 74, Transduction and Cellular Mechanisms in Sensory Receptors," W. Hamann and A. Iggo, eds., Elsevier Science Pub., Amsterdam

Kürten, L., and Schmidt, U., 1982, Thermoperception in the common vampire bat *(Desmodus rotundus)*, *J. Comp. Physiol.*, 146:223

Kürten, L., 1984, Vergleichende Untersuchungen zur Anatomie und Physiologie des Tast- und Wärmesinnes im Nasenaufsatz der Vampirfledermaus *Desmodus rotundus*, Dissertation, Bonn

Kürten, L., Schmidt, U., and Schäfer, K., 1984, Warm and cold receptors in the nose of the vampire bat *Desmodus rotundus*, *Naturwissenschaften*, 71:327

Lindblom, U., 1965, Properties of touch receptors in distal glabrous skin of the monkey, *J. Neurophysiol.*, 28:965

Mense, S., and Craig, A. D., 1988, Spinal and supraspinal terminations of primary afferent fibers from the gastrocnemius-soleus muscle in the cat, *Neurosci.*, 26:1023

Müller, J., 1844, "Handbuch der Physiologie des Menschen," Coblenz Hölscher

Poulton, E. B., 1885, On the tactile terminal organs and other structures in the bill of *Ornithorhnychus*, *J. Physiol. (Lond.)*, 5:15

Poulton, E. B., 1894, The structure of the bill and hairs of *Ornithorhynchus paradoxus* with a discussion of the homologies and origin of mammalian hair, *Quart. J. micr. Sci.*, 36:143

Proske, U., Gregory, J. E., Iggo, A., and McIntyre, A. K., 1988, An electrical sense in echidnas, *Proc. Aust. physiol. pharmacol. Soc.*, 19:190

Santini, M., 1968, Noradrenergic fibres in Pacinian corpuscles, *Anat. Rec.*, 160:494

Scheich, H., Langner, G., Tidemann, C., Coles, R. B., and Guppy, A., 1986, Electroreception and electrolocation in platypus, *Nature*, 319:401

Schwarztkopff, J., 1973, Mechanoreception, in: "Avian Biology Vol 3," D. S. Fahrner, J. R., King, B., and K. C. Parkes, eds., Academic Press, London

Vallbo, A. B., and Hagbarth, K-E., 1968, Activity from skin mechanoreceptors recorded percutaneously in awake human subjects, *Exp. Neurol.*, 21:270

Wilson, J. T., and Martin, C. J., 1893, On the peculiar rod-like tactile organs in the integument and mucous membrane of the muzzle of *Ornithorhynchus*, *Linn. Soc. NSW.*, MacLeay Memorial Vol.

2

SENSORY INNERVATION OF THE HAIRLESS AND HAIRY SKIN IN

MAMMALS INCLUDING HUMANS

Zdenek Halata

Anatomisches Institut der Universität Hamburg
Abteilung für funktionelle Anatomie, Hamburg
F.R.G.

INTRODUCTION

The sense of touch developed early in phylogeny. One of the reasons might be that this sense is highly essential for survival: using this sense, the newborn animal explores its new surroundings, and by doing this, it might come across sources of food and support. For this purpose the skin contains complex nervous structures, named "touch organs" or "mechanoreceptive complexes" (Halata, 1975). These organs develop at different stages before birth depending on their significance for the animal., In animals which leave their parents at an early point they are fully developed immediately after birth (Walter, 1960/61), whereas with those animals which stay with their parents for a longer period of time, the process of maturation takes longer.

Like other sense organs, touch organs consist of a mechanic transductive component transmitting and guiding forces toward the second component, the sensory nerve ending. These mechanic transductive components are, e.g., the epidermal papillae of the cone skin or the epidermal ridges of the ridged skin. In the case of hairy skin, examples of these mechanic transductive components can be seen in the epithelium of hair follicles including the sebaceous gland and the connective tissue with blood vessels of the papillary layer.

Different regions of the surface of the body are equipped with touch organs of differing degrees. Thus, the skin surrounding the body entrances and the glabrous skin of the tip of the nose is rich in mechanoreceptive complexes, such as can be seen in the finger tip skin of primates and a number of small animals. By contrast, the skin of the back only reveals low touch-organ density. Furthermore, the representation of these sensory organs in the central nervous system is of considerable importance: the representation area in the primary sensory cortex of the skin of the lips or of the thumb is much larger than the representation area of the back skin.

In this study the touch organs of the hairy and the glabrous skin of several marsupials and eutherians including humans are investigated using light and electron microscopy. Characteristic features of these sense organs will be described and compared.

MATERIAL AND METHODS

Specimens for light and electron microscopic purposes were prepared from following sources: the glabrous skin of the tip of the nose *(planum nasale)*, the hairy skin of the upper lip and of the surrounding skin. Marsupials from America and Australia (American opossum - *Didelphis virginiana*, bandicoot - *Isoodon macrourus*, Kowari - *Dasyuroides byrnei*, gray short-tailed opossum - *Monodelphis domestica)* as well as eutherians (mole - *Talpa europaea*, cat - *Felis silvestris*, mini-pig Göttingen, long-tailed monkey - *Macaca mulata* and *Macaca cynomolgus)* were used. Human skin was taken from the edge of the eye lid, from the skin of the lips, the fingers, the thigh and the upper arm.

For light microscopy, the animals were perfused with Buoin's solution (after deep anaesthesia with 80 mg/kg Nembutal i.p.). Following this, the pieces of skin were dissected, embedded in Paraffin and sectioned into slices of 6 μm to perform silver impregnation according to Spaethe (1984). Sections of 1 μm from the electron microscopic specimens were stained according to Ito and Winchester (1963) or Laczko and Levai (1975).

For electron microscopy, perfusion was carried out with a 6% glutaraldehyde solution (pH 7.2 in 0.05 M Millonig-sodium-phosphate-buffer) for 5 to 30 minutes (with regard of the animals body size). After being cut into small pieces of about 3 mm³, the tissue specimens were immersed for a maximum of 1 hour in the same fixans at 40°C. Postfixation was performed with 1% OsO$_4$ (in 0.1 M Millionig-sodium-phosphate-buffer containing 1% sucrose). The specimens were dehydrated and embedded in EPON 812 (Serva) according to Luft (1961). Thin sections were obtained (Ultramicrotome OmU2, "Ultracut" Reichert and Porter Blum) and contrasted (Reynolds, 1963). Visualization was performed on a Philips 300 and a Zeiss EM 109 electron microscope.

RESULTS AND DISCUSSION

1. Mechanoreceptive complexes of the glabrous skin

1. 1 The cone skin (Fig. 1-7)

The cone skin was firstly described by Mojsisovics (1876) in the mole, and then by Botezat (1903), Quilliam (1966a, b) and more recently by Halata (1971a, b), Andres and von Düring (1973) and Buisseret et al., (1976) with the use of the electron microscope. The arrangement of the epidermal cones in cat nose was described in detail (Halata, 1970), in pig nose (Andres and von Düring, 1973; Halata, 1975), in several marsupials (Loo and Halata, 1985a, b) and in the platypus (Bohringer, 1981; Andres, 1984). These investigations suggest that blind or nocturnal animals are primarily equipped with this type of glabrous skin at the tip of the nose.

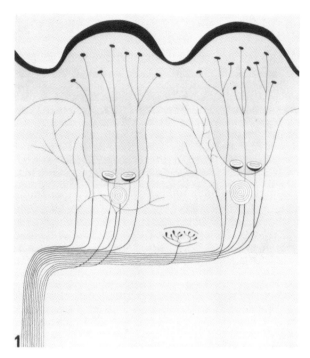

Fig. 1. Innervation of the cone skin. In the epidermis there are free nerve endings in the stratum spinosum and the stratum granulosum, and Merkel nerve endings in the basal layer. In the papillary layer of the dermis there are free nerve endings, and below the epithelial cones Pacini corpuscles. In the reticular layer of the dermis lies a Ruffini corpuscle.

In the mechanoreceptive complex of the cone skin (Fig. 1) three different parts can be distinguished: an epidermal and a dermal part as well as nerve endings. The epidermal part consists of a cone of the superficial epidermal layer and an epidermal cone embedded in the corium. The dermal part is composed of the connective tissue of the papillary layer surrounding the cones and ring-like capillaries emerging from blood sinus of the nose (Boeke, 1933). An example of this type of glabrous skin is the tip of the nose in moles, in which mechanoreceptive complexes can be found. First described by Eimer (1871) and for this reason known as *Eimer's organs*, they consist of so-called free nerve endings and of Merkel endings innervating the epidermal cone. Below these cones, the papillary layer shows small lamellated corpuscles and also free nerve endings (Halata, 1971a). Eimer (1871) himself estimated the number of such touch organs present in the nose of the mole to be about 5,000.

In the *epidermis*, the number of *free nerve endings* (Fig. 1, 3 and 4) in the stratum spinosum and granulosum of each cone varies from 15 to 25 depending on the thickness of the cone. The non-myelinated, afferent axons display an average diameter of 1-2 µm and are situated in the dermis immediately adjacent to the epidermis. Entering the epidermis they lose their Schwann cell sheath and run perpendicularly to the surface, between the keratinocytes, often enveloped by cytoplasmatic pockets of the keratinocytes. More superficially, in the spinous and granular layers, these free nerve endings reveal axonal swellings, which contain mitochondria, light vesicles and glycogen granules. Axoplasmatic

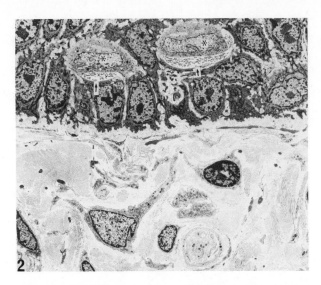

Fig. 2. Epithelial cone from didelphis snout in section perpendicular to skin surface. Merkel nerve endings (*) are situated between the basal and spinous layers of the epidermis. (arrows) - Nerve terminals. Magnification 2,500 x.

protrusions (spiculae) are interposed between the surrounding keratinocytes. These spikes enlarge the axonal membrane, which seems to be the region of mechano-electric transformation.

The base of each cone displays 3-5 *Merkel* endings (Fig. 1 and 2) in the basal layer. A Merkel ending consists of a Merkel cell (Merkel, 1875) and a sensory nerve ending which clings to the Merkel cell from the dermal aspect, forming synapse-like contacts with the cytomembrane of the Merkel cell (Hartschuh et al., 1986). The afferent axons (1-2 per cone) are myelinated and reveal an average diameter of 2-4 µm. After loosing their myelin sheath at a distance of about 10 - 50 µm below the basal layer, they enter the epidermis. They can branch dichotomically into unmyelinated axons both in the dermis and the epidermis.

In the *dermis, free* nerve endings are situated in the papillary layer just beneath the basal lamina of the cones and between the cones (Fig. 1 and 5). They display axonal swellings and often abut directly to the basal lamina (Hensel, 1973). Small *lamellated* corpuscles are situated below the epidermal cones just

Fig. 3. Epithelial cone from didelphis snout in section perpendicular to skin surface. Terminal thickening of the free nerve ending in the spinous layer of the epidermis contains accumulations of mitochondria and clear vesicles. Magnification 11,000 x.

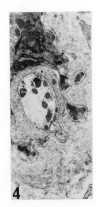

Fig. 4. Epithelial cone from didelphis snout in section perpendicular to skin surface. Nerve terminal is situated in the basal layer of the epidermis. Magnification 10,000 x.

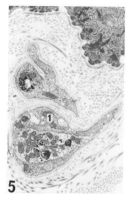

5

Fig. 5. Cone skin of the planum nasale of the cat. Section perpendicular to the skin surface. The free nerve ending reveals numerous mitochnondria and clear vesicles and is ensheathed by a terminal Schwann cell (1). Magnification 10,000 x.

beneath the center of the cone (Fig. 6 and 7). They consist of the terminal axon, the inner core and the capsule. The myelinated afferent axon has an average diameter of about 4-5 μm. In the inner core the afferent axon loses the myelin sheath and terminates in an axonal thickening. The inner core is composed of thin cytoplasmic lamellae of terminal Schwann cells. The capsule surrounding the corpuscle comprises 1-3 layers of flat perineural cells.

The mechanical function of the epidermal cones (i.e., conducting mechanical

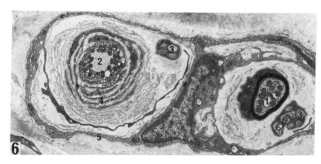

6

Fig. 6. Small lamellated corpuscle in a nerve fascicle from the cone skin of the nose of the kowari in section perpendicular to the skin surface. 1 myelinated afferent axon, 2 nerve terminal of the lamellated corpuscle, 3 non-myelinated axons in a nerve fascicle, 4 cytoplasmic lamellae of the terminal Schwann cell, 5 perineural capsule cells. Magnification 4,000 x.

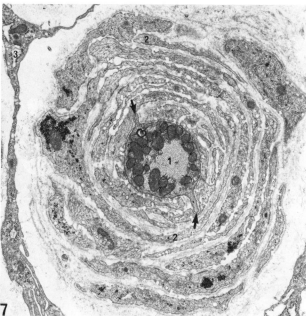

7

Fig. 7. Small lamellated corpuscle from the cone skin of the nose of the kowari in section perpendicular to the skin surface. 1 nerve terminal with accumulations of mitochondria, (arrows) finger-like protrusions of the nerve terminal (spikes), 2 cytoplasmic lamellae of the terminal Schwann cell, 3 fibroblast from the subcapsular space. Magnification 20,000x.

stimuli) resembles that of the hair follicles of touch hairs. The ring-like capillaries isolate neighbouring cones from each other so that each cone can be stimulated separately. The arrangement and total number of about 5,000 cones associated with nerve endings (about 150,000 epidermal and dermal free nerve endings, 20,000 Merkel endings and 5,000 - 10,000 small lamellated corpuscles) enables the blind or nocturnal animal to create a reliable image of its surroundings, based solely on the sense of touch.

1. 2 The ridged skin (Fig. 8)

This type of glabrous skin is present in the fingers, palms, toes and feet of humans and other primates as well as some marsupials. Furthermore, human and primate lips are covered in part by ridged skin.

A characteristic example of the mechanoreceptive complex of ridged skin can be seen in the tips of the fingers (Fig. 8). The superficial relief of the epidermal layer reveals ridges and grooves. The continuation of the superficial ridges are glandular ridges extending deep into the reticular layer of the dermis. Underneath the superficial grooves, smaller adhesive ridges run parallel to the glandular ones. The latter two deep ridges of the epidermal layer are connected by cross ridges, perpendicular to both (Horstmann, 1957). In the skin of the finger in rhesus monkeys (Macacus rhesus) the total height of the glandular ridge including the superficial ridge is approximately 2,000 µm and its width about 500 µm. As it passes through the epidermis, the sweat gland duct is twisted helically, with each loop of the duct being followed by the next at a constant distance. From the basal apex to the superficial groove, the adhesive ridge has a total height of about 1,000 µm and a width of 200 µm. The cross ridges are smaller, i.e., only about 300 µm in height and about 100 µm in width. Thus, the basal or dermal aspect of the epidermis is divided by these ridges into crypts, and the surface of the papillary layer of the dermis mirrors this aspect again. Thus, the papillary layer is arranged in rows of papillae. Each dermal papilla is separated by the epidermal ridges and by a ring-like capillary. These epithelial ridges and crypts, the corresponding dermal papillae and their blood vessels and nervous elements together make up the touch organs of the ridged skin. In the basal layer of the epidermal glandular ridges accumulations of Merkel cells are frequently visible (up to 10).

In the connective papilla of each crypt, formed by the lateral wall of one glandular ridge (crista intermedia), of one adhesive ridge (crista limitans) and of two cross ridges, one Meissner corpuscle (Wagner and Meissner, 1852; Meissner, 1853) is suspended from collagen fibres which are fixed into the basal lamina. These corpuscles are oval in shape (40 - 70 µm in diameter and 100 - 150 µm in length) and their longitudinal axis is oriented towards the surface of the skin (Halata, 1975; Chouchkov, 1978). They are positioned immediately beneath the basal lamina of the top of the crypt or the apex of the papilla. In the skin of the monkey, this apex extends to the bottom of neighbouring superficial grooves. Thus, Meissner corpuscle is the most superficially located touch organ. It consists of 1-5 myelinated afferent axons of an average diameter of 3-5 µm. In the corpuscle the axons no longer have a myelin sheath. Rather, they become flatter and assume the shape of oval discs, lying parallel to the epidermal surface. These terminal axons are separated by cytoplasmatic lamellae of

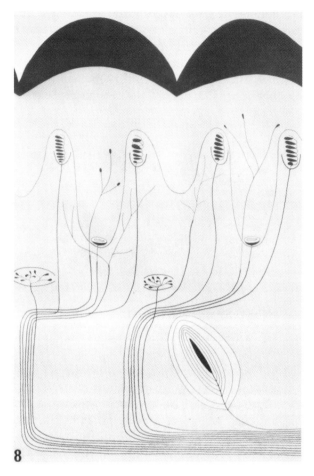

Fig. 8. Diagrammatic representation of the ridged skin of the finger of rhesus monkey in cross section. Free nerve endings are present in the spinous layer of the epidermis and in the papillary layer of the dermis. Merkel nerve endings are present in the basal layer of glandular ridges. Meissner corpuscles are situated in the papillary layer of the dermis, Ruffini corpuscles are situated in the reticular layer of the dermis and Pacini corpuscles in the subcutis.

Schwann cells. A perineural capsule envelopes the corpuscle in a cup-like fashion only in the lowest quarter. The collagen fibrils, suspending the corpuscle beneath the top of the papilla, run in the fissures of Schwann cell lamellae. Thus, deformations of the surface of the skin are transmitted and led towards the terminal afferent axons of the Meissner corpuscle.

Free nerve endings are distributed in the connective tissue of the papillary layer as well as in the connective tissue adjacent to the basal parts of the glandular and adhaesive ridges (Cauna, 1980). Often they come into contact with the basal lamina. It was only by using light microscopy that free nerve endings were described in the spinous layers of the epidermis (Kadanoff, 1928; Novotny and Gommert-Novotny, 1988).

Large *Pacini* corpuscles are situated deep in the dermis, without any topographic relation to the surface. The same applies to the small lamellated corpuscles, which consist of an afferent axon, an inner core and a perineural capsule. The diameter of the myelinated afferent axon varies between 5 - 8 µm. The terminal axon is unmyelinated and slightly thickened and is surrounded concentrically by cytoplasmatic lamellae of Schwann cells. This system of lamellae is termed the inner core (Pease and Quilliam, 1956). The number of lamellae varies between 20 - 60, according to the corpuscle's size.

Like the cone skin, the ridged skin is adapted for the function of transmitting tactile stimuli (Halata, 1975). The capillaries separate each crypt containing Meissner corpuscles from the neigbouring crypt. The glandular ridges transmit surface pressure to Meissner corpuscles and to the Merkel endings. The large Pacini corpuscles in the depth of the dermis are stimulated by vibration. The innervation pattern of the skin of the lip is similar to that of the finger tips (Halata and Munger, 1983).

2. Mechanoreceptive complexes of the hairy skin

The touch organs of hairy skin include the so-called "Tastscheiben" (or Pinkus corpuscles) and hair follicles with their nerve endings.

2. 1 The "Tastscheibe" of the hairy skin

In hairy skin the so-called "Tastscheibe" was primarily described by Pinkus (1902, 1905), and for this reason it is also referred to as *Pinkus* corpuscle. With the assistance of light and electron microscopy, it has been described by several authors (Straile, 1960, 1961; Mann, 1968; Smith, 1967; Iggo and Muir, 1969; Straile and Mann, 1972; Fitzgerald et al., 1984). The "Tastscheibe" is composed of an epithelial rise, close to groups of hair follicles, and of *Merkel* cells and terminal axonal thickenings. The total number of Merkel cells corresponds to the size of the epithelial rise, i.e., large Pinkus corpuscles reveal up to 50 Merkel endings. By means of the electron microscope, "Tastscheiben" were studied in the ventral abdominal skin (Smith, 1967) and in the skin of the edge of the eye lid (Munger and Halata, 1984), also in humans. Apart from the epithelial rise, the epithelium of the hair funnel may conduct mechanical stimuli to the "Tastscheibe". Hairs of this kind are often longer and thicker than those of other groups (tylotrichic hairs - Straile 1960, 1961; Straile and Mann, 1972).

2. 2 The sinus hair as a mechanoreceptive complex (Fig. 9)

An additional example of typical touch organs are sinus hairs or vibrissae (Vincent, 1913; Melaragno and Montagna, 1953; van Horn, 1970). The hair follicle of a sinus hair is nearly completely embedded in a blood sinus (Fig. 9). In the region of the hair bulb, this sinus is cavernous. On the other hand, the sinus in the upper regions of the hair follicle is an undivided ring-like cavern. Often, fibres of striated mimic muscles originate in the thick connective tissue sheath of the sinus' adventitia. Due to this arrangement, sinus hairs are voluntarily movable. Obviously, the deep cavernous sinus is a kind of erectile organ, in contrast to the upper lacunar sinus, which separates the well innervated hair follicle from the surrounding tissue of the dermis (Andres, 1966).

Longitudinal sections of sinus hairs reveal two distinct compartments: one upper region, continuous with the epidermis with a prominent rete ridge collar

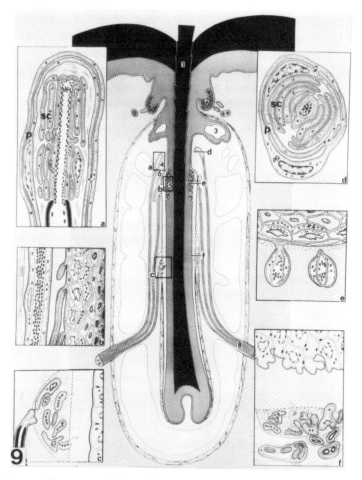

Fig. 9. Diagrammatic representation of the innervation of sinus hair from the lip of rhesus monkey in longitudinal section. 1 Hair shaft, 2 rete ridge collar of the epidermis with Merkel nerve endings, 3 sebaceous gland, 4 small lamellated corpuscles, 5 thickening of the hair follicle with Merkel nerve endings, 6 lanceolate nerve endings, 7 Ruffini-like nerve endings, 8 nerve fascicle. A. Detail of a lamellated corpuscle in longitudinal, and D in cross section. (*) Nerve terminal, sc - cytoplasmic lamellae of the terminal Schwann cell, p - perineurial capsule. B. Detail of lanceolate and Merkel nerve endings in longitudinal, and E in cross section. C. Detail of pilo-Ruffini corpuscle in longitudinal, and F in cross section. Modified from Halata and Munger (1980b).

surrounding the hair shaft (Halata and Munger, 1980a), and a lower region, which is completely surrounded by a sheath of connective tissue. This lower region contains the excretory duct opening of the sebaceous gland, and, in the depths of the sinus, a characteristic reduction in diameter of the hair follicle just below the gland can be seen. At a deeper level, the diameter widens again. This thickening is called "Haarwulst" (Halata, 1975). Sensory nerve endings in the connective tissue are present at the angle between the sebaceous gland and the follicle epithelium. By contrast, intraepithelial nerve endings are present in the region of the "Haarwulst". All kinds of mechanoreceptors are present, from the slowly adapting type to the rapidly adapting ones (Gottschaldt et al., 1973). The electron microscopic appearance of sinus hairs has been described in the rat (Andres, 1966), the sea lion (Stephens et al, 1973), the cat, the mole, the mini-pig (Halata, 1975) and the rhesus monkey (Halata and Munger, 1980a) in detail.

Immediately beneath the sebaceous gland free nerve endings can be found. In the connective tissue, close to the "Haarwulst" and below the sebaceous gland, there are small *lamellated* corpuscles and *lanceolate* nerve endings. The latter are rapidly adapting receptors. In the same region, but in the epithelium of the hair follicle, a collar is formed by up to 2,000 *Merkel* endings. The opposing side of the hair follicle epithelium reveals lanceolate nerve endings, arranged in a pallisade manner (Andres, 1966). The deepest third of the hair follicle is innervated by *Ruffini* corpuscles, intertwined with the glassy membrane of the dermal connective tissue, which surrounds the sinus hair. These Ruffini endings are also named *pilo-Ruffini complexes* (Biemesderfer et al., 1978; Halata and Munger, 1980a).

Thus, each sinus hair is equipped with more than 2,000 different mechano-receptors of rapid or slow adaptation characteristics which register pressure, tension or pain. It therefore comes as no surprise that these sinus hairs are distributed at exposed body parts. In animals they are present around the mouth, emerging from the lips, and at the volar aspect of the wrist and the ankle. All mammals are equipped with these highly differentiated mechanoreceptive complexes, only humans have no sinus hairs.

2. 3. Guard (contour) and vellus hairs as mechanoreceptive complexes (Fig. 10-12)

All the guard (contour) and many vellus hairs are touch organs (Fig. 10). Like the sinus hairs, the region just below the sebaceous gland is of importance for the function as a tactile sense organ. The majority of mechanoreceptors are arranged in such a way that they touch the epithelium of the hair follicle and its connective tissue sheath (Halata, 1979, 1980; Hashimoto, 1979). Like the blood sinus in sinus hairs, the hair follicle of guard and vellus hairs is separated from the surrounding connective tissue of the dermis by one or two spirally twisted arteries. Often, the sebaceous gland is bell-shaped and girdles the hair follicle. Guard hairs are mainly present in the cranial body part (head, neck, upper extremities and shoulders).

Most *guard* hairs (Fig. 11) are equipped with Merkel endings, free nerve endings, palisades of lanceolate nerve endings and pilo-Ruffini complexes (Biemesderfer et al., 1978; Munger and Halata, 1983). Unlike sinus hairs, the total number of nerve endings is small: it varies between 50 - 200, depending on the size of the follicles.

The *Merkel* endings are distributed in the thickening of the follicle's epithelium, the so-called "Stehkragen" (Horstmann, 1957). A minimum of 4 and, in large follicles, a maximum of 12 Merkel endings were counted. In human guard hairs, Merkel endings are often situated in the connective tissue sheath of the dermis (Fig. 12; Orfanos, 1967; Halata, 1979).

Free nerve endings are arranged spirally, running in the connective tissue just below the sebaceous gland. Their afferent axons are predominantly myelinated and measure 1 - 2 µm in diameter. Palisades of *lanceolate* endings are orientated radially to the epithelium of the hair follicle and contact its basal lamina. Each terminal lancette is covered by lamellae of Schwann cells in a

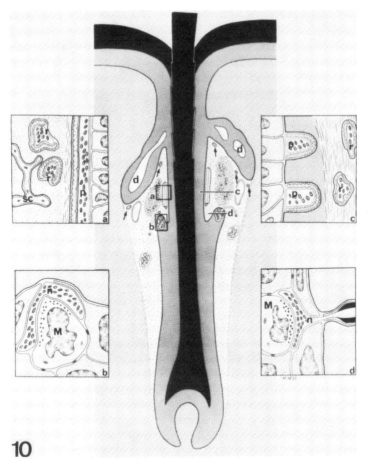

Fig. 10. Diagrammatic representation of the innervation of guard hair from the lip of didelphis in longituinal section. In the angle between the sebaceous gland (d) and hair follicle there are free nerve endings (arrows), pilo-Ruffini corpuscles (a) and lanceolate nerve endings (p). The protrusion of the hair follicle contains Merkel nerve endings (M). A. Detail of Ruffini terminals and lanceolate nerve endings in longitudinal, and C in cross section. B. Detail of Merkel nerve endings in longitudinal, and D in cross section. n - Nerve terminal of Merkel nerve ending, sc - cytoplasmic lamellae of Schwann cell in pilo-Ruffini corpuscle, r - Ruffini nerve terminal.

sandwich-like fashion. Between 20 and 200 lancettes were counted depending on the size of the follicle. The myelinated afferent axons (diameter about 3-5 μm) branches into up to 12 terminal lancettes.

The so-called *pilo-Ruffini complexes* (Biemesderfer et al., 1978) are also situated in the connective tissue below the sebaceous gland. The terminal axons are anchored between the collagen fibres, which girdle the follicle. The myelinated afferent axons, 3 - 8 per hair follicle, reveal a diameter of about 3 - 5 μm.

Some *vellus* hairs are associated with free nerve endings, pilo-Ruffini complexes and lanceolate nerve endings (Munger and Halata, 1983, 1984). However, the most are only equipped with one or two different mechanoreceptors, and only a few lack any sensory innervation.

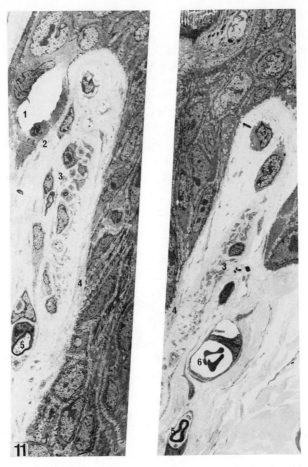

Fig. 11. Guard hair from the skin of the lip of kowari in longitudinal section. 1 Sebaceous gland, 2 free nerve endigs, 3 pilo-Ruffini corpuscle, 4 lanceolate nerve ending, 5 myelinated afferent axon of the pilo-Ruffini corpuscle, 6 capillary. Magnification 2,000x.

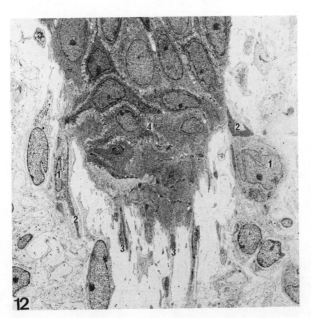

Fig. 12. Merkel nerve endings in the dermis of the hairy skin of human arm. 1 Merkel cell, 2 nerve terminal of Merkel nerve ending, 3 lanceolate nerve endings, 4 hair follicle. Magnification 2,500 x.

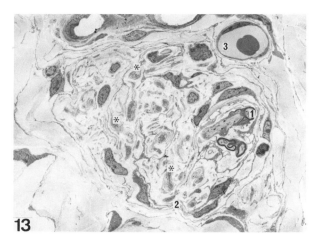

Fig . 13. Ruffini corpuscle from the mini-pig snout in cross section. 1 Myelinated afferent axon, (*) nerve terminals, 2 perineural cells, 3 capillary. Magnification 1,700 x.

3. Ruffini corpuscles (Fig. 13 and 14)

Ruffini corpuscles are mechanoreceptors of the connective tissue. They are present in the dermis and in the joint capsules (Halata, 1988). Ruffini corpuscles have nothing to do with the mechanoreceptive complexes of the skin, except the pilo-Ruffini complex. It is only described in this text for systematic reasons.

In the glabrous skin of the pig, Ruffini corpuscles are present in the dermis often close to the insertion zones of the mimic muscles. In the corpus and glans penis of man and goat, Ruffini corpuscles are distributed in the dense connective tissue, and in the prepuce close to the smooth muscle cells of the tunica dartos (Halata, 1988).

In hairy skin, Ruffini corpuscles are revealed in the reticular layer of human scalp irrespective of whether hair follicles are present or not (Halata and Munger,

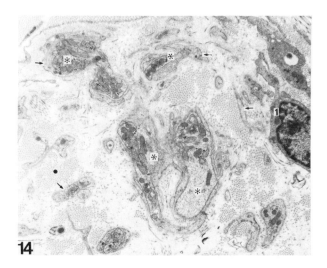

Fig . 14. Detail from Ruffini corpuscle from the mini-pig snout in cross-section. (*) Nerve terminals, (arrows) lamellae of the terminal Schwann cell, 1 perineural cell. Magnification 10,000 x.

1981a). Other parts of hairy skin show Ruffini corpuscles as pilo-Ruffini complexes (Biemesderfer et al., 1978) adjacent to hair follicles of guard hairs and eye lashes (Halata and Munger, 1980a, 1983; Munger and Halata, 1983, 1984). In the sinus hairs of monkeys (Halata and Munger, 1980b) they are present at the connective tissue sheath of the hair follicle.

Typical Ruffini corpuscles (Fig. 13 and 14) are ensheathed by a perineural capsule, which reveals the shape of a cylinder open at both poles. Through these openings collagen fibers of the surrounding dermis enter the corpuscle at one pole and leave it at the other. Mostly, two or even more cylinders fuse at different angles to form a single Ruffini corpuscle. The terminal nerve endings are anchored between the fibrils. These terminal axons are unmyelinated branches of the myelinated afferent axon (diameter about 5 µm) which enters the corpuscles either from one pole (Chambers et al., 1972; Chouchkov, 1978) or from a longitudinal side. As a rule, one cylinder is supplied by one myelinated afferent axon, but it may branch in one cylinder to form two or more myelinated daughter axons.

The structure of Ruffini corpuscles is correlated with the structure of the surrounding connective tissue. On the one hand, corpuscles of the hair follicle lack a capsule, on the other hand, Ruffini corpuscles present in joint capsules reveal a capsule which is the more differentiated the more the surrounding connective tissue is organized (Strasmann et al., 1987).

Ruffini corpuscles are slowly adapting tension-receptors (Chambers et al., 1972; Iggo and Gottschaldt, 1974; Biemesderfer et al., 1978).

Acknowledgements: The author thanks Ms B. Knuts, Mr St. Schillemeit and Ms T. Coellen for technical assistance, and Mr St. Fletcher and Dr T. Strasmann for translating the manuscript. Supported by DFG (HA 1194/3-1)

REFERENCES

Andres, K. H., 1966, Ueber die Feinstruktur der Rezeptoren an Sinushaaren, *Z. Zellforsch.*, 75:339-365

Andres, K. H., and von Düring, M., 1973, Morphology of cutaneous receptors, in: "Handbook of sensory physiology," H. Autrum, R. Jung, W.R. Loewenstein, and D. M. MacKay, eds., Springer, Berlin, New York

Andres, K. H., and von Düring, M., 1984, The platypus bill. A structural and functional model of a pattern-like arrangement of different cutaneous sensory receptors., in: "Sensory Receptor Mechanisms," W. Hamann and A. Iggo, eds., World Scientific Publ.Co., Singapore

Biemesderfer, D., Munger, B. L., Binck, J,. and Dubner, R., 1978, The pilo-Ruffini complex: a non-sinus hair and associated slowly-adapting mechanoreceptor in primate facial skin, *Brain Res.*, 142:197-222

Boeke, J., 1933, Innervationsstudien: II.Ueber Bau und Entwicklung des Eimerschen Organs in der Schnauze des Maulwurfs *(Talpa europaea), Zschrft. mikr. anat. Forsch.*, 33:47-90

Bohringer, R. C., 1981, Cutaneous receptors in the bill of the platypus *(Ornithorhynchus anatinus), Aust. Mammal.*, 4:93-105

Botezat, E., 1903, Ueber epidermoidalen Tastapparate in der Schnauze des Maulwurfs und anderer Saeugetiere mit besonderer Beruecksichtigung derselben fuer die Phylogenie der Haare, *Arch. mikrosk. Anat.*, 61:730-764

Buisseret, C., Bauchot, R. and Allizard, F., 1976, L'equipement sensoriel de la trompe du Desman des Pyrenees, *(Galemys pyrenaicus, Insectivores Talpidae)* - Etude en microscopie electronique, *J. Microsc. Biol. Cell.*, 25:259-264

Cauna, N., 1980, Fine morphological characteristics and microtopography of the free nerve endings of the human digital skin, *Anat. Rec.*, 198:643-656

Chambers, M. R., Andres, K. H., von Düring, M., and Iggo, A., 1972, The structure and function of the slowly adapting type II mechanoreceptor in hairy skin, *Quart. J. Experiment. Physiol.*, 57:417-445

Chouchkov, N. C., 1978, Cutaneous Receptors, *Adv. Anat. Embryol.*, 54:1-86

Eimer, T., 1871, Die Schnauze des Maulwurfes als Tastwerkzeug, *Arch. mikrosk. Anat.*, 7:181-201

Fitzgerald, M. J. T., Saaid, A. H., and Jennings, S. G., 1984, Experimental behaviour of touch domes (Haarscheiben), *Acta anat. (Basel)*, 120:25

Gottschaldt, K.-M., Iggo, A., and Young, D. W., 1973, Functional characteristics of mechano-receptors in sinus hair follicles of the cat, *J.Physiol. (Lond.)*, 235:287-315

Halata, Z., 1970, Zu den Nervenendigungen (Merkelsche Endigungen) in der haarlosen Nasenhaut der Katze, *Z. Zellforsch.*, 106:51-60

Halata, Z., 1971a, Innervation der unbehaarten Nasenhaut des Maulwurfs *(Talpa europaea)*. I. Intraepidermale Nervenendigungen, *Z. Zellforsch.*, 125:108-120

Halata, Z., 1971b, Innervation der unbeahaarten Nasenhaut des Maulwurfs *(Talpa europaea)*. II. Innervation der Dermis (einfache eingekapselte Koerperchen), *Z. Zellforsch.*, 125:108-120

Halata, Z., 1975, The mechanoreceptors of the mammalian skin. Ultrastructure and morphological classification, *Adv. Anat. Embryol.*, 50:1-77

Halata, Z., 1979, Spezifische Innervation, in: "Haar und Haarkrankheiten," C. E. Orfanos, ed., Gustav Fischer Verlag, Stuttgart, New York

Halata, Z., 1980, Sensory innervation of various hair follicles, in: "Linnean Society Symposium Skin of Vertebrates," R. I. C. Spearman and P. A. Riley, eds. , Acad. Press, London

Halata, Z., 1988, Ruffini corpuscle - a stretch receptor in the connective tissue of the skin and locomotion apparatus, in: "Progress in Brain Research, Vol 74, Transduction and Cellular Mechanisms in Sensory Receptors," W. Hamann and A. Iggo, eds., Elsevier, Amsterdam, New York

Halata, Z., and Munger, B.L., 1980a, The sensory innervation of primate eyelid, *Anat. Rec.*, 198:657-670

Halata, Z., and Munger, B.L., 1980b, Sensory nerve endings in Rhesus monkey sinus hairs, *J. Comp. Neurol.*, 192:645-663

Halata, Z., and Munger, B.L., 1983, The sensory innervation of the primate facial skin. II.Vermilion border, *Brain Res. Rev.*, 5:81-107

Hartschuh, W., Weihe, E., and Reinecke, M., 1986, The Merkel cell, in: "Biology of the Integument, Vol. 2, Vertebrates," J. Bereiter-Hahn, A. G. Matoltsy, and K. S. Richards, eds., Springer Verlag, Berlin, Heidelberg

Hashimoto, K., 1979, Nerven- und Blutversorgung des Haarfollikels, in: "Haar und Haarkrankheiten," C. E. Orfanos, ed., Gustav Fischer Verlag, Stuttgart, New York

Hensel, H., 1973, Cutaneous Thermoreceptors, in: "Hanbook of Sensory Physiology. Somatosensory system," A. Iggo, ed., Springer-Verlag, Berlin, Heidelberg

Horstmann, E., 1957, Die Haut, in: "Handbuch der mikroskopischen Anatomie des Menschen," von Moellendorf, ed., Springer, Berlin, Heidelberg

Iggo, A., and Gottschaldt, K.-.M., 1974, Cutaneous mechanoreceptors in simple and complex sensory structures, in: "Symposium Mechanoreception," Rheinisch-Westfaelische Akademie der Wissen-schaften, ed., Westdeutscher Verlag

Iggo, A., and Muir, A. R., 1969, The structure and function of a slowly adapting touch corpuscle in hairy skin, *J. Physiol. (Lond.)*, 200:763-796

Ito, S., and Winchester, R. J., 1963, The fine structure of the gastric mucosa in the bat, *J. Cell Biol.*, 16:541-578

Kadanoff, D., 1928, Ueber die intraepithelialen Nerven und ihre Endigungen beim Menschen und bei den Säugetieren, *Z. Zellforsch.*, 7:553-576

Laczko, J., and Levai, G., 1975, A simple differential staining method for semi-thin sections of ossifying cartilage and bone tissue embedded in epoxy resin, *Mikroskopie*, 31:1-4

Loo, S. K., and Halata, Z., 1985a, The sensory innervation of the nasal glabrous skin in the short-nosed bandicoot *(Isodoon macrourus)* and the opossum *(Didelphis virginiana)*, *J. Anat. (Lond.)*, 143:167-180

Loo, S. K., and Halata, Z., 1985b, Die Ultrastruktur sensibler Nervenendigungen in der unbehaarten Haut des Possums und des amerikanischen Opossums, *Verh. Anat. Ges.*, 79:507-508

Luft, J. H., 1961, Improvements in epoxy resin embedding methods, *J. biophys. biochem. Cytol,* 9:409-414

Mann, S. J., 1968, The tylotrich (hair) follicle of the American opossum, *Anat. Rec.*, 160:171-180

Meissner, G., 1853, "Beitraege zur Anatomie und Physiologie der Haut," Leopold Voss, Leipzig

Melaragno, H. P., and Montagna, W., 1953, The tactile hair follicles in the mouse, *Anat. Rec.*, 115:129-150

Merkel, F., 1875, Tastzellen und Tastkoerperchen bei den Hausthieren und beim Menschen, Arch. mikr. Anat., 11:636-652

Mojsisovics, 1876, Ueber die Nervenendigungen in der Epidermis der Saeuger, II. S. B. Akad. wiss. Wien, math-nat. Kl., 73

Munger, B. L., and Halata, Z., 1983, The sensory innervation of the primate facial skin. I.Hairy skin, Brain Res. Rev., 5:45-80

Munger, B. L., and Halata, Z., 1984, The sensorineural apparatus of the human eyelid, Am. J. Anat., 170:181-204

Novotny, G. E. K., and Gommert-Novotny, E., 1988, Intraepidermal nerves in human digital skin, Cell Tiss. Res., 254:111-117

Orfanos, C., 1967, Elektronenmikroskopischer Nachweis epithelioneuraler Verbindungen am Haarfollikelepithel des Menschen, Arch. klin. Derm., 228:421-429

Pease, D. C., and Quilliam, T. A., 1956, Electron microscopy of the Pacinian corpuscle, J. biophys. biochem. Cytol., 3:331-342

Pinkus, F., 1902, Ueber einen bisher unbekannten Nervenapparat am Haarsystem des Menschen: Haarscheiben, Derm. Z., 9:465-469

Pinkus, F., 1905, Ueber Hautsinnesorgane neben dem menschlichen Haar (Haarscheiben) und ihre vergleichend-anatomische Bedeutung, Arch. mikr. Anat., 65:121-179

Quilliam, T. A., 1966a, The moles sensory apparatus, J. Zool. (Lond.), 149:76-88

Quilliam, T. A., 1966b, Unit design and array patterns in receptor organs, in: "Touch, Heat and Pain. A Ciba Foundation Symposium," A. V. S. de Reuck and J. Knight, eds., Churchill, London

Reynolds, E. S., 1963, The use of lead citrate at high pH as an electron-opaque staining in electron microscopy, J. Cell Biol., 17:208-212

Smith, K. R., 1967, The structure and function of Haarscheibe, J. Comp. Neurol., 131:459-474

Spaethe, A., 1984, Eine Modifikation der Silbermethode nach Richardson für die Axonfärbung in Paraffinschnitten, Verh. Anat. Ges., 78:101-102

Stephens, R. J., Beebe, J. J., and Poulter, T. C., 1973, Innervation of the vibrissae of the california sea lion, Zalophus californianus, Anat. Rec., 176: 421-442

Straile, W. E., 1960, Sensory hair follicles in mammalian skin: the tylotrich follicle, Amer. J. Anat., 106:133-148

Straile, W. E., 1961, The morphology of tylotrich follicles in the skin of the rabbit, Am. J. Anat., 109:1-13

Straile, W. E., and Mann, S. M., 1972, Discharges in neurons that innervate specific groups of tactile receptors in mice, Brain Res., 42:89-102

Strasmann, T., Halata, Z., and Loo, S. K., 1987, Topography and ultrastructure of sensory nerve endings in the joint capsules of the Kowari (Dasyuroides byrnei), an Australian marsupial, Anat. Embryol., 176:1-12

van Horn, R. N., 1970, Vibrissae structure in the rhesus monkey, Folia primat., 13:241-285

Vincent, S. B., 1913, The tactile hair of the white rat, J. Comp. Neurol., 23: 1-36

Wagner, R., and Meissner, G., 1852, Ueber das Vorhandensein bisher unbekannter eigenthümlicher Tastkörperchen (corpuscula tactus) in den Gefühlswärzchen der menschlichen Haut und über die Endausbreitung sensitiver Nerven, Nachr. G.-A.-Univers. königl. Ges. Wiss. Göttingen, 2:17-30

Walter, P., 1960/61, Die sensible Innervation des Lippen- und Nasenbereiches von Rind, Schaf, Ziege, Schwein, Hund und Katze, Z. Zellforsch., 53:394-410

3

TOPOGRAPHY AND ULTRASTRUCTURE OF GROUP III AND IV NERVE TERMINALS OF THE CAT'S GASTROCNEMIUS-SOLEUS MUSCLE

Monika von Düring and Karl Hermann Andres

Ruhr-Universität Bochum, Institut für Anatomie/
Neuroanatomie, Bochum, F.R.G.

INTRODUCTION

Topography, structure and values of muscle spindles and Golgi tendon organs within the gastrocnemius-soleus muscle of the cat are well documented (Swett et al., 1960; Chin et al., 1962; Boyd and Davey, 1968; Ariano et al., 1973). More than two thirds of thick myelinated nerve fibers in a muscle nerve innervate these proprioreceptors (Boyd and Davey, 1968; Boyd and Smith, 1984). There is much evidence that the receptor rich portion of the striated muscle is rich in oxidative muscle fibers (Gonyea and Ericson, 1978; Botterman et al., 1978; Richmond and Stuart, 1985). In physiological studies of slowly conducting fine afferent units (group III and group IV afferents) in the gastrocnemius-soleus muscle different receptor types are classified: 1. low threshold mechanoreceptors (LTM), 2. receptors that are stimulated by contraction only (ergoreceptors), 3. thermoreceptors and 4. receptors stimulated by ischemic contraction (nociceptors) (Mense and Stahnke, 1983; Kaufman et al., 1984; Mense and Meyer, 1985). Other investigators describe the thin-fiber receptors in the skeletal muscle as polymodal receptors responding to mechanical, chemical and thermal stimulation (Kumazawa and Mizumura, 1977; Kumazawa and Tadaki, 1983; Kumazawa et al., 1983). The functional significance of fine afferent fibers of skeletal muscle receptors for the control of cardiovascular reflexes has been reviewed by Mitchell and Schmidt (1983). Little is known about the morphology of these fine afferents. Light microscopical observations on the sensory innervation of the cat skeletal muscle by thin myelinated (group III) and unmyelinated fibers (group IV) has been reported with silver methods (Stacey, 1969) demonstrating the gross distribution of the fibers. In an ultrastructural study terminals of group III and group IV afferent fibers are described in relation to the surrounding tissue (von Düring et al., 1984). In the following presentation a classification of group III and group IV afferent terminals in the skeletal muscle is proposed.

METHODS

In two cats a complete sympathectomy including the rami communicantes was carried out from L2 to S1 on the left side. The cats were anesthetized with Ketamin hydrochlorid, initial dose 10 -20 mg/kg i.m. After a survival time of 40 days which is sufficient for a complete disappearance of all postganglionic sympathetic fibers (Andres et al., 1985) the animals were anesthetized with pentobartital (40 mg/kg i.p.) for an in situ perfusion via the thoracic aorta with warm Tyrode solution (39°C, pH 7.4) for 30 sec followed by a fixation solution of the temperature (2.5% glutaraldehyde and 1.5% formaldehyde in 0.1 M phosphate buffer pH 7.4, Karnovsky, 1965). The gastrocnemius-soleus muscles of both sides were dissected, the muscle nerves were cut from the tibialis nerve 1 cm before their entrance into the muscle, shifted to phosphate buffer (0.22 M saccharose in 0.1 M phosphate buffer, pH 7.4), surrounded by 4% agarose, and cut in 2.5 mm slices (Andres and von Düring, 1981). Nerves and muscle slices of the right side served as control. The slices were postfixed in 4% OsO₄, photographed in 70% ethanol from both sides for documentation, dehydrated in graded ethanol and embedded in araldite. Blocks taken from the slices were cut in series of alternate semithin and ultrathin sections with a Reichert Om UIII and a LKB Ultrotom III. Semithin sections were stained with toluidine blue (pH 9) and photographed with a Zeiss Photomicroscope II. The ultrathin sections were stained with uranyle-acetate and lead citrate and examined with a Philips 300 electron microscope. Location of nerve fiber counting was done 1 cm before the entrance of the muscle nerves into the muscle. Ultrathin sections of the nerves were taken on aperture grids to allow counting of all unmyelinated fibers at a magnification of 20.000 x in the electron microscope.

RESULTS

Two branches of the tibialis nerve supply the gastrocnemius-soleus muscles, a medial nerve branch enters the proximal part of the medial gastrocnemius muscle and a lateral nerve branch runs to the proximal part of the lateral gastrocnemius muscle and supplies both, the lateral gastrocnemius- and the soleus muscle. The number of myelinated axons in the medial nerve branch contains 800 myelinated fibers and 2.800 unmyelinated axons. 46% (1.250) of the unmyelinated axons (group IV) are sensory. The nerve branch to the lateral gastrocnemius muscle contains 1.200 myelinated axons and 3.000 unmyelinated axons. 43% (1.300) of the unmyelinated axons are afferent. Boyd and Davy (1968) calculated 9% group III afferent nerve fibers in the lateral branch of the lateral gastrocnemius muscle. Together with our results the fine myelinated and unmyelinated afferent nerve fibers represent a quantitative significant sensory component in the skeletal muscle.

The location of the nerve terminals of group III and IV afferent fibers is restricted to the perimysium externum surrounding the whole muscle and the perimysium internum surrounding smaller muscle fiber bundles in contrast to the efferent somatic nerve fiber bundles which enter the endomysium to terminate in motor end-plates at the extrafusal muscle fibers. This fact explains the different composition of most peripheral nerve fiber fascicles (Fig. 1) except those supplying the muscle spindles. The types of group III and of group IV

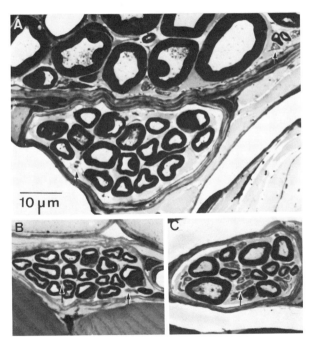

Fig. 1. Semithin sections of 3 nerve fiber bundles in the medial gastrocnemius muscle 56 days after sympathectomy. **A.** Nerve fiber bundle composed of afferent and efferent somatic axons. A small nerve fiber bundle of motor fibers exhibiting few afferent C fibers (arrow) has just separated. **B.** The same bundle some microns deeper. The myelin sheath and the axon diameter have reduced. Afferent C fibers (nervi nervorum) are indicated (arrow). **C.** Nerve fiber bundle composed of afferent nerve fibers different in diameter and myelination. Numerous C fibers (arrow).

endings could be characterized by serial sections. The following criteria are used for the characterization (Andres, 1973; von Düring, 1978): 1. the classification of the afferent axon as group III or group IV fiber, 2. the localization of the nerve ending in relation to the connective tissue compartments of the muscle, 3. the topographical relations between the nerve terminals and their surrounding tissue including blood- and lymphatic vessels, 4. the structural specifications of the terminal axon such as its membrane, the arrangement of the cytoskeletal structures, the presence of clear and dense core vesicles and the receptor matrix. The axon membrane which lack Schwann cell covering is undulated (Fig. 4).

To classify the fine afferent endings we designate terminals of group III fibers as M III (M for muscle) and group IV as M IV. The different location of termination is indicated, e.g., M IV/EN = ending in the endoneurium. Terminations of group III and group IV fibers are observed in three major locations within the muscle: 1. in the adventitia of arteries (M III/A; Fig. 2), arterioles and precapillary segments (M IV/PC; Fig. 3), veins and lymphatic vessels (M IV/V, M III IV/L; Fig. 2), 2. in the endoneurium of larger and smaller nerve fiber bundles (M IV/EN; Fig. 4), and 3. in the connective tissue of the perimysium proper which are often in close relationsship to "pseudo mast cells" (M IV/PM; Andres et al., 1985). All types of muscle receptors observed in the gastrocnemius-soleus muscles represent unencapsulated free terminals. The morphological differences bet-

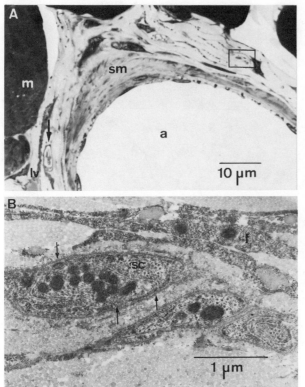

Fig. 2. Example of an M III/A receptor in the soleus muscle of the cat. The inset in **A** represents the receptor in **B**. Large arrow indicates a Remak bundle with nerve terminals of type M IV /LV. m (skeletal muscle fiber), sm (smooth muscle of an artery), lv (lymphatic vessel) starting in the perimysium internum, f (fibrocyte), sc (Schwann cell). Exposed axonal membrane with receptor matrix (small arrows).

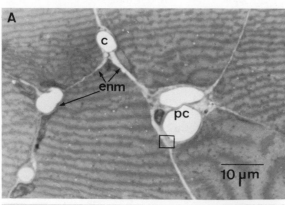

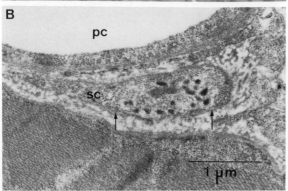

Fig. 3. Example of an M IV/ PC receptor in the soleus muscle of the cat. Inset in **A** represents the receptor in **B**. The terminal axon contains numerous dense core and clear vesicles. pc (precapillary segment), c (capillary), enm (endomysium), sc (Schwann cell), arrows indicate exposed axonal membrane of the terminal.

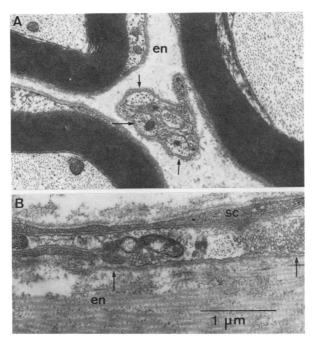

Fig. 4. Example of type M IV /EN terminating in the endoneural connective tissue of nerve fiber fascicles. **A.** Electron micrograph of Fig. 1B shows afferent C fiber terminals in a fascicle with motor nerve fibers. **B.** Type M IV/EN in a perpendicular section of the nerve branch to the medial gastrocnemius muscle just after entering the muscle.

ween group III and group IV afferents are the size of the ending which in case of group III terminals is larger than group IV afferent terminals. The group III terminals contain more mitochondria and a more distinct receptor matrix.

DISCUSSION

The present light- and electron microscopic study confirms our former results that the endomysium of the gastrocnemius-soleus muscle of the cat has no sensory innervation (von Düring et al., 1984). The sensory innervation is restricted to the perimysium internum and externum, to the connective tissue compartment of the vascular bed except the capillary and postcapillary venules, and to the endoneurium of the nerve fiber fascicles *(nervi nervorum)*. All described endings are unencapsulated. The axonal parts which lack the Schwann cell covering may represent the receptive area. They are characterized by an undulated membrane, and the axonal cytoskeletal structures differentiate the receptormatrix, vesicles and mitochondria. Morphological characteristics of the axon terminal, its location and the associated surrounding tissue of the receptive area argue for a specific sensory function. For example the polarized arrangement of the receptive structures may indicate that, depending on the configuration, either stretch or pressure may be transmitted to the transductive areas (Chambers et al., 1972; von Düring et al., 1974; Andres et al., 1987; von Düring and Andres, 1988). In this context the M III/A terminal forming lanceolate endings in the collagenous and elastic fiber system of the vascular adventitia is a good candidate controlling the tension of the vascular wall. Nerve

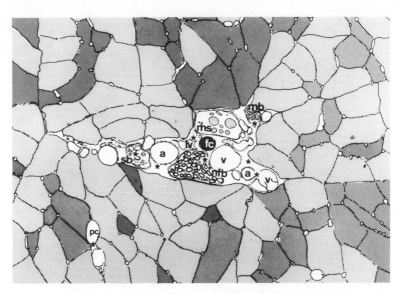

Fig. 5. Schematic diagram of the medial gastrocnemius muscle with the locations of the group III and group IV afferent nerve terminals (asteriks). The perimysium containing arterial (a), venous (v) and lymphatic vessels (lv), precapillary segments (pc), fat cell (fc), different nerve fiber bundles in the perimysium: large mixed fiber bundle (nfb), motor branch (mb), sensory branch (sb), mixed branch (mib) to the muscle spindle (ms). Capillary plexus is situated in the endomysium surrounding single muscle fibers. Skeletal muscle fibers differ in size, staining contrast and capillary density.

terminals which lack these morphological characteristics for mechanoreception or which exhibit a special relation to "pseudo mast cells" containing serotonin or histamine possibly are candidates for nociceptive functions (M IV/PC, Fig 3; M IV/EN, Fig. 4).

A main distribution of the thin myelinated and unmyelinated afferent terminals to the red portion (oxidative compartment) of the rat's sternomastoid muscle was observed by Zenker et al. (1988). In our study we are not able to confirm these observations which may depend on the different function of the triceps surae muscles of the cat.

Acknowledgements: This work was supported by the Deutsche Forschungsgemeinschaft. We thank Prof. Dr. W. Jänig, Kiel, who performed the sympathectomies and we extend our appreciation to Mrs. Luzie Augustinowski and Ms. Kerstin Lukas for their skillful technical assistance.

REFERENCES

Andres, K. H., 1973, Morphological criteria for the differentiation of mechanoreceptors in vertebrates, *in:* "Symposium Mechanorezeption," Schwartzkopff, ed., Abhdlg. Rhein.-Westf. Akad. Wiss., Westdeutscher Verlag, 53:135-152

Andres, K. H., and von Düring, M., 1981, General methods for characterization of brain regions, *in:* "Techniques in neuroanatomical research," Ch. Heym and W. G. Forssmann, eds., Springer, Berlin, Heidelberg, New York: 100-108

Andres, K. H., von Düring, M., Jänig, W., and Schmidt, R. F., 1985, Degeneration patterns of postganglionic fibers following sympathectomy, *Anat. Embryol.*, 172:133-143

Andres, K. H, von Düring, M., and Schmidt, R. F., 1985, Sensory innervation of the Achilles tendon by group III and IV afferent fibers, *Anat. Embryol.*, 172:145-156

Andres, K. H., von Düring, M., Muszynski, K., and Schmidt, R. F., 1987, Nerve fibres and their terminals of the dura mater encephali of the rat, *Anat. Embryol.*, 175:289-301

Ariano, M. A., Armstrong, R. B., Edgerton, V. R., 1973, Hind limb muscle fiber populations of five mammals, *J. Histochem. Cytochem.*, 21:51-55

Botterman, B., Binder, M. D., and Stuart, D. G., 1978, Functional anatomy of the association between motor units and muscle receptors, *Amer. Zool.*, 18:135-152

Boyd, I. A., and Davey, M. R., 1968, "Composition of peripheral nerves," Livingston, Edinburgh

Boyd, A., and Smith, R. S., 1984, The muscle spindle, *in:* "Peripheral Neuropathy, 2nd ed.," P. J. Dyck, P. K. Thomas, E. H. Lombert, and R. Bunge, eds., W. B. Saunders Comp.

Chambers, M. R., Andres, K. H., von Düring, M., and Iggo, A., 1972, Structure and function of the slowly adapting type II mechanoreceptor in hairy skin, *Quart. J. Exp. Physiol.*, 57:417-445

Chin, N. K., Cope, M., and Pang, M., 1962, Number and distribution of spindle capsules in seven hindlimb muscles of the cat, *in:* "Muscle receptors," D. Barker, ed., Hong Kong University Press, Hong Kong: 241-248

Düring, M. von, and Andres, K. H., 1988, Structure and functional anatomy of visceroreceptors in the mammalian respiratory system, *in:* "Transduction and cellular mechanisms in sensory receptors, Progr. Brain Res., Vol. 74," W. Hamann and A. Iggo, eds., Elsevier, Amsterdam, New York, Oxford: 139-154

Düring, M. von, Andres, K. H., and Iravani, J., 1974, The fine structure of pulmonary stretch receptor in the rat, *Z. Anat. Entwickl. Gesch.,* 143:215-222

Düring, M. von, Andres, K. H., and Schmidt, R. F., 1984, Ultrastructure of fine afferent fibre terminations in muscle and tendon of the cat, *in:* "Sensory receptor mechanisms," W. Hamann and A. Iggo, eds., World Scien. Publ. Co., Singapore: 15-23

Düring, M. von, and Miller, M. R., 1978, Sensory nerve endings of the skin and deeper structures, *in:* "Biology of the reptilia, Vol. 9, Neurology," C. Gans, ed., Academic Press, London, New York: 407-441

Gonyea, W. J., and Ericson, G. C., 1978, Morphological and histochemical organization of the flexor carpi radialis muscle in the cat, *Am. J. Anat.*, 148:329-344

Karnovsky, M. F., 1965, A formaldehyde-glutaraldehyde fixative of high osmolarity for use in electron microscopy, *J. Cell Biol.*, 27:137A

Kaufman, M. P., Rybicki, K. J., Waldrop, T. G., and Ordway, G. A., 1984, Effect of ischemia on responses of group III and IV afferents to contraction, *J. Appl. Physiol.*, 57:644-650

Kumazawa, T., and Mizumura, K., 1977, Thin-fibre receptors responding to mechanical, chemical, and thermal stimulation in the skeletal muscle of the dog, *J. Physiol. (Lond.)*, 273:179-194

Kumazawa, T., and Takadi, E., 1983, Two different inhibitory effects on respiration by thin-fiber muscular afferents in cats, *Brain Res.*, 272:364-367

Kumazawa, T., Takaki, E., Mizumura, K., and Kim, K., 1983, Post-stimulus facilitatory and inhibitory effects on respiration induced by chemical and electrical stimulation of thin-fiber muscular afferents in dogs, *Neurosci. Lett.*, 35:283-287

Mense, S., and Stahnke, M., 1983, Responses in muscle afferent fibres of slow conduction velocity to contractions and ischaemia in the cat, *J. Physiol. (Lond.)*, 342:383-397

Mense, S., and Meyer, H., 1984, Different types of slowly conducting afferent units in cat skeletal muscle and tendon, *J. Physiol. (Lond.)*, 363:403-417

Mitchell, J. H., and Schmidt, R. F., 1983, Cardiovascular reflex control by afferent fibers from skeletal muscle receptors, *in:* "Handbook of Physiology, Cardiovascular system III," Williams and Wilkins Comp., Baltimore, Maryland: 623-658

Richmond, F. J. R., and Stuart, D. G., 1985, Distribution of sensory receptors in the flexor carpi radialis muscle of the cat, *J. Morph.*, 183:1-13

Stacey, M. J., 1969, Free nerve endings in skeletal muscle of the cat, *J. Anat.*, 105:231-254

Swett, J. E., and Eldred, E., 1960, Distribution and numbers of stretch receptors in medial gastrocnemius and soleus muscles of the cat, *Anat. Rec.*, 137:461-473

Zenker, W., Sandoz, P. A., and Neuhuber, W., 1988, The distribution of anterogradely labeled I-IV primary afferents in histochemically defined compartments of the rat's sternomastoid muscle, *Anat. Embryol.*, 177:235-243

4

NECK MUSCLE AFFERENTS: THE SENSORY INNERVATION OF THE

STERNOMASTOID MUSCLE OF THE RAT

Wolfgang Zenker

Anatomisches Institut, Universität Zürich-Irchel
Switzerland

INTRODUCTION

The proprioceptive information from the neck obviously plays an important role for the coordination of body, head and eye movements. The contribution of muscle afferents to the sensory input from the neck gained increasing attention in recent years. The question whether neck muscles, in the light of their potentially particular afferent role, differ in their sensory innervation from the majority of other muscles, is being discussed lively. A clear answer will remain pending until sufficient data from all neck muscles will be available and can be compared with those from other muscles. Certainly, it has been shown that there is an exceptionally high density of muscle spindles in different neck muscles (Voss, 1958; Richmond and Abrahams, 1975; Bakker and Richmond, 1982; Pfister and Zenker, 1984). However, the fact cannot be neglected that, e.g., in the rat, there are considerable differencees between different neck muscles: we calculated 200 spindles per gram muscle weight for rectus capitis minor, but only 53/g for rectus capitis major and 86/g for sternomastoid muscles (Bank-oul, unpublished results). Moreover, the content of other types of sensory receptors has to be taken into consideration.

I shall now present anatomical data on the content and regional distribution of afferent fibers and sensory receptors in the sternomastoid muscle of the rat. These data may yield a basis for chapter 14 of this volume, where the central projections of neck muscle afferents are described in detail.

FIBER TYPE COMPARTMENTS OF THE STERNOMASTOID MUSCLE

The anatomical location of this muscle makes it evident that it plays an important role within the ensemble of muscles concerned with the precise adjustment of head position. For two reasons the sternomastoid muscle represents a particularly suitable model for studying the afferent innervation of a neck muscle: a) near its entrance into the muscle, the sternomastoid nerve is

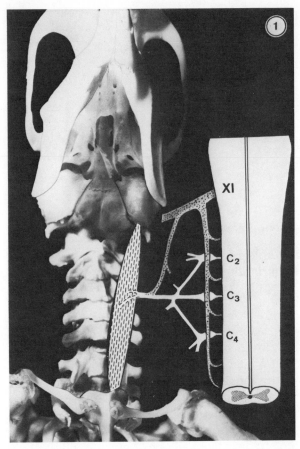

Fig. 1. The anatomical situation of the sternomastoid muscle in the rat and its innervation. Dotted: motor root; white: sensory root of the sternomastoid nerve.

formed by two well separable roots, a motor one (from the accessory nerve) and a sensory one (from the cervical plexus; Fig. 1). Thus, separate analyses of the two roots are possible and retrograde labeling experiments can be performed separatly in the motor and in the sensory root; b) already macroscopically, this muscle can be subdivided into a white and a red portion. Histochemically, the white portion of the muscle consists predominantly of IIb (fast glycolytic twitch) fibers, whereas the red portion, besides Ia (slow oxidative twitch) fibers contains mainly IIa (fast oxidative-glycolytic twitch) fibers (Dulhunty and Dlutowski, 1979; Gottschall et al. 1980; Luff, 1985). (Fig. 2). This conspicuous compartmentalisation of the muscle enables us to search for potential differences in the afferent and efferent innervation possibly occuring in these two histochemically different muscle portions.

THE EXTRAMUSCULAR SECTION OF THE STERNOMASTOID NERVE

Before entering the muscle, the sternomastoid nerve consists of myelinated and of unmyelinated fibres. Electronmicroscopically, about 260 myelinated and 320 unmyelinated axons have been counted. Using acetylcholinesterase (AChE) histochemistry, (Zenker and Hohberg, 1973; Gottschall et al. 1980), 59% of the

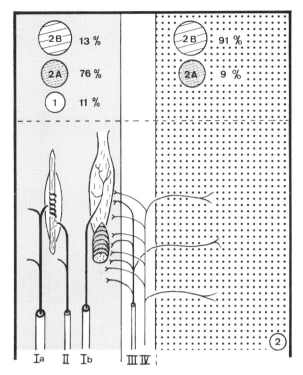

Fig. 2. Upper section of the figure: Distribution of muscle fiber types in the red (finely dotted) and white (coarsely dotted) portions of the sternomastoid muscle. Lower section of the figure: Distribution of receptors and afferent fiber types in the red versus white portion.

myelinated axons showed high AChE activity and therefore were regarded as motor fibers, whereas 41% of the fibers showed low AChE levels and were classified as sensory fibers (type I to III).The differentiation of myelinated fibers into motor and sensory elements on the basis of AChE-histochemistry is especially well supported in our case as almost all myelinated fibers of the motor root (accessory nerve) showed high enzyme activity, whereas those of the sensory root (cervical plexus) at the same time displayed only very low enzyme activity.

The population of *unmyelinated fibers* contains postganglionic sympathetic axons as well as sensory type IV axons. We tried to find out the proportion of unmyelinated sensory fibers by comparison of the number of unmyelinated fibers in cross sections of nerves of four normal rats with those of three animals after extensive cervical sympathectomy. 42 days after surgery, the nerves of the sympathectomized animals showed a highly significant 40% loss of unmyelinated axons. If it is justified to conclude that the remaining unmyelinated fibers are to be regarded as sensory (type IV) axons, this population amounts to 60% of the total number of C-fibers. (Fig. 3)

However, I would not like to withhold that I have some reservations about these calculations: it cannot be excluded that our surgical sympathectomy may be associated with unforeseeable biological effects. I refer to findings of several authors (Varon and Bunge, 1978; Richardson and Ebendal, 1982; Kessler et al.

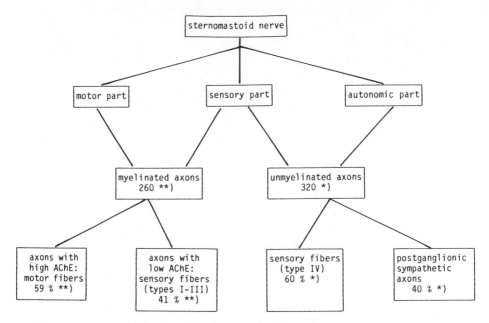

Fig. 3. Axon populations in the rat sternomastoid nerve. Exemplary counts on electronmicroscopic sections (asterisk) and on acetylcholinesterase preparations (double asterisks) where sensory and motor axons could be differentiated (Gottschall et al, 1980). Some of the smallest myelinated fibers probably escaped light microscopic detection, since counts by electronmicroscopy showed slightly higher numbers (from Sandoz and Zenker, 1986, with permission).

1983; Rush, 1984; Shelton and Reichardt, 1984; Abrahamson et al. 1985), from which it can be inferred that denervation may induce sprouting of remaining fibers caused by some growth factors produced by the target tissue or by activated Schwann cells in the nerve. Such an induction of sprouting processes could in fact falsify our calculations (Sandoz and Zenker, 1986).

SPINAL GANGLIA AND SENSORY UNITS

The afferent innervation of the sternomastoid muscle is provided by spinal ganglia C2 - C4 with a main contribution from C3 (Gottschall et al., 1980; Mysicka and Zenker, 1981). A total of about 70 ganglion cells has been found by retrograde tracing (Gottschall et al., 1980). This might be an underestimate, since these data are based on material processed with diaminobenzidine. Given this number of afferent cell bodies, and a number of nearly 300 myelinated and unmyelinated afferent axons in the sternomastoid nerve, the ratio of cell bodies to axons is about 1:4, or probably even less. It has to be left open whether this reflects axonal branching in the sternomastoid nerve, or simply insufficient retrograde labeling of cell bodies. The spinal ganglion cells innervating the sternomastoid muscle are significantly smaller than those innervating proximal leg muscles. A similar relation has been shown for afferents from the splenius muscle of the rat (Pfister and Zenker, 1984). The size of a sensory perikaryon supposedly is determined by several factors (see chapter 16), one of which might be the size of the sensory unit which is probably small in neck muscle afferents.

THE INTRAMUSCULAR DISTRIBUTION OF THE STERNOMASTOID NERVE

After the sternomastoid nerve has entered the muscle, it splits up into two or more branches, distributing to the red ("red nerve") and white ("white nerve") portion of the muscle. Already macroscopically, it can be observed that the overall nerve supply to the red portion is considerably larger than that to the white portion. In AChE-preparations of the "white nerve" almost all myelinated axons display high enzyme activity, the axon diameters showing an unimodal distribution (most probably only alpha-axons) with a mean of 3,3 μm. On the other hand, in the "red nerve" the myelinated fibers show two populations of axons, those with high, and those with low enzyme activity (Zenker et al., 1988). The latter were regarded as sensory ones and we concluded that sensory myelinated fibers innervate exclusively the red portion of the muscle. Moreover, the population of AChE-positive fibers of the "red nerve", as compared with the "white nerve" show a broader distribution with a bimodal shape, suggesting the content of both, alpha and gamma fibers. These observations are in accordance with the presence of muscle spindles exclusively in the red part, first reported by Bottermann et al.(1978) and later confirmed by Gottschall et al. (1980) and Brichta et al. (1987).

MUSCLE SPINDLES AND MUSCLE COMPARTMENTALISATION

The topographic relation of muscle spindles to those regions of muscles containing mainly type IIa and type I muscle fibers has been shown also in other muscles in rats and cats (Yellin, 1969; Richmond and Abrahams, 1975; Gonyea and Ericson, 1977; Maier, 1979; Brichta et al. 1987). The accumulation of muscle spindles in red muscle compartments fits into the concept of a particularly precise reflex regulation in muscle regions rich in I and IIa muscle fibers, which generally belong to smaller motor units. Such units are being most readily recruited and therefore are being most frequently used during less than maximal effort endeavours (Henneman and Olsen, 1965; Yellin, 1969). In muscles showing an even distribution of the various fiber types, no preferential location of muscle spindles was found.

GOLGI TENDON ORGANS

In a previous study (Zenker et al., 1988), it has been demonstrated that in the sternomastoid muscle of the rat Golgi tendon organs are attributed to muscle fibers of the red part only. This is in accordance with observations of Hinsey (1927), who found an association between spindles and tendon organs in skeletal muscles in general. In our case, this functional linkage between muscular tension detectors clearly is restricted to the red fiber compartment.

FINE MUSCLE AFFERENTS

In recent years, growing attention has been paid to the innervation of skeletal muscles by fine afferents (III and IV-fibers). I am referring to the contribution of von Düring and Andres in this volume, and to previous reports of (Hinsey, 1927; Stacey, 1969; Andres et al., 1985; for review see Mense, 1986). According to these studies, many unmyelinated fibers (IV fibers) as well as nonmyelinated

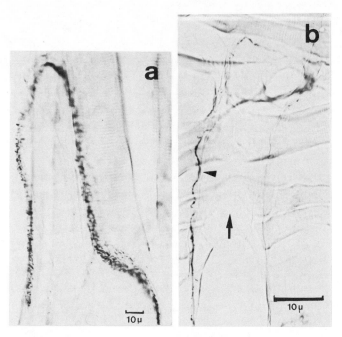

Fig. 4. Anterogradely HRP-labeled afferent fibers within the sternomastoid muscle. **A.** I-fiber; **B.** thin calibre, probably IV-fiber in the perivascular tissue of a venule (B from Zenker et al., 1988, with permission).

branches of myelinated stem fibers (III fibers) are distributed preferentially to the adventitia of blood vessels and to the connective tissue between the muscle fibers. This location is identical with that of sympathetic efferent nerve fibers of the same caliber class. To overcome the difficulty of morphological separation of these different modalities, several authors studying fine muscle afferents performed their investigations in sympathectomized animals (Stacey, 1969; Andres et al., 1985). We used a different method and tried to label selectively muscle afferents by injection of wheatgerm agglutinin-horseradish peroxidase conjugate (WGA-HRP) into spinal ganglia C2 - C4; this was followed by anterograde transport of tracer into the terminals of sensory nerves. After histochemical processing of muscle sections, performed after survival times of 1 - 3 days, reaction product appeared in all types of sensory muscle nerves: in thick axons (I and II afferents), granules were distributed over a band much larger than the size of individual granules (Fig. 4a), whereas in thin axons they were arranged like beads in single lines (Fig. 4b). Thin afferent axons were usually detected running individually or in small groups, sometimes forming plexus around small blood vessels and in the connective tissue of the muscle. There was an uneven distribution of these nerve fibers to the red and to the white portion of the muscle: the maximal labeling of the fibers was found in the red portion, whereas the number of labeled fine fibers was rather sparse in the white portion.

The question must be left open whether a complete labeling of the whole population of fine afferents was achieved in our experiments. The fact that the number of labeled muscle spindles (20 - 22) corresponded to previous counts of muscle spindles in conventionally stained serial sections of the sternomastoid muscle, speaks in favour of a complete labeling of all muscle afferents.

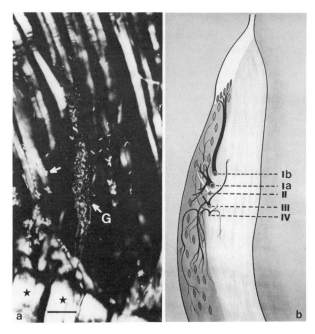

Fig. 5. A. Labeled Golgi tendon organ (G) in the proximal red transition zone, where striated muscle fibers (arrows) of the red portion insert into collagen bundles (asterisks, no cross-striation) of their respective portion of the proximal (mastoid) tendon. Bar: 100 μm. **B.** Schematic representation of the afferent innervation of the sternomastoid muscle of the rat. Dark grey: red portion. Pale grey:white portion. White:cranial tendon. All muscle spindles are located in the red portion. Golgi tendon organs are found in the transition zone only between the red part of the muscle and the cranial tendon. Muscle fibers of the white portion are attached to the tendon cranially of the region where tendon organs occur. The majority of III and IV afferents are also found within the red portion (from Zenker et al., 1988, with permission).

Thus, our results suggest not only muscle spindles and tendon organs being exclusively present in the red portion, but also the majority of III- and IV-afferents (Fig. 5).

The finding of a higher content of substance P in the red as compared to the white portion of the sternomastoid muscle as determined by radioimmuno assay (Zenker et al., 1988), leads us to the assumption that there might be a number of substance P-reactive fibers within the population of IV-axons (cf., Molander et al., 1987).

CONCLUSIONS

The exclusive localisation of muscle spindles and tendon organs, the almost exclusive location of fine calibre afferents and the higher content of substance P within the red portion, indicate this portion to be the "sensory compartment" of the muscle. That means that, besides proprioceptors, also nociceptors, receptors for deep pressure sensations, metabolic and thermoreceptors should be concentrated in this special "sensory" region of the muscle.

REFERENCES

Abrahamson, I. K., Wilson, P. A., and Rush, R. A., 1985, Origin of trophic activity in peripheral nerve, *Neurosci. Lett.*, (Suppl.) 22:334

Andres, K. H., von Düring, M., and Schmidt, R. F., 1985, Sensory innervation of the Achilles tendon by group III and IV afferent fibers, *Anat. Embryol.*, 172:145-156

Bakker, D. A., and Richmond, F. J. R., 1982, Muscle spindle complexes in muscles around upper cervical vertebrae in the cat, *J. Neurophysiol.*, 48:62-74

Bottermann, B. R., Binder, M. B., and Stuart, D. G., 1978, Functional anatomy of the association between motor units and muscle receptors, *Am. Zool.*, 18:135-152

Brichta, A. M., Callister, R. J., and Peterson, E. H., 1987, Quantitative analysis of cervical musculature in rats: Histochemical composition and motor pool organization, *J. Comp. Neurol.*, 255:351-368

Dulhunty, A. F., and Dlutowski, M., 1979, Fiber types in red and white segments of rat sternomastoid muscle, *Am. J. Anat.*, 156:51-62

Gonyea, W. J., and Ericson, G. C., 1977, Morphology and histochemical organization of the flexor carpi radialis muscle in the cat, *Am. J. Anat.*, 148:329-344

Gottschall, J., Zenker, W., Neuhuber, W., Mysicka, W., and Muentener, M., 1980, The sternomastoid muscle of the rat and its innervation. Muscle fiber composition, perikarya and axons of efferent and afferent neurons, *Anat. Embryol.*, 160:285-300

Henneman, E., and Olsen, C. B., 1965, Relations between structure and function in the design of sceletal muscles, *J. Neurophysiol.*, 289:581-598

Hinsey, J. C., 1927, Some observations on the innervation of skeletal muscle of the cat, *J. Comp. Neurol.*, 44:87-195

Kessler, J. A., Bell, W. O., and Black, I. B., 1983, Interactions between the sympathetic and sensory innervation of the iris, *J. Neurosci.*, 3:1301-1307

Luff, A. R., 1985, Dynamic properties of fiber bundles from the rat sternomastoid muscle, *Exp. Neurol.*, 89:491-502

Maier, A., 1979, Occurrence and distribution of muscle spindles in masticatory and suprahyoid muscles of the rat, *Am. J. Anat.*, 155:483-506

Mense, S., 1986, Slowly conducting afferent fibers from deep tissues: neurobiological properties and central nervous actions, *Progr. Sensory Physiol.*, 6:139-219

Molander, C., Ygge, J., and Dalsgaard, C.-J., 1987, Substance P-, somatostatin-, and calcitonin gene-related peptide-like immunoreactivity and fluoride resistant acid phosphatase-activity in relation to retrogradely labeled cutaneous, muscular and visceral primary sensory neurons in the rat, *Neurosci. Lett.*, 74:37-42

Pfister, J., and Zenker, W., 1984, The splenius capitis muscle of the rat, architecture and histochemistry, afferent and efferent innervation as compared with that of the quadriceps muscle, *Anat. Embryol.*, 169:79-89

Richardson, P. M., and Ebendal, T., 1982, Nerve growth activities in rat peripheral nerve, *Brain Res.*, 246:57-64

Richmond, F. J. R., and Abrahams, V. C., 1975, Morphology and distribution of muscle spindles in dorsal muscles of the cat neck, *J. Neurophysiol.*, 38:1322-1339

Rush, R. A., 1984, Immunohistochemical localization of endogenous nerve growth factor, *Nature*, 312:364-367

Sandoz, P. A., and Zenker, W., 1986, Unmyelinated axons in a muscle nerve: Electron microscopic morphometry of the sternomastoid nerve in normal and sympathectomized rats, *Anat. Embryol.*, 174:207-213

Shelton, D. L., and Reichardt, L. F., 1984, Expression of the beta-nerve growth factor gene correlates with the density of sympathetic innervation in effector organs, *Proc. Natl. Acad. Sci. USA*, 81:7951-7955

Stacey, M. J., 1969, Free nerve endings in skeletal muscle of the cat, *J. Anat.*, 105:231-254

Varon, S. S., and Bunge, R. P., 1978, Trophic mechanisms in the peripheral nervous system, *Annu. Rev. Neurosci.*, 1:327-361

Voss, H., 1958, Zahl und Anordnung der Muskelspindeln in den unteren Zungenbeinmuskeln, dem M. sternocleidomastoideus und den Bauch- und tiefen Nackenmuskeln, *Anat. Anz.*, 105:265-275

Yellin, H., 1969, A histochemical study of muscle spindles and their relationship to extrafusal fiber types extrafusal fiber types in the rat, *Am. J. Anat.*, 125:31-45

Zenker, W., and Hohberg, E., 1973, Alphamotorische Nervenfasern, Axonquerschnittsfläche von Stammfaser und Endästen, *Z. Anat. Entw. Gesch.*, 139:163-172

Zenker, W., Sandoz, P.A., and Neuhuber, W., 1988, The distribution of anterogradely labeled I-IV primary afferents in histochemically defined compartments of the rat's sternomastoid muscle, *Anat. Embryol.*, 177:235-243

5

THE ULTRASTRUCTURE OF MECHANORECEPTORS IN THE MUSCULO-SKELETAL SYSTEM OF MAMMALS

Z. Halata and T. Strasmann

Anatomisches Institut der Universität Hamburg
Abteilung für funktionelle Anatomie, Hamburg
F.R.G.

INTRODUCTION

At the insertion zone of muscles, free nerve endings together with small lamellated corpuscles, large Pacini corpuscles, Golgi-tendon organs and muscle spindles form mechanoreceptive complexes which register the sensation of tension of muscle fasciae, aponeuroses and muscle fibers themselves (Barker, 1974). Tendons and fasciae originate in joint capsules as well as in ligaments and in the periosteum. Like in the skin, receptive fields (Halata, 1975) are also present in the musculoskeletal system (Polacek, 1966), controlled by free nerve endings, Ruffini corpuscles and lamellated corpuscles.

All types of sensory nerve endings, those in the muscles and their tendons and aponeuroses as well as those in joint capsules and their ligaments, are sense organs for kinesthesia, the sense of position and movement of the body, and are necessary for the equilibration of movements (Proske et al., 1988).

1. Free nerve endings (Fig. 1, 2 and 3)

Recent investigations have shown that the number of free nerve endings (FNEs) of the joint capsules is remarkably large. At present, there is no method of labelling all FNEs of the joint capsules exactly. However, Langford and co-workers counted sensory axons (C or A ∂-fibers) after sympathectomy (Langford, 1983; Langford and Schmidt, 1983; Langford et al., 1984). They found that about 34% of the articular nerves of the knee joint of the cat are C or A ∂-fibers (Langford, 1983; Langford et al., 1984). Using electron microscopy, hitherto only one type of FNE belonging to the A ∂-fiber group (Fig. 1 and 2) in the joint capsules has been described (Halata, 1977; Halata and Munger, 1980; Halata et al., 1984, 1985). It is present either in the vicinity of blood vessels in the subsynovial layer or periarticularly. Its structure resembles that of nerve terminals in the Ruffini corpuscle. The axons are myelinated and display diameters of about 1 to 3 μm (A ∂-fibers). The nerve terminal is slightly thickened and contains mitochondria, clear vesicles, net-like arranged microfilaments and some microtubuli.

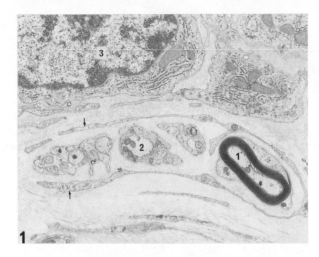

Fig. 1. Free nerve ending from the subsynovial part of the joint capsule of the jaw joint of the mouse. 1, myelinated axon (A-delta fiber); 2, free nerve ending, (arrows) cytoplasmic processes of fibroblasts; 3 B-cells of synovia. From Dreessen et al. (1988), magnification 15,000 x.

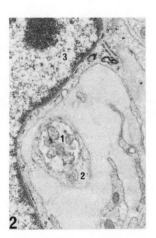

Fig. 2. Free nerve ending from the synovial layer of the joint capsule of the jaw joint of the mouse. 1, Free nerve ending; 2 terminal Schwann cell; 3 fibroblasts. Magnification 20,000 x.

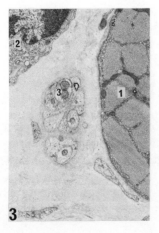

Fig. 3. Free nerve endigs form the epimysium of m. pterygoideus lateralis of the mouse. 1 striated muscle, 2 fibroblast, 3 free nerve endings. Magn. 12,000 x.

Within the connective tissue of muscles (epimysium, perimysium and intramuscular tendons) FNEs have been investigated by von Düring and Andres using electron microscopy (von Düring et al., 1984; Andres et al., 1985). Again, they observed nerve endings belonging to the A ∂-fiber group (diameters about 1 to 4.5 µm) and to the C-fiber group (Fig. 3). Tracing these nerve endings over long distances (several millimeters) von Düring and Andres described five morphological types of A ∂-fibers, located within the adventitia of small vessels or within the peritenonium and terminating as lanceolate or bulboid endings. As these terminals are covered by Schwann cell processes and are not "free" or "naked", Andres et al. (1985) proposed the term "unencapsulated endings".

Sensory nerve endings of the C-fiber group are solely present in connective tissue surrounding blood or lymphatic vessels. Their terminals have been termed "penicillate nerve endings" (Cauna, 1969) and contain not only clear but also granulated vesicles.

The function of these different types of FNEs of muscle tendons and fasciae is still subject to speculation. Based on physiological investigations performed by Kumazawa and Mizumura (1977, 1984), Kumazawa and Perl (1977a, b) and Mense and Meyer (1981), it is reasonable to assume that terminals with close contact to connective tissue might be mechanoreceptors, whereas those in the vicinity of vessels might be nocireceptors, thermoreceptors or chemoreceptors (Andres et al., 1985) or even polymodal receptors (Kamuzawa and Mizumura, 1977, 1984).

2. Ruffini corpuscle (Fig. 4, 5 and 6)

In the musculoskeletal system Ruffini corpuscles are only present in joint capsules. Polacek extensively studied Ruffini corpuscles using light microscopy (Polacek, 1965, 1966) and called them "spray-like endings". He grouped these kinds of corpuscles into five morphological types. The dimension and the presence or absence of a perineural capsule are the criteria for his classification system. With the aid of electron microscopy, it was possible to simplify Polacek's classification. On the basis of ultrastructural studies of joint capsules of several different species (eutherians: rabbit - Goglia and Sklenska, 1969; cat - Halata, 1977; monkey - Halata et al., 1984; humans - Halata et al., 1985; marsupial: different species - Halata et al., 1987; Strasmann et al., 1987; aves: Halata and Munger, 1980) three morphological types of Ruffini corpuscles can be described: corpuscles without any capsule, corpuscles with a capsule formed by connective tissue elements and by perineural cells (Fig. 4), and corpuscles resembling Golgi-tendon organs, with a well-developed perineural capsule (Fig. 5 and 6).

All three types are present in the fibrous layer of the joint capsule. Areas of the capsule which are reinforced by ligaments or muscle tendons, reveal Ruffini corpuscles resembling Golgi-tendon organs. This type displays a perineural capsule of a cylindric shape. Through gaps of the capsule on each pole, collagen fibers run into the corpuscles leaving it at the other pole. The afferent axon is myelinated and shows a diameter of about 4-6 µm. Entering the cylinder, the axon loses its myelin sheath and may branch dichotomally. The nerve terminals are covered incompletely by cytoplasmic lamellae of the terminal Schwann cell and contain accumulations of mitochondria and clear vesicles. Schwann cells anchor the terminals between the collagen fibers by long processes.

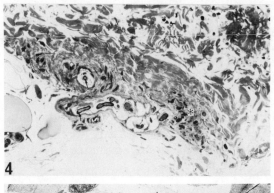

Fig. 4. Ruffini corpuscle from the fibrous layer of the joint capsule of the human knee joint in an oblique section. The afferent axon is myelinated (arrows), the corpuscle (*) is embedded between bundles of collagen fibrils. Semithin section, magnification 350 x.

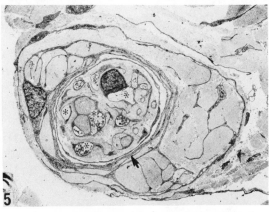

Fig. 5. Cylinder of a Ruffini corpuscle from the fibrous layer of the joint capsule of the knee joint of the macacus rhesus in cross section. The cylinder is ensheathed by a perineural capsule (arrow). Nerve terminals (*) with terminal Schwann cells are anchored between bundles of collagen. Magnification 3,000 x.

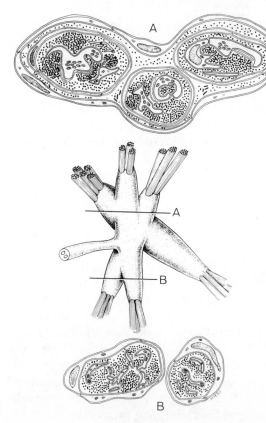

Fig. 6. Diagrammatic representation of a Ruffini corpuscle from the joint capsule of a shoulder joint of the pigeon. The cylinders longitudinal axis is orientated paralelly to the course of the collagen fibrils of the fibrous layer. **A** and **B** represent cross-sectional images through the planes of the diagram of the corpuscle as indicated.

In the fibrous layer of the joint capsule, Ruffini corpuscles with or without a partly perineural and partly fibrocytic capsule are situated superficially as well as deeply. The terminals show a structure similar to those with a well-developed perineural capsule. Apparently, the capsule has only a mechanic function, providing protection against surrounding undirected forces. It is likely, that in case of injuries, iatrogen joint-position alterations, scar formation or ageing of joint capsules a remodelling of Ruffini corpuscles in accordance with the present tissue texture takes place.

Ruffini corpuscles of the joint capsules are slowly adapting tension receptors (Boyd and Roberts, 1953; Boyd, 1954; Eklund and Skoglund, 1960; Skoglund, 1956, 1973; Freeman and Wyke, 1967; McCloskey, 1978; Grigg and Hoffman, 1982, 1984).

3. Golgi-tendon organs (Fig. 7 and 8)

94% of the Golgi-tendon organs (GTOs, Fig. 7 and 8) are present at the muscle-tendon junction, the remaining 6% are situated within the main tendon (Barker, 1974). All tendons investigated reveal these organs, even in the tendinous centre of the diaphragm and at the tendinous intersections of the rectus abdominis muscle (Stoehr jr., 1928). The number of GTOs varies according to the species and the muscle investigated, e.g., slow muscles such as the soleus contain more GTOs than fast muscles such as the m. flexor digitorum brevis or the gastrocnemius muscle (Eldred et al., 1962; Zelena and Hnik, 1963).

Golgi-tendon organs are spindle-shaped corpuscles. The size of the corpuscle correlates with the body size of the species. In humans, the maximum length is 1.6 mm, the maximum diameter about 120 μm (Bridgman, 1970), in the rat the max. length/max. diameter is 580/52 μm (Soukup, 1983; Zelena and Soukup, 1983) and in the cat 500/100 μm (Barker, 1974). Both the muscular

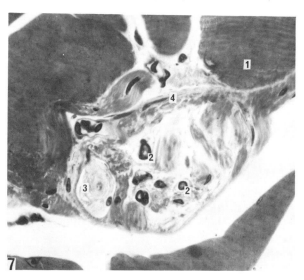

Fig. 7. Golgi tendon organ from the tendon of the m.peroneus longus of the mouse. 1 Muscle cells, 2 myelinated axons, 3 lamellated corpuscle of Pacini type, 4 perineural capsule. Semithin section, Magnification 800 x.

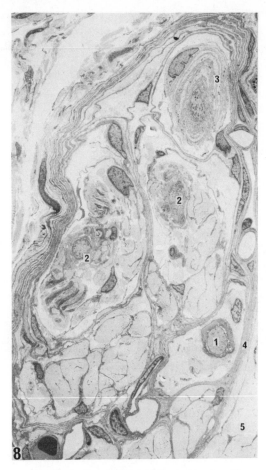

Fig. 8. Golgi tendon organ from the tendon of the m.semitendineus from the kowari. 1 myelinated axon, 2 nerve terminals, 3 small lamelated corpuscle of Pacini type, 4 perineural capsule, 5 tendon. Magnification 2,000 x.

tip and the tendinous tip are pointed. The muscular tip mostly reveals a 25% larger diameter than the tendinous tip (Bridgman, 1968, 1970). Most of the GTOs have a simple spindle-shape (63%), but there are also GTOs with two or three tips (Barker, 1974; Soukup, 1983). The muscular tip contains 3 to 25 muscle fibers, which are fixed to the capsule of the GTO or run into the inner core (Regaud, 1907; Bridgman, 1970; Barker, 1974). Since Cattaneo (1888) and Ruffini (1893a, b), we know that small lamellated corpuscles may be present within the GTOs (Fig. 7 and 8). In the cat 7-15%, in the rat 5-10% of the GTOs reveal these small lamellated corpuscles. Mostly, they are enveloped by the capsule of GTO, but they also are situated on the surface of the tendon organ (Barker, 1974). Apart from these lamellated corpuscles, also free nerve endings are contained in about 6% of the GTOs (Stacey, 1967).

GTOs are composed of several afferent axons and their terminals as well as of terminal Schwann cells, fibrocytes and other connective tissue components, and of perineural cells which form the perineural capsule. The following description is based upon electron microscopic studies of Merrillees (1960, 1962), Schoultz and Swett (1972, 1976), Sklenska (1973), Strasmann and

Halata (1988d) and Nitatori (1988). Three compartments of the GTO are discernible: (1) the perineural capsule and the similarly formed septa emerging from the capsule, (2) the sensory compartment with collagen fibres, fibrocytes, Schwann cells and axons and (3) the connective tissue compartment, filled with collagen fibres, vessels and other connective tissue.

The afferent axons are myelinated and enter the corpuscle from its longitudinal side. The perineurium of the supplying nerve fuses with that of the GTO (Sklenska, 1973). Bridgman (1968) observed two classes of nerve fibers within the GTOs: axons with a diameter of 5-7 µm and those with a diameter of 12-15 µm. Mostly, more than three myelinated axons are present, whereas even six myelinated axons have been described by Sklenska (1973) in the GTOs of the lumbricales muscles in the cat.

After entering the corpuscle, the myelinated axons may branch into myelinated axons again. These daughter axons are surrounded by perineural cells exclusively or by perineural cells, Schwann cells and often also fibroblasts, forming a complete sheath around the axons. These sheaths are part of the capsule or of the perineural septa emerging from the inner side of the capsule. Leaving these sheaths, the axons lose their myelin sheath and enter the sensory compartment. The non-myelinated branches of these afferent axons run between bundles of collagen fibers, branching profusely and anchored by axoplasmatic processes between these fiber bundles. Like the Ruffini corpuscle, the terminal axons contain clusters of mitochondria, clear vesicles and display all other features of terminal axons. Terminal axons are incompletely covered by terminal Schwann cell processes and are often in contact with adjacent collagen fibers.

Myelinated afferent axons of the small lamellated corpuscles in GTOs display a diameter of about 3 to 5 µm, and axons of free nerve endings in the GTO belong to the A ∂- or the C-fiber class.

As is the case with Ruffini corpuscles, the terminal Schwann cells reveal flat cytoplasmatic sheaths, covering the axon terminals. The cell membrane of these Schwann cell processes shows numerous micropinocytotic vesicles. Like the axon terminals, Schwann cell processes encircle collagen fiber bundles, forming "collagen pockets" (Fig. 8).

The capsule of the GTO is formed by several (up to 12) sheaths of flat cells, displaying the same features as perineural cells of nerve fascicles do. Their processes are fused by intermediate and tight junctions (Nitatori, 1988). The cell membrane reveals numerous micropinocytotic vesicles and is covered by a basal lamina. Thin collagen fibers run predominantly longitudinally through the gaps between the cellular sheaths. The capsule cells form septa which are often strengthened by Schwann cells and by fibroblasts. The latter are also termed "septal cells" (Schoultz and Swett, 1972) and resemble the "inner septal cells" of muscle spindles (Ovalle and Dow, 1983). The septa divide the corpuscle into several pockets. Some pockets only contain collagen fibers and cells of connective tissue, others contain collagen fibers and sensory axons terminating between them. Thus, two compartments are discernible: the connective tissue compartment and the sensory compartment respectively. Collagen fibers as well as muscle fibers enter and penetrate the capsule at the pits to reach either the sensory or the connective tissue compartment. The latter seems only to fix the GTO at the muscle-tendon transition zone.

Blood vessels enter the corpuscle from one pole and often leave it from another. They curve through the corpuscles by running in the septa surrounded by perineural cells, Schwann cells and fibrocytes. Recently, fenestrations have been observed in capillaries of GTOs in several species (duck - Haiden and Awad, 1981; rat - Dubovy and Malinovsky, 1986; humans - Nitatori, 1988; marsupials - Strasmann et al., 1989). In all cases, capillaries never reach the sensory compartments irrespective of whether they are fenestrated or not.

Like the Ruffini corpuscle, the GTO is a slowly adapting mechanoreceptor (Iggo and Gottschaldt, 1974). In contrast to muscle spindles, which respond to elongation of relaxing muscles fibres, GTOs respond to longitudinal stress of tendons in case of muscle strain. One GTO registers one or multiple muscle units (Houk et al., 1980; Proske and Gregory, 1980). The total number of such proprioceptors varies according to the particular muscle. Together with muscle spindles (Peck et al., 1984, 1988; van der Wal and Drukker, 1988), GTOs are distributed in distinct regions of the muscle (Strasmann et al., 1989; Stuart et al., 1972; Soukup, 1983; Zelena and Soukup, 1983).

4. Lamellated corpuscles (Fig. 9-12)

Lamellated corpuscles reveal a large spectrum of morphological variation. Based on the number of axons, the branching pattern of axons and inner cores, Polacek (1966) described 12 morphological variations. According to him, this variation occurs not only among different species, but also within a single one. He also took the size of the corpuscles into account, e.g., lamellated corpuscles with an inner core of the joint capsules in rabbit and in guinea pig are larger than those in hamster and rat (Polacek and Sklenska, 1964).

By using electron microscopy, it was possible to simplify the classification of lamellated corpuscles (Halata, 1975). Thus today we differentiate between two types of lamellated corpuscles: (1) simple encapsulated corpuscles with an inner core and (2) Pacini corpuscles.

4.1. Simple encapsulated lamellated corpuscles with an inner core (Fig. 9)

These corpuscles are located in the musculoskeletal system firstly in the fibrous layer of the joint capsules, secondly in the connective tissue of muscles, surrounding tendons and near the periosteum and rarely in conjunction with GTOs, as described above. In contrast to the large Pacini corpuscles, simple encapsulated corpuscles (sLCs) are often present in dense connective tissue (Strasmann et al., 1988a), even in ligaments (Halata et al., 1985) or in the insertion zone of muscles (Strasmann et al., 1988b, c, 1989; van der Wal et al., 1987, 1988).

The ultrastructure of these corpuscles observed in joint capsules was described in the cat (Halata and Groth, 1976; Halata, 1977), in the bat (Malinovsky and Novotny, 1980), in the monkey (Halata et al., 1984), in humans (Halata et al., 1985), in the rat (Strasmann et al., 1989) and in marsupials (Strasmann et al., 1987).

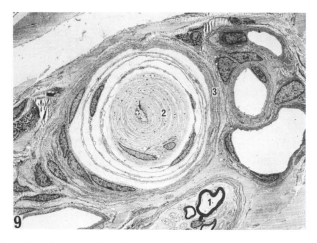

Fig. 9. Small lamellated corpuscle of Pacini type from the stratum fibrosum of the joint capsule of the knee joint of the cat. 1 myelinated fibres, (arrow) nerve terminal, 2 inner core, 3 perineural capsule. Magnifiaction 1,700 x.

The shape of the corpuscles is mostly oval (Fig. 9), but often branched or spiral shaped corpuscles can be seen. Three constituents form a sLC: (1) the afferent axon with its terminal, (2) terminal Schwann cells forming the inner core and (3) the perineural capsule. Branching corpuscles may show a divided inner core as well as two complete corpuscles pending on two branches of one afferent axon.

The afferent axon (Fig. 9) is myelinated and reveals a diameter of about 3 to 6 microns. Losing the myelin sheath after entering the capsule, the axon may branch into non-myelinated axon terminals. In its course the axon balloons intermittently as well as at its tip. These enlargements contain clusters of mitochondria, clear vesicles (diameter of about 60 nm) and occasionally glycogen granules.

Neurotubuli are visible in the axon center, whereas neurofilaments are situated at its periphery, close to the axolemma. Both along the length and at the tip of the terminal axon finger-like processes (spiculae - Strasmann et al., 1987) emerge from it and project about 1 micron over the surface of the terminal.

The terminal axon is surrounded by terminal lamellated Schwann cells which form concentric cytoplasmatic lamellae forming the inner core of the LC in an onion-like way. 3 to 20 lamellae in one inner core can be counted. The thinnest lamellae envelop the axon, and the more the lamella is at the surface of the core, the thicker it is. The outermost Schwann cell cytoplasmic lamella contains the nucleus of the cell. Both sides of the lamellae display abundant micropinocytotic vesicles, microfilaments, some microtubuli and, rarely, coated pits. The lamellae are connected by desmosome-like contacts, gap junctions (predominantly reflexive - Herr, 1976; Strasmann et al., 1989) and are fixed by *maculae densae* in the basal lamina, which covers the cytoplasm all around. Thin collagen fibrils run longitudinally and twisted within the circular gaps between the lamellae. The larger corpuscles (about 25 - 100 microns in length) also reveal two radially symmetric longitudinal clefts, extending from the periaxonal space to the sub-capsular space. The smaller corpuscles lack these symmetric clefts, only showing a cleft on one side or none at all.

The capsule (Fig. 9) of the small encapsulated corpuscles with an inner core is formed by flat perineural cells. Each cell is very thin, connected by desmosomes and tight junctions with neighbouring perineural cells and, in this way, these cells form several concentric layers. Each layer is covered by a basal lamina. Small collagen fibrils run twisted in the concentric gaps between the layers of perineural cells. The diameter of these fibrils is smaller than those of the epineurium or endoneurium. According to Breathnach (1977) they are synthesised by the perineural cells. The larger the corpuscle, the more perineural lamellae are present. The capsule is continuous with the perineurium of the afferent nerve. In very rare cases, these small encapsulated corpuscles can be observed in a peripheral nerve (Halata, 1977).

As for the function of these small encapsulated and lamellated corpuscles of the joint capsules, several physiologists have observed that these are fast adapting pressure receptors (Burgess and Clark, 1969; Clark, 1975; Clark and Burgess, 1975; Grigg, 1975; Grigg et al., 1982; McCloskey, 1978; Rossi and Rossi, 1985; Skoglund, 1956).

4.2. Pacini corpuscle (Fig. 10-12)

The largest mechanoreceptor in mammals is the Pacini corpuscle (PC), also termed as "Vater-Pacini corpuscle" (Halata, 1975). It is located in the musculoskeletal system in muscle septa (Ruffini, 1893a, b; Regaud, 1907), at the interosseous membrane of the lower leg (Rauber, 1867; Tello, 1922) and of the lower foreleg (Barker, 1974), in the periosteum (Rauber, 1867; Pansini, 1889) and in the periarticular connective tissue (Rauber, 1867).

Like the simple encapsulated corpuscle, the Pacini corpuscle consists of three parts: (1) the afferent axon with its terminal, (2) the inner core built by terminal Schwann cells and (3) the perineural capsule.

The afferent axon of the corpuscle is myelinated and reveals a diameter of about 6 to 10 microns. Only one single axon is present in a PC as a rule. Large lamellated corpuscles with two or even more axons and correspondingly more inner cores are termed "Golgi-Mazzoni corpuscles". Such corpuscles are predominantly found in the human knee joint capsule (Fig. 12; Halata et al., 1985) and the human anterior cruciate ligament (Fig. 11; Halata and Haus, 1989). In contrast to these, Pacini corpuscles always reveal one inner core and a total diameter ranging between 400 and 1.000 microns or even more.

The axon loses its myelin sheath a few microns after entering the inner core. The terminal axon is now only in part ensheathed by lamellae of terminal Schwann cells. Spiculae emerge from parts which are only covered by basal lamina and protrude into a cleft resembling the cleft of small lamellated corpuscles. Axonal balloons are also present and contain the same organelles as in the small lamellated corpuscle.

In contrast to the inner core of a LC , the inner core of a PC is exactly symmetrical: in the periphery the nuclei of the Schwann cells are arranged in

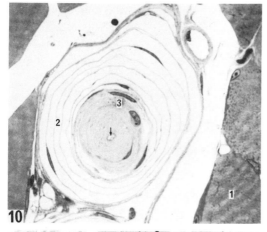

Fig. 10. Pacini corpuscle from the membrana interossea of the mouse. 1 muscle fibers, 2 layers of the perineural capsule, 3 inner core, (arrow) nerve terminal. Semithin section, magnification 800 x.

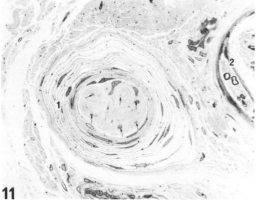

Fig. 11. Lamellated corpuscle from the subsynovial layer of cruciate ligament of the human knee joint. The corpuscle contains several inner cores (Golgi-Mazzoni type). (arrows) Nerve terminals, 1 perineural capsule, 2 nerve with myelinated and unmyelinated axons. Semithin section, magnification 800 x.

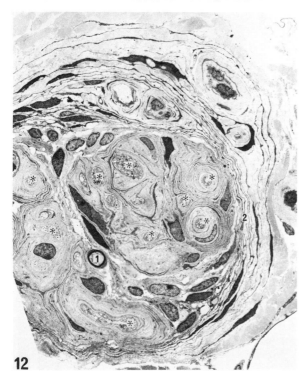

Fig. 12. Lamellated corpuscle from the human knee joint capsule. The corpuscle contains several inner cores (Golgi-Mazzoni type). 1 myelinated afferent axon, * nerve terminals, 2 perineural capsule. Magnification 1,600 x.

a chain-like fashion. Cytoplasmatic lamellae of Schwann cells, covered with basal lamina and with a gap between the adjacing lamellae, form one half of the inner core, separated from the other half by two symmetrically longitudinal clefts. The gaps are bridged by desmosome-like contacts and by gap junctions (Sakada and Sasaki, 1984; Strasmann et al., 1989). Collagen fibrills curve through the central as well as through the lamellar clefts. The axonal spikes protrude into the central clefts. Cross sections of the Pacini corpuscles display their symmetric structure (Fig. 10).

The perineural capsule consists of flat perineural cells which are continous with the perineurium of the afferent nerve (Shanthaveerappa and Bourne, 1964). Their structure resembles that of perineural cells of LCs, but the gaps between the cellular layers are wider (up to 10 microns) than in LCs. The total number of perineural layers varies, reaching 70 in the largest PCs.

The physiology of Pacini corpuscles has been investigated in the skin and in the mesenterium by several authors (Adrian and Umrath, 1929; Gray and Matthews, 1951; Loewenstein and Skalak, 1966; Loewenstein, 1971; Jänig, 1971; Gottschaldt, 1973). They are fast adapting acceleration-receptors (Gottschaldt, 1973; Schmidt, 1976), registering vibration. According to Gottschaldt (1973), the sheaths of the capsule and the inner core work like antishock pads which filter out slow pressure changes. Only fast pressure changes can pass the two lamellar systems (capsule and inner core) so that frequencies between 40 and 1.000 Hz are registered as physiological stimuli.

Acknowledgements: The authors whish to thank Ms T. Coellen, Ms B. Knuts, and Mr St. Schillemeit for technical assistance, and Mr St. Fletcher for assistance in translating the manuscript. Supported by DFG (HA 1194/3-1) and the "Verein fuer die Erforschung und Bekaempfung rheumathischer Krankheiten e.V.", Bad Bramstedt.

REFERENCES

Adrian, E. D., and Umrath, K., 1929, The impulse discharge from the Pacinian corpuscle, *J. Physiol. (Lond.)*, 68:139-154

Andres, K. H., von Düring, M., and Schmidt, R.F., 1985, Sensory innervation of the Achilles tendon by group III and IV afferent fibers, *Anat. Embryol.*, 172:145-156

Barker, D., 1974, The morphology of muscle receptors, in: "Handbook of Sensory Physiology. Muscle Receptors," C. C. Hunt, ed., Springer, Heidelberg, New York

Boyd, I. A., 1954, The histological structure of the receptors in the knee-joint of the cat correlated with their physiological response, *J. Physiol. (Lond.)*, 124:476-488

Boyd, I. A., and Roberts, T. D., 1953, Proprioceptive discharges from stretch receptors in the knee joint of the cat, *J. Physiol. (Lond.)*, 122:38-58

Breathnach, A., 1977, Freeze-fracture replication of peripheral nerve, in: "Struktur und Pathologie des Hautnervensystems," G. Lassmann, W. Jurecka, and G. Niebauer, eds., Verb. wiss. Ges., Wien

Bridgman, C. F., 1968, The structure of tendon organs in the cat: a proposed mechanism for responding to muscle tension, *Anat. Rec.*, 162:209-220

Bridgman, C. F., 1970, Comparison in structure of tendon organs in the cat, rat and man, *J. Comp. Neurol.*, 138:369-372

Burgess, P. R., and Clark, F.J., 1969, Characteristics of knee joint receptors in the cat, *J. Physiol. (Lond.)*, 203:317-335

Cattaneo, A., 1888, Sugli organi nervosi terminali muscolo-tendinei in condizioni normal e sul loro modo di comportarsi in sequito al taglio delle radici nervose e dei nervi spinali, *Memorie della R. Accademia delle Science di Torino*, 38:237-256

Cauna, N., 1969, The fine morphology of the sensory receptor organs in the auricula of the rat, *J. Comp. Neurol.*, 136:81-98

Clark, F. J., 1975, Information signalled by sensory fibers in medial articular nerve, *J. Neurophysiol.*, 38:1464-1472

Clark, F. J., and Burgess, P.,R., 1975, Slowly adapting receptors in the cat knee joint: can they signal joint angle?, *J. Neurophysiol.*, 38:1448-1463

Dreessen, D., Halata, Z., and Strasmann, T., 1988, Distribution and structure of mechanoreceptors in the mandibular joint of STR/1N mice, *in:* "Mechanoreceptors. Development, Structure and Function.," P. Hnik, T. Soukup, R. Vejsada, and J. Zelena, eds., Plenum Press, New York and London

Dubovy, P., and Malinovsky, L., 1986, The topography of capillaries in two different types of sensory nerve endings and some ultrastructural features of their walls, *Z. mikr. - anat. Forschung*, 100:577-587

Düring, M. von, Andres, K. H., and Schmidt, R. F., 1984, Ultrastucture of fine afferent fibre terminations in muscle and tendon of the cat, *in:* "Proceedings of the International Symposium on Sensory Receptor Mechanisms (Mechanoreceptors, Thermoreceptors and Nocireceptors)," W. Hamann and A. Iggo, eds., World Sci. Publ. Corp., Singapore

Eklund, G., and Skoglund, S., 1960, On the specificity of the Ruffini like joint receptors, *Acta Physiol. Scand.*, 49:184-191

Eldred, E., Bridgman, C. F., Swett, J. E., and Eldret, B., 1962, Quantitative comparison of muscle receptors of the cat's medial gastrocnemius, soleus and extensor digitorum brevis muscle, *in:* "Symposium on Muscle Receptors," D. Barker, ed., Hong Kong University Press, Hong Kong

Freeman, M. A. R., and Wyke, B., 1967, The innervation of the knee joint. An anatomical and histological study in the cat, *J. Anat. (Lond.)*, 101:505-532

Goglia, G., and Sklenska, A., 1969, Ricerche ultrastrutturali sopra i corpuscoli di Ruffini delle capsule articolari nel coniglio, *Quad. Anat. Prat.*, 25:14-27

Gottschaldt, K.-M., 1973, Mechanorezeptoren als Grundlage des Tastsinnes, *Biologie in unserer Zeit*, 3:184-190

Gray, J. A. B., and Matthews, P. B. C., 1951, Response of Pacinian corpuscles in the cat's toe, *J. Physiol. (Lond.)*, 113:475-482

Grigg, P., 1975, Mechanical factors influencing response of joint afferent neurons from cat knee, *J. Neurophysiol.*, 38:1473-1484

Grigg, P., and Hoffman, A. H., 1982, Properties of Ruffini afferents revealed by stress analysis of isolated sections of knee capsule, *J. Neurophysiol.*, 47:41-54

Grigg, P., and Hoffman, A. H., 1984, Ruffini mechanoreceptors in isolated joint capsule: response correlated with strain energy density, *Somatosens. Res.*, 2:149-162

Grigg, P., Hoffman, A. H., and Forgaty, K. E., 1982, Properties of Ruffini afferents revealed by stress analysis of isolated sections of knee capsule, *J. Neurophysiol.*, 47:41-54

Haiden, G. J., and Awad, E. A., 1981, The ultrastructure of the avian Golgi tendon organ, *Anat. Rec.*, 200:153-161

Halata, Z., 1975, The mechanoreceptors of the mammalian skin. Ultrastructure and morphological classification, *Adv. Anat. Embryol.*, 50:1-77

Halata, Z., 1977, The ultrastructure of the sensory nerve endings in the articular capsule of the knee joint of the domestic cat (Ruffini corpuscles and Pacinian corpuscles), *J. Anat. (Lond.)*, 124:717-729

Halata, Z., Badalamente, M. A., Dee, R., and Propper, M., 1984, Ultrastructure of sensory nerve endings in monkey *(Macaca fascicularis)* knee joint capsule, *J. Orthopedic Res.*, 2:169-176

Halata, Z., and Groth, H.-P., 1976, Innervation of the synovial membrane of cat knee joint capsule, *Cell Tiss. Res.*, 169:415-418

Halata, Z., and Haus, J., 1989, The ultrastructure of sensory nerve endings in human anterior cruciate ligament, *Anat. Embryol.*, in press

Halata, Z., and Munger, B. L., 1980, The ultrastructure of the Ruffini and Herbst corpuscles in the articular capsule of domestic pigeon, *Anat. Rec.*, 198:681-692

Halata, Z., Rettig, T., and Schulze, W., 1985, The ultrastructure of sensory nerve endings in the human knee joint capsule, *Anat. Embryol.*, 172:265-275

Halata, Z., Strasmann, T., and Loo, S. K., 1987, Sensible Innervation der Gelenkkapseln eines australischen Marsupials *(Dasyuroides byrnei)*. Licht- und elektronenmikroskopische Untersuchung, *Verh. Anat. Ges.*, 81:979-980

Herr, J. C., 1976, Reflexive gap junctions, *J. Cell Biol.*, 69:495-501

Houk, J. C., Crago, P. E., and Rymer, W. Z., 1980, Functional properties of Golgi tendon organs, *in:* "Spinal and supraspinal mechanisms of voluntary motor control and locomotion. Progress in Clinical Neurophysiology," J. E. Desmedt, ed., Karger, Basel

Iggo, A., and Gottschaldt, K.-M., 1974, Cutaneous mechanoreceptors in simple and complex sensory structures, *in:* "Symposium Mechanoreception. Rheinisch-Westfälische Akademie der Wissenschaften," ed., Westdeutscher Verlag

Jänig, W., 1971, The afferent innervation of the central pad of the cats hind foot, *Brain Res.*, 28:103-216

Kumazawa, T., and Mizumura, K., 1977, Thin-fiber receptors responding to mechanical, chemical and thermal stimulation in the skeletal muscle of the dog., *J. Physiol. (Lond.)*, 273:179-194

Kumazawa, T., and Mizumura, K., 1984, Functional properties of the polymodal receptors in the deep tissues, *in:* "Sensory Receptor Mechanisms," W. Hamann and A. Iggo, eds., World Scientific, Singapore

Kumazawa, T., and Perl, E. R., 1977a, Primate cutaneous sensory units with unmyelinated (C) afferent fibers, *J. Neurophysiol.*, 40:1325-1338

Kumazawa, T., and Perl, E. R., 1977b, Primate cutaneous receptors with unmyelinated (C) fibres and their projection to the substantia gelatinosa, *J. Physiol. (Paris)*, 73:287-304

Langford, L. A., 1983, Unmyelinated axon ratios in cat motor cutaneous and articular nerves, *Neurosci. Lett.*, 40:19-22

Langford, L. A., and Schmidt, R. F., 1983, Afferent and efferent axons in the medial and posterior articular nerves of the cat, *Anat. Rec.*, 206:71-78

Langford, L. A., Schaible, H.-G., and Schmidt, R. F., 1984, Structure and function of fine joint afferents: Observations and speculations, *in:* "Sensory Receptor Mechanisms," W. Hamann and A. Iggo, eds., World Scientific, Singapore

Loewenstein, W. R., 1971, Mechano-electric transduction in the Pacinian corpuscle. Initiation of sensory impulses in mechanoreceptors, *in:* "Principles of Sensory Physiology," W. R. Loewenstein, ed., Springer, Berlin

Loewenstein, W. R., and Skalak, R., 1966, Mechanical transmission in a Pacinian corpuscle. An analysis and theory, *J. Physiol. (Lond.)*, 182:246-278

Malinovsky, L., and Novotny, V., 1980, Ultrastructure of sensory corpuscles in joint capsules of the bat *(Myotis myotis)*, *Folia morph. (Prague)*, 28:230-232

McCloskey, D. I., 1978, Kinesthetic sensibility, *Physiol. Rev.*, 58:763-820

Mense, S., and Meyer, H., 1981, Response properties of group III and IV receptors in the Achilles tendon of the cat, *Pflügers Arch.*, 389:R25

Merrillees, N. C. R., 1960, The fine structure of muscle spindles in the lumbrical muscles of the rat, *J. Biophys. Biochem. Cytol.*, 7:725-742

Merrillees, N. C. R., 1962, Some observations on the fine structure of a Golgi tendon organ of a rat, *in:* "Symposium on Muscle Receptors," D. Barker, ed., Hong Kong Univ. Press, Hong Kong

Nitatori, T., 1988, The fine structure of human Golgi tendon organs as studied by three-dimensional reconstruction, *J. Neurocytol.*, 17:27-41

Ovalle, W. K., and Dow, P. R., 1983, Comparative ultrastructure of the inner capsule of the muscle spindle and the tendon organ, *Am. J. Anat.*, 166:343-357

Pansini, S., 1889, Des terminaisons des nerfs sur les tendons des vertebres, *Arch. ital. Biol.*, 11:225-228

Peck, D., Buxton, D. F., and Nitz, A. J., 1984, A comparison of spindle concentrations in large and small muscle acting in parallel combinations, *J. Morph.*, 180:243-252

Peck, D., Buxton, D. F., and Nitz, A. J., 1988, A proposed mechanoreceptor role for the small redundant muscles which act in parallel with large primemovers, *in:* "Mechanoreceptors. Development, Structure and Function," P. Hnik, T. Soukup, R. Vejsada, and J. Zelena, eds., Plenum Press, New York and London

Polacek, P., 1965, Differences in the structure and variability of spray-like nerve endings in the joint of some mammals, *Acta anat. (Basel)*, 62:568-583

Polacek, P., 1966, Receptors of the joints. Their structure, variability and classification, *Acta Fac. Med. Univ. Brunensis*, 23:1-107

Polacek, P., and Sklenska, A., 1964, Druhove rozdily v usporadani kloubnich receptoru u nekterych hlodavcu, *Sbornik ved.praci LK UK Hradec Kralove*, 7:617-623

Proske, U., and Gregory, J. E., 1980, The discharge rate:tension relation of Golgi tendon organs, *Neurosci. Lett.*, 16:287-290

Proske, U., Schaible, H. G., and Schmidt, R. F., 1988, Joint receptors and kinaestesia, *Exp. Brain Res.*, 72:219-224

Rauber, 1867, "Untersuchungen über das Vorkommen und die Bedeutung der Vaterschen Koerperchen," München

Regaud, C., 1907, Les terminaisons nerveuses et les organes nerveux sensitifs de l'appareil locomoteur, *Rev. gen. Histol.*, 2:583-689

Rossi, A., and Rossi, B., 1985, Characteristics of the receptors in the isolated capsule of the hip in the cat, *Int. Orthop.*, 9:123-127

Ruffini, A., 1893a, Sur un reticule nerveux special, et sur quelques corpuscules de Pacini qui se trouvent en connexion avec les organes musculo-tendineux du chat, *Arch. ital. Biol.*, 18:101-105

Ruffini, A., 1893b, Sur la terminaison nerveuse dans les faisceaux musculaires et leur signification physiologique, *Arch. ital. Biol.*, 18:106-114

Sakada, S., and Sasaki, T., 1984, Blood-nerve barrier in the Vater-Pacini corpuscle of cat mesentery, *Anat. Embryol.*, 169:237-247

Schmidt, R. F., 1976, Somato-viscerale Sensibilität: Hautsinne, Tiefensensibilität, Schmerz, *in:* "Physiologie des Menschen," R. F. Schmidt and G. Thews, eds., Springer, Berlin

Schoultz, T. W., and Swett, J. E., 1972, The fine structure of the Golgi tendon organs, *J. Neurocytol.*, 1:1-26

Schoultz, T. W., and Swett, J. E., 1976, Ultrastructural organisation of the sensory fibres innervating the Golgi tendon organ, *Anat. Rec.*, 179:147-162

Shanthaveerappa, T. R., and Bourne, G. H., 1964, New observations on the structure of the Pacinian corpuscle and its relation to the pia-arachnoid of the central nervous system and peripheral epithelium of the peripheral nervous system, *Z. Zellforsch.*, 61:742-753

Sklenska, A., 1973, Die Ultrastruktur des Golgischen Sehnenorgans bei der Katze, *Acta anat. (Basel)*, 86:205-221

Skoglund, S., 1956, Anatomical and physiological studies of knee joint innervation in the cat, *Acta Physiol. Scand. Suppl. 124*, 36:1-101

Skoglund, S., 1973, Joint receptors and kinaesthesis, *in:* "Hanbook of Sensory Physiology," A. Iggo, ed., Springer, Berlin and New York

Soukup, T., 1983, The number, distribution and size of Golgi tendon organs in developing and adult rat muscles, *Physiol. Bohemoslovaca*, 32:211-224

Stacey, M. J., 1967, "The innervation of mammalian skeletal muscle by medium and small diameter afferent nerve fibres Ph.D. Thesis," University Durham, Durham

Stoehr, P., jr., 1928, Das periphere Nervensystem, *in:* "Handbuch der mikroskopischen Anatomie des Menschen," von Moellendorf, ed., Springer, Berlin

Strasmann, T., and Halata, Z., 1988a, Topography of mechanoreceptors in the connective tissue of the elbow joint region in *Monodelphis domestica*, a laboratory animal, *in:* "Mechanoreceptors. Development, Structure and Function," P. Hnik, P. Soukup, R. Vejsada, and J. Zelena, eds., Plenum Press, New York and London

Strasmann, T., and Halata, Z., 1988b, Verteilung von Mechanorezeptoren im Muskuloskelettal-System des Vorderlaufes von *Monodelphis domestica*, *Verh. Anat. Ges.*, in press

Strasmann, T., and Halata, Z., 1988c, Applications for 3-D-image processing in functional anatomy - reconstruction of the cubital joint region and spatial distribution of mechanoreceptors surrounding this joint in *Monodelphis domestica*, a laboratory marsupial, *Suppl. Europ. J. Cell Biol.*, (in press)

Strasmann, T., and Halata, Z., 1988d, Vergleichende Untersuchung der Ultrastruktur der bindegewebigen Kompartimente von Golgi-Sehnen-Organen, Lamellen- und Ruffini-Koerperchen, *Verh. Anat. Ges.*, (in press)

Strasmann, T., Halata, Z., and Loo, S. K., 1987, Topography and ultrastructure of sensory nerve endings in the joint capsules of the kowari *(Dasyuroides Byrnei)*, an australian marsupial, *Anat. Embryol.*, 176:1-12

Strasmann, T., Van der Wal, J. C., Halata, Z., and Drukker, J., 1989, Functional topography and ultrastructure of periarticular mechanoreceptors in the lateral elbow region of the rat, *Acta anat. (Basel)*, (in press)

Stuart, D. G., Mosher, C. G., Gerlach, R. L., and Reinking, R. M., 1972, Mechanical arrangement and transducing properties of Golgi tendon organs, *Exp. Brain Res.*, 14:274-292

Tello, J. F., 1922, Die Entstehung der motorischen und sensiblen Nervenendigungen, *Z. Anat. Entwickl. Gesch.*, 64:348-440

Van der Wal, J. C., and Drukker, J., 1988, The occurence of muscle spindles in relation to the architecture of the connective tissue in the lateral cubital region of the rat, *in:* "Mechanoreceptors. Development, Structure and Function," P. Hnik, T. Soukup, R. Vejsada, and J. Zelena, eds., Plenum Press, New York and London

Van der Wal, J. C., Strasmann, T., Drukker, J., and Halata, Z., 1987, The occurrence of sensory nerve endings in the lateral cubital region of the rat in relation to the architecture of the connective tissue, *Acta anat. (Basel)*, 130:94

Van der Wal, J. C., Strasmann, T., Drukker, J., and Halata, Z., 1988, Sensory nerve endings in the deep lateral cubital region: A topographical and ultrastructural study, *in:* "Mechanoreceptors. Development, Structure and Function," P. Hnik, T. Soukup, R. Vejsada, and J. Zelena, eds., Plenum Press, New York and London

Zelena, J., and Hnik, P., 1963, A new approach to muscle deafferentation, *in:* "Central and Peripheral Mechanisms of Motor Functions. Proceedings of the Conference," E. Gutmann and P. Hnik, eds., Academia, Prague

Zelena, J., and Soukup, T., 1983, The in-series and in-parallel components in rat hindlimb tendon organs, *Neuroscience*, 9:899-910

6

SENSORY INNERVATION PATTERNS OF THE ORIGINS OF THE SUPINATOR MUSCLES IN THE RAT AND THE GRAY SHORT-TAILED OPOSSUM IN RELATION TO FUNCTION

T. Strasmann[1], J. C. van der Wal[2], Z. Halata[1], and J. Drukker[2]

[1]Anatomisches Institut der Universität Hamburg
Abteilung für funktionelle Anatomie, Hamburg, F.R.G.

[2]Capaciteitsgroep Anatomy / Embryology
Rijksuniversiteit Limburg, Biomedische Centrum
Maastricht, The Netherlands

INTRODUCTION

Mechanoreceptors of the locomotive system are to be found in joint capsules and ligaments, in the connective tissue sheaths of muscles, as well as in aponeuroses, tendons and muscles (see chapter 5, this volume). For kinesthesia, which is the "sense of position as well as movements of body parts" (Proske et al., 1988), all these receptors even including receptors of the skin are necessary, however, different joints and different kinds of mechanoreceptors may play different roles in these sensations (Ferrell, 1988; Ferrell et al., 1987; Ferrell and Smith 1988; Ferrell and Baxendale 1988; Matthews, 1982; McClosey et al., 1987). The controversial discussion of physiologists as to the significance of individual receptors or receptor groups for kinesthesia has not yet been decisively concluded. As anatomists, all we can do is speculate on the physiology of mechanoreceptors, but we can at least describe their distribution patterns in relation to functional joint anatomy. To this end, we compared two similar joint regions (i.e., the deep lateral elbow joint with the origin of the supinator muscle) of different species and the distribution patterns of mechanoreceptors within these. "Similar" means that both regions are composed of the same constituents, but that they have differing mechanoreceptor distribution patterns and that the joints are used in a different way by the animal in question.

MATERIAL AND METHODS

Two adult wistar rats and two adult gray short-tailed opossums *(Monodelphis domestica)* were used for silver staining of peripheral nerves and light microscopy. The animals were perfused with Bouin's solution, the right elbow joint with surrounding muscles and skin dissected and, after decalcification, processed for serial silver impregnation of 7,5 µm paraffin sections according to Spaethe (1984). Using light microscopy, each slice was investigated and the positions of diagnosed mechanoreceptors transferred to camera lucida drawings. With a VAX 11-780 computer, these drawings were reconstructed and standard image processing was performed (Strasmann and Halata, 1988c).

Additional animals (5 and 16, respectively) were used for total foreleg-reconstructions as well as for ultrastructural purposes. The specimens were fixed with formalin or with glutaraldehyde. This material was used for scrutinising either the reconstruction or the diagnosis of the respective mechanoreceptor.

RESULTS

Anatomy of the joint regions compared

In the *rat,* the supinator muscle originates from a broad aponeurotic tendon which is fixed at the lateral humeral epicondyle. This aponeurosis is composed of dense regular longitudinal collagen fiber bundles and reveals a sesamoid bone, whose deep aspect is covered with cartilage and faces the joint cavity. The joint capsule beneath the aponeurosis is very thin. Only towards the ventral and the dorsal aspect of the joint does the fibrous layer of the capsule become discernible. The supinator muscle itself is slender and covers nearly three-fourths of the radial bone.

In the *short-tailed opossum,* the supinator originates from a smaller aponeurosis which lacks a sesamoid bone. The joint capsule surrounds the aponeurosis by embracing it from beneath or from above with the dense irregular connective tissue of the fibrous layer. The muscle itself is thick, shorter than in the rat and covers only the proximal third of the bone.

Distribution patterns of mechanoreceptors

In the deep lateral elbow joint region, where the joint capsule, the aponeurosis already mentioned and the origin of the supinator muscle are situated, only small *lamellated corpuscles (LCs)* and a few *free nerve endings* are present. For technical reasons (Strasmann et al., 1989), we counted only the LCs.

In the *rat,* 10 to 13 LCs can be observed at the distal edge of the sasamoid bone (Fig. 1b). They lie immediately adjacent to muscle fibers of the supinator, which originate here in the aponeurosis. Superficially to this muscle-aponeurosis transition zone and in the joint capsule adjacent to the aponeurosis, LCs cannot be found. It is only further away, within the capsule or within periarticular connective tissue that LCs can be seen again.

In the *opossum,* the distribution is completely different (Fig. 1a). In the transition zone from the muscle to the aponeurosis there are no LCs. But in the

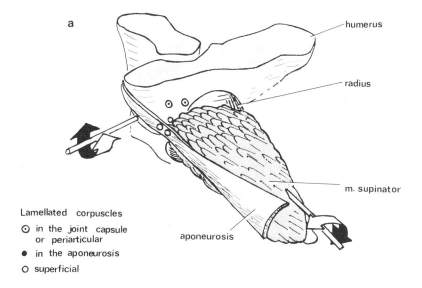

Lamellated corpuscles

⊙ in the joint capsule
  or periarticular

● in the aponeurosis

○ superficial

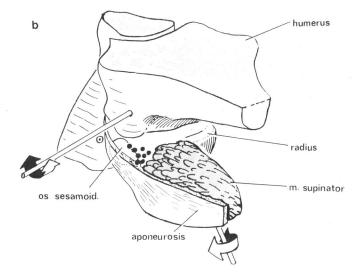

Fig. 1. Comparative anatomy of elbow joint

A. Ventro-lateral aspect of the elbow joint of the gray short-tailed opossum *(Monodelphis domestica)*, inflected about 10 degrees. The aponeurosis, partly dissected in this diagram, forms the origin of the supinator muscle and is itself a part of the joint capsule. Superficial to the aponeurosis, small lamellated corpuscles (open circles) have been added. The joint capsule adjacent to the aponeurosis reveals two small lamellated corpuscles in its fibrous layer (encircled dots). Ruffini corpuscles (not displayed) are regularly observable in the joint capsule at its flexion side, adjacent to the radial head. Drawing based on two reconstructed section series.

B. Ventro-lateral aspect of the elbow joint of the rat, extremely inflected. The aponeurosis, partly dissected in this diagram too, forms the origin of the supinator muscle, as is the case in the opossum. The aponeurosis contains a small sesamoid bone which faces the joint cavity with a cartilaginous surface. The originating muscle fibers of the supinator abut directly to the bone. For display reasons, the gap between muscle and sesamoid bone has been drawn too wide. In this gap, 10 or more small lamellated corpuscles (large filled dots) are regularly present. Only one corpuscle, lying superficially and further apart within loose periarticular tissue, is shown (encircled dot). Drawing based on two reconstructed section series, slightly reduced in comparison to Fig 1A.

fibrous layer covering the aponeurosis, LCs are present (3 in each series). A few LCs are also situated beneath the muscle: either within the fibrous layer of the capsule (1 to 2), where the proximal part of the muscle runs a short way over it, or between the periosteum of the radial bone and the muscle (1 to 2). In addition, the joint capsule reveals a cluster of *Ruffini corpuscles* at its flexion side adjacent to the aponeurosis.

DISCUSSION

The gross as well as the microscopic anatomy of the elbow joint in the rat and in *Monodelphis domestica*, the opossum used here, have been described before (Strasmann et al., 1988a; Van der Wal et al., 1984, 1988b). Likewise, the distribution patterns of different mechanoreceptors of this joint have been published (Strasmann et al., 1988a; Strasmann and Halata, 1988b; Van der Wal et al., 1987, 1988a). The subject of this paper is the comparison of the respective patterns.

It may be argued that such small numbers of corpuscles at each locus may lead to considerable errors if only two joints are investigated in each animal. We counted and reconstructed two joints of each species in the same manner, but we verified the data by performing additional investigations on 21 animals. 16 animals (8 of each species) were used for electron microscopy of the LCs, which always were observed at the loci mentioned and not at places which lack LCs in the reconstructed series.

The *comparison* of these two species shows two different distribution patterns of small lamellated corpuscles (LCs), which are known as to be rapidly adapting pressor receptors (Burgess and Clark, 1969; Clark, 1975; Clark and Burgess, 1975; Grigg, 1975; Grigg et al., 1982; McCloskey, 1978; Rossi and Rossi, 1985; Skoglund, 1956).

In the *rat*, the supinator muscle at its origin in the aponeurotic tendon reveals 10 or more LCs as a rule. From the gross and the microscopic anatomy it would be reasonable to assume that at this position of the corpuscles, shearing forces pass from the muscles fibers to the aponeurosis. The LCs are arranged in such a way that a single LC might register forces which are generated by only a few muscle fibers (Strasmann et al., 1989). Supposing that these corpuscles are predominantly supplied by separate afferent axons, or, in other words, the convergence from peripheral to central is minimal, this arrangement could provide exact information on the changing activities of muscle fibers.

In the *opossum*, only a few LCs surrounding the origin of the supinator muscle can be found as a rule. These corpuscles may be pressed by the muscle itself as well as by neighbouring muscles (mm. brachialis, brachioradialis, extensores) or by whole-joint movements. As far as we know from gross or microscopic anatomy of this joint, these corpuscles do not record actions of a small group of muscle fibers of the supinator. Close to the aponeurosis, a cluster of Ruffini corpuscles is always present in the flexion side of the joint capsule. Ruffini corpuscles (RCs) are known to be slowly adapting tension receptors (Boyd and Roberts, 1953; Boyd, 1954; Eklund and Skoglund, 1960; Freeman and Wyke, 1967; Grigg and Hoffman, 1982, 1984; McCloskey, 1978; Skoglund,

1973). Taking this into account, it might be possible that in the opossum LCs together with RCs record joint movements (probably predominantly LCs) or contribute to the detection of joint positions near the end of range (probably predominantly RCs).

Recently, Van der Wal and Drukker (1988) demonstrated that the supinator muscle in the rat has to be labeled "monitor muscle" (Peck et al., 1984, 1988). In brief, monitor muscles reveal a high mechanoreceptor content for movement perception in contrast to muscles with low mechanoreceptor content, meaning that these muscles predomintly generate power and movement. For the elbow joint of the gray short-tailed opossum, we were able to demonstrate that only the cluster of Ruffini corpuscles mentioned above can be observed, whereas other parts of the joint capsule lack these RCs (Strasmann and Halata 1988a, b, c).

The two animals, rat and opossum, are nearly of the same size, but they differ extremely in behavior: the rat is predominantly quadrupedal, the opossum is, of course, quadrupedal too but uses its forefeet more manipulatively than the rat would ever do. Monodelphis is an excellent climber in contrast to the rat. Therefore, the supination in the rat may be predominantly used for balancing in quadrupedal locomotion, whereas supination in the opossum may be used for forefoot rotation during manipulation. If this is true, the sensory innervation of the supinator muscle and its aponeurosis in the rat registers or "monitors" alterations in muscle tension during quadrupedal locomotion, whereas in the opossum the sensory innervation of the joint capsule adjacent to the origin of the supinator muscle is necessary to prevent harmful distortions in the cubital joint during climbing.

Acknowledgements: The autors whish to thank Mr T. Feilscher, Ms B. Knuts, Mr St. Schillemeit and Ms B. Luecht for technical assistance, Mr St. Fletcher for assistance in translating the manuscript. Supported by DFG (HA 1194/3-1) and Neth.Org.Adv.Pure Res. (ZWO R 92-97/107)

REFERENCES

Boyd, I. A., 1954, The histological structure of the receptors in the knee joint of the cat correlated with their physiological response, *J. Physiol. (Lond.)*, 124:476-488

Boyd, I. A., and Roberts, T. D., 1953, Proprioceptive discharges from stretch receptors in the knee joint of the cat, *J.Physiol. (Lond.)*, 122:38-58

Burgess, P. R., and Clark, F. J., 1969, Characteristics of knee joint receptors in the cat, *J.Physiol. (Lond.)*, 203:317-335

Clark, F. J., 1975, Information signalled by sensory fibers in medial articular nerve, *J. Neurophysiol.*, 38:1464-1472

Clark, F. J., and Burgess, P. R., 1975, Slowly adapting receptors in the cat knee joint: can they signal joint angle?, *J. Neurophysiol.*, 38:1448-1463

Eklund, G., and Skoglund, S., 1960, On the specificity of the Ruffini-like joint receptors, *Acta Physiol. Scand.*, 49:184-191

Ferrell, W. R., 1988, Discharge characteristics of joint receptors in relation to their proprioceptive role, *in:* "Mechanoreceptors. Development, Structure and Function," P. Hnik, T. Soukup, R. Vejsada, and J. Zelena, eds., Plenum Press, New York and London

Ferrell, W. R., and Baxendale, R. H., 1988, The effect of acute joint inflammation on flexion reflex excitability in the decerebrate, low-spinal cat, *Quart. J. Exp. Physiol.*, 73:95-102.

Ferrell, W. R., Gandevia, S. C., and McCloskey, D. I., 1987, Role of joint receptors in human kinaesthesis when intramuscular receptors cannot contribute, *J. Physiol. (Lond.)*, 386:63-71.

Ferrell, W. R., and Smith, A., 1988, Position sense at the proximal interphalangeal joint of the human index finger, *J. Physiol. (Lond.)*, 399:49-61

Freeman, M. A. R., and Wyke, B., 1967, The innervation of the knee joint. An anatomical and histological study in the cat, *J. Anat. (Lond.)*, 101:505-532

Grigg, P., 1975, Mechanical factors influencing response of joint afferent neurons from cat knee, *J. Neurophysiol.*, 38:1473-1484

Grigg, P., and Hoffman, A. H., 1982, Properties of Ruffini afferents revealed by stress analysis of isolated sections of knee capsule, *J. Neurophysiol.*, 47:41-54

Grigg, P., and Hoffman, A. H., 1984, Ruffini mechanoreceptors in isolated joint capsule: response correlated with strain energy density, *Somatosens. Res.*, 2:149-162

Grigg, P., Hoffman, A. H., and Forgaty, K. E., 1982, Properties of Ruffini afferents revealed by stress analysis of isolated sections of knee capsule, *J. Neurophysiol.*, 47:41-54

Matthews, P. B. C., 1982, Where does Sherrington's "muscular sense" originate? Muscles, joints, corollary discharges?, *Ann. Rev. Neurosci.*, 5:189-218

McCloskey, D. I., 1978, Kinesthetic sensibility, *Physiol. Rev.*, 58:763-820

McCloskey, D. I., Macefield, G., Gandevia, S. C., and Burke, D., 1987, Sensing position and movements of the fingers, *NIPS*, 2:226-230

Peck, D., Buxton, D. F., and Nitz, A. J., 1984, A comparison of spindle concentrations in large and small muscle acting in parallel combinations, *J. Morph.*, 180:243-252

Peck, D., Buxton, D. F., and Nitz, A. J., 1988, A proposed mechanoreceptor role for the small redundant muscles which act in parallel with large primemovers, *in:* "Mechanoreceptors. Development, Structure and Function," P. Hnik, T. Soukup, R. Vejsada, and J. Zelena, eds., Plenum Press, New York and London

Proske, U., Schaible, H. G., and F, Schmidt, R., 1988, Joint receptors and kinaestesia, *Exp. Brain Res.*, 72:219-224

Rossi, A., and Rossi, B., 1985, Characteristics of the receptors in the isolated capsule of the hip in the cat, *Int. Orthop.*, 9:123-127

Skoglund, S., 1956, Anatomical and physiological studies of knee joint innervation in the cat, *Acta Physiol. Scand. Suppl.124*, 36:1-101

Skoglund, S., 1973, Joint receptors and kinaesthesis, *in:* "Handbook of Sensory Physiology," A. Iggo, ed., Springer, Berlin and New York

Spaethe, A., 1984, Eine Modifikation der Silbermethode nach Richardson für die Axonfärbung in Paraffinschnitten, *Verh. Anat. Ges.*, 78:101 - 102

Strasmann, T., and Halata, Z., 1988a, Topography of mechanoreceptors in the connective tissue of the elbow joint region in *Monodelphis domestica*, a laboratory animal, *in:* "Mechanoreceptors. Development, Structure and Function," P. Hnik, P. Soukup, R. Vejsada, and J. Zelena, eds., Plenum Press, New York and London

Strasmann, T., and Halata, Z., 1988b, Verteilung von Mechanorezeptoren im Muskuloskelettal-System des Vorderlaufes von *Monodelphis domestica*, *Verh. Anat. Ges.*, (in press)

Strasmann, T., and Halata, Z., 1988c, Applications for 3-D-image processing in functional anatomy - reconstruction of the cubital joint region and spatial distribution of mechanoreceptors surrounding this joint in Monodelphis domestica, a laboratory marsupial, *Suppl. Europ. J. Cell Biol.*, (in press)

Strasmann, T., Van der Wal, J. C., Halata, Z., and Drukker, J., 1989, Functional topography and ultrastructure of periarticular mechanoreceptors in the lateral elbow region of the rat, *Acta anat. (Basel)*, (in press)

Van der Wal, J. C., Drukker, J., and Van Mameren, H., 1984, The organisation of the connective tissue in relation with muscle and nerve tissue in the lateral cubital region in the rat, *Verh. Anat. Ges.*, 78:631-632

Van der Wal, J. C., and Drukker, J., 1988, The occurence of muscle spindles in relation to the architecture of the connective tissue in the lateral cubital region of the rat, *in:* "Mechanoreceptors. Development, Structure and Function," P. Hnik, T. Soukup, R. Vejsada, and J. Zelena, eds., Plenum Press, New York and London

Van der Wal, J. C., Strasmann, T., Drukker, J., and Halata, Z., 1987, The occurrence of sensory nerve endings in the lateral cubital region of the rat in relation to the architecture of the connective tissue, *Acta anat. (Basel)*, 130:94

Van der Wal, J. C., Strasmann, T., Drukker, J., and Halata, Z., 1988a, Sensory nerve endings in the deep lateral cubital region: A topographical and ultrastructural study, *in:* "Mechanoreceptors. Development, Structure and Function," P. Hnik, T. Soukup, R. Vejsada, and J. Zelena, eds., Plenum Press, New York and London

Van der Wal, J. C., Strasmann, T., Drukker, J., and Halata, Z., 1988b, The connective tissue apparatus in the lateral elbow region of the rat as instrument in the quality of centripetal information, *Verh. Anat. Ges.*, (in press)

7

THE SENSORY TERMINAL TREE OF "FREE NERVE ENDINGS" IN THE ARTICULAR CAPSULE OF THE KNEE

B. Heppelmann[1], K. Meßlinger[1], W.F. Neiss[2*], and
R.F. Schmidt[1§]

[1]Physiologisches Institut der Universität Würzburg
Würzburg, F.R.G., [2]Anatomisches Institut der
Universität zu Köln, Köln, F.R.G.

INTRODUCTION

In recent years the normal and artificially inflamed knee joint of the cat has become an important model for neurophysiological studies on arthritic pain. The noxious stimuli causing pain are presumably perceived via "free nerve endings", i.e., non-corpuscular endings of thinly myelinated fine afferent nerve fibers (group III or A∂-fibers) and of non-myelinated fine afferent group IV or C-fibers (Schaible and Schmidt, 1983a, b, 1985; Grigg et al., 1986). But in comparison with neurophysiological knowledge, little is known about the morphological characteristics of articular "free nerve endings". This is why we have studied the ultrastructure of "free nerve endings" and fine afferent nerve fibers in the normal knee joint capsule of the cat by serial sectioning and with three-dimensional reconstruction of nerve fibers for up to 300 μm length. These reconstructions yielded the following major results:

Firstly, our findings provide evidence that the transduction areas (i.e., the sensory areas proper) of afferent "free nerve endings" are not confined to the tips of the nerve fibers; instead, the "free nerve ending" in connective tissue forms a *sensory terminal tree* ("*sensible Endstrecke*"), extending over several hundred microns in size. The branches of this sensory terminal tree have no myelin sheath (in group III endings), are not enclosed by perineurium (in group III and IV endings), and carry *multiple receptive sites* evenly spaced along their entire lengths.

---

[*] Author for correspondence

[§] The authors appear in alphabetical order.

Secondly, our study turned out that "free nerve endings" of thinly myelinated group III and non-myelinated group IV nerve fibers differ in 3 parameters, namely in mean fiber diameter of the terminal tree (1.0 - 1.5 µm in group III vs. 0.3 - 0.6 µm in group IV), mean number of mitochondria (≥10/fiber profile vs. ≤7/fiber profile), and - most remarkably - in their cytoskeleton: Only the sensory terminal tree of group III fibers contains a conspicuous bundle of microfilaments that extends throughout all branches of the tree close up to their tips. This bundle appears in cross-cut nerve fibers as a distinct *neurofilament core* that is free from cell organelles and microtubules.

MATERIALS AND METHODS

In 6 cats the sympathetic trunk including ganglia L3-S1 was removed in order to cause a total loss of sympathetic fibers and to get an exclusively afferent innervation of the knee joint. Eight to 12 weeks later 3 animals were perfused with a mixture of 2.5% glutaraldehyde and 1.5% paraformaldehyde in 0.1 M phosphate buffer (Andres et al., 1985a). The dissected knee joints were decalcified for 2 months, sectioned in 2 mm thick slices, postfixed with osmium tetroxide, dehydrated and embedded in araldite according to Andres and von Düring (1981). Three other cats were perfused with 6% glutaraldehyde with an initial bolus of 20% glutaraldehyde in 0.1 M cacodylate buffer. Parts of the capsule were dissected, photographed and cut into triangular pieces. After osmium postfixation and dehydration they were embedded in ULV-resin.

Examination and reconstruction of the nerve fibers were based on 2 different serial section techniques (Heppelmann et al., in press). For the *re-embedding technique*, the tissue blocks were cut into serial semithin sections (1 µm thick). After light microscopic examination and reconstruction, selected sections were re-embedded and cut ultrathin for an electron microscopic investigation. In the *serial section-ESI technique* the blocks were cut into series of alternate semi- and ultrathin sections (1 x 500 nm, 5 x 100 nm, and so on). All sections were examined with a Zeiss EM 902 transmission electron microscope at 80 kV using conventional microscopy and - in the case of 500 nm sections - the inelastic mode of electron spectroscopic imaging (ESI; Bauer, 1987).

Three-dimensional reconstructions were performed with the 3D-REC program on the image processing system (IPS, Kontron) or with the HISTOL program (List-electronic, D-6100 Darmstadt 13, FRG) on an Olivetti M24 personal computer.

RESULTS

Most neurophysiological data have been collected from fine afferent nerve fibers of the medial articular nerve. This nerve contains 91% afferent fibers belonging to groups III and IV (Heppelmann et al. 1988) and supplies the antero-medial part of the knee joint capsule (Freeman and Wyke, 1967). To allow future correlations of morphological and physiological data, our study was confined to the sensory area of the medial articular nerve and included 11 reconstructions of group III and 32 of group IV "free nerve endings" situated in the patellar ligament, the medial patellar retinaculum, the medial collateral ligament, and in the thin parts of the articular capsule between these ligaments.

Histotopography

Within each fiber group (III and IV) all endings showed the same ultrastructure, regardless of their histotopographical localization in the structures mentioned above. The majority of small nerve fiber bundles and their endings are found in the vicinity of or in close contact to venules, arterioles and lymphatic vessels; no nerve fibers have been observed around capillaries.

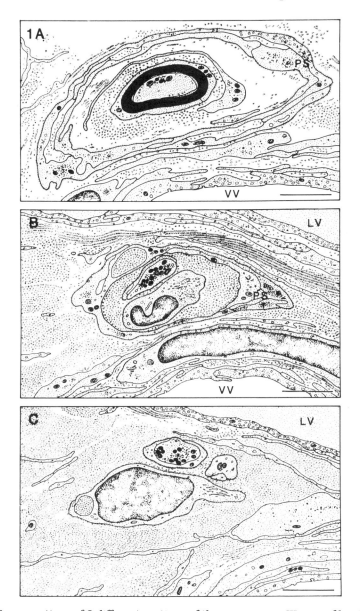

Figs. 1. Cross sections of 3 different portions of the same group III nerve fiber that runs between a venous (VV) and a lymphatic vessel (LV) in the medial patellar retinaculum of the knee joint. Drawings from original electron micrographs; scale bar: 2 μm.
A. The proximal portion of the nerve fiber is myelinated and enclosed by a perineurial sheath (PS). **B.** The intermediate portion of the nerve fiber has lost its myelin sheath, but is still surrounded by perineurium (PS). Section 40 μm distal to that of Fig. 1 A. **C.** The distal portion comprising the *sensory terminal tree* of the nerve fiber lacks myelin and perineurial sheath. Section 59 μm distal to that of Fig. 1 A.

In the patellar and medial collateral ligament, fine afferent nerve fibers mainly spread within a narrow layer at the inner and outer surface of the dense connective tissue, also accompanying blood vessels. Likewise, in the region of the patellar retinaculum, the nerve fibers run rather within vascular layers than within strands of dense collagenous tissue.

Transition from peripheral nerve to "free nerve endings"

Small peripheral nerves. The medial articular nerve branches several times within the articular capsule. As expected each successive branch has a smaller diameter and contains a lower number of nerve fibers, but interestingly these reductions are also linked with a decrease in thickness of the perineurial sheath. This perineurium which separates as blood-nerve-barrier (for reviews see Low, 1976; Montes et al., 1984) the nerve fibers from the interstitial fluid of the connective tissue, consists of 4-6 layers of flat perineurial cells around the mainstem of the medial articular nerve, but only of one cell layer in the smallest, most distal peripheral nerves (Fig. 1a; cf., Fig. 150 in: Breathnach, 1971, for small nerve in human skin).

The smallest peripheral nerves in the knee joint capsule contain only one myelinated group III fiber (Fig. 1a) and/or one Remak bundle enclosing up to 13 group IV fibers. The single perineurial cell layer surrounding these nerves is coated both on its inner (i.e., endoneurial) and outer side with a basal lamina. Such basal laminae are regarded as a characteristic criterion for perineurial cells elsewhere (Low, 1976).

Ending of perineurium. In all reconstructions the single-layered perineurium, despite of its thinness, still appears as a gapless, completely continuous sheath of cells, and there is no morphological evidence for any leaks of the blood-nerve-barrier along the course of the peripheral nerve. At the end of the peripheral nerve, however, the perineurium ceases to exist (compare Figs. 1b and 1c; *arrowheads* in Fig. 2a). All reconstructions show that the perineurium of the branching peripheral nerves has the three-dimensional shape of a branched system of tubing, the ends of which are open. Here, where peripheral nerve and blood-nerve-barrier end, the group III and IV nerve fibers leave the perineurial tube (cf., Fig. 7 in Andres et al., 1985b for perineurium in Achilles tendon) and immediately enter the general interstitial space of the connective tissue. At this point the sensory terminal tree (see below) starts.

Transition of group III nerve fibers to endings. Along its course within the peripheral nerve up to the very tips of its fiber endings, 3 different portions of a group III nerve fiber can be discriminated in cross-sections and reconstructions.

The *proximal* portion of group III fibers (Fig. 1a) is thinly myelinated and runs in the endoneurial space of peripheral nerves that is separated from the surrounding tissue by the perineurium. Presumably this portion of the nerve fiber has purely conductive function, there are no structural signs of receptive sites.

The *intermediate* portion of group III fibers (Fig. 1b) has no myelin sheath, but is still surrounded by perineurium. It starts where the myelin sheath ends with

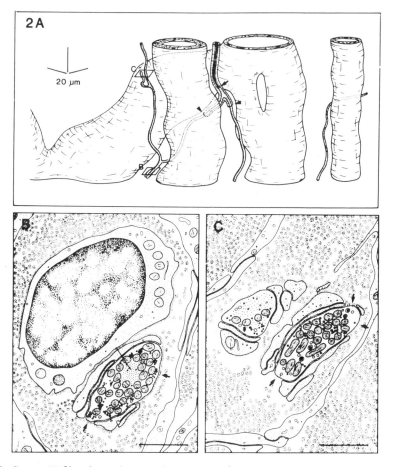

Figs. 2. Group III fiber branching in the vicinity of venous vessels within the innermost layer of the patellar ligament.
A. Gross reconstruction of the group III fiber and its terminal tree, based on serial semi- and ultrathin sections examined with the electron microscope. Arrow, end of myelin sheath; arrowheads, end of perineurial sheath.
B and C. Drawings of original electron micrographs from sections of the nerve fiber marked in A; scale bar: 1 µm. **B.** Cross section of a bulged segment of the nerve fiber with accumulations of parallely arranged mitochondria and glycogen particles. The *neurofilament core* with the central axoplasm (thin arrow) contains bundles of microfilaments but is devoid of organelles. Areas of axolemma not covered by Schwann cell lamellae (bare areas) are marked by thick arrows. **C.** Cross section of the final thickening of a short side branch of the terminal tree. The final thickening has an otherwise most similar ultrastructure as bulged segments, but does not contain a neurofilament core, and its mitochondria appear piled up irregularly. Thick arrows indicate same structures as in B.

a half-node of Ranvier (*arrow* in Fig. 2a), and changes into the distal portion at the open ending of the perineurial cell-tube (see above; *arrowheads* in Fig. 2a). In our reconstructions the length of the intermediate portion varied between 7 and 98 µm, but in all cases the myelin sheath ended before the perineurial sheath.

The *distal* portion of group III fibers (Figs. 1c, 2b, 2c) starts at the end of the peripheral nerve. It has neither myelin nor perineurial sheath. Based on the conclusions drawn from the ultrastructural findings presented below, we have termed this part of the fiber sensory terminal portion or *sensory terminal tree* (*sensible Endstrecke*).

Both intermediate and distal portions of group III fibers have the same ultrastructure. They do not possess a myelin sheath but are nevertheless accompanied and partly enclosed by a string of Schwann cells (Figs. 1B, 1C, 2B, 2C), the common basal lamina of which is wrapped completely around the nerve fiber up to its very tip.

Terminal portion of group IV fibers. In the unmyelinated group IV nerve fibers only a proximal portion inside and a distal portion outside of the perineurium can be discriminated. As far as the smallest peripheral nerves are concerned, both portions have the same ultrastructure, and some group IV nerve fibers even end within the peripheral nerve. Because of the branching, however, and of the long course of group IV fibers outside of the blood-nerve-barrier (see below), we have also used the term "sensory terminal tree" for that part of the group IV nerve fiber without perineurium.

Ultrastructure of the sensory terminal tree

Group III fibers

After the loss of myelin and perineurial sheath, group III fibers split up into several (about 4) branches with variable lengths of about 10 to 200 µm (Fig. 2a), and a mean diameter ranging from 1.00 to 1.50 µm. The branches run as single fibers or associated with some group IV fibers in Remak bundles forming a more or less two-dimensional termination field within an area of about 150 µm x 200 µm (Fig. 2a).

All branches of this terminal tree consist of pearl-string-like series of thick and thin segments, the diameters of which vary between 0.4 µm and 2.5 µm. The thick segments (bulges) appear roughly spindle-shaped and are 5 to 12 µm long. The thin (waist-like) segments between the bulges are somewhat shorter.

"Bare" axolemma. The sensory terminal tree of the nerve fiber is not completely covered by Schwann cells: Nearly all of the cross-section-profiles from investigated group III fibers show gaps between the Schwann cell processes. At these sites the axolemma of the nerve fiber is separated from the surrounding connective tissue only by the basal lamina of the Schwann cells (*thick arrows* in Fig. 2b). The reconstructions reveal that the areas of bare axolemma form long continuous streaks on one or two sides of the nerve fiber and can comprise as much as 37% of the total fiber surface. The widths of these streaks varies, but usually it is greatest in bulged segments. Moreover, at these sites the axolemma tends to protrude through the gaps of the Schwann cell processes towards the adjacent tissue (left *thick arrow* in Fig. 2b). A preferential orientation of the bare areas towards specific structures, e.g., vessels or bundles of connective tissue, cannot be recognized.

Axoplasmatic bulges. The bulged segments of the axoplasm always contain in their cytoplasm accumulations of mitochondria orientated in parallel, and clusters of glycogen particles (Fig. 2b). In addition some clear or dense core vesicles are found. The axoplasm underlying the bare axolemma shows a faint filamentous substructure strongly resembling the classical "receptor matrix"

(Andres, 1969; Andres and von Düring, 1973). Microtubules and microfilaments extend over the total length of the nerve fiber. These microfilaments are bundled in a "*neurofilament core*" which is completely devoid of cell organelles, but within the bulges closely surrounded by mitochondria and glycogen particles (*thin arrow* in Fig. 2b). Microtubules rarely occur within this core but are localized inbetween and in close contact to the peripherally arranged mitochondria.

Final thickening. The final thickenings are the most distal bulges at the tips of the nerve fibers (Fig. 2a, 2c). In three-dimensional ultrastructure they closely resemble the bulges within the course of the fiber. Both final thickenings and bulges have identical size, are covered by equal areas of bare axolemma, and contain the same amount of mitochondria, glycogen particles and vesicles in their cytoplasm (compare Figs. 2b and 2c). They strikingly differ, however, in two parameters, and this allows a clear discrimination between bulges and final thickenings of group III fibers even in single sections: 1) The mitochondria are piled up at the end of the final thickenings and thus loose their parallel arrangement. 2) Microtubules and the "neurofilament core" are lacking.

Group IV fibers

Having left the perineurial sheath, group IV fibers run over a distance of up to several hundred (more than 300) microns through the connective tissue. As group III fibers, they may also give rise to several long branches during their course, but in addition they frequently show multiple branching at their very distal portion forming several short end-branches. The orientation of terminal group IV fibers is nearly always parallel to the course of the blood vessels (Fig. 3a). The mean diameter of group IV fibers ranges from about 0.30 to 0.55 μm. This diameter distribution is nearly identical to that of group IV fibers in the mainstem of the medial articular nerve (Heppelmann et al., 1988).

"Bare" axolemma. The axolemma of the terminal portion is not completely covered by Schwann cell processes but shows discontinuous patches or short streaks of bare areas. At bulged segments the bare nerve fiber membrane can extend up to half of the whole circumference. At these sites the membrane is separated from the adjacent tissue only by a basal lamina (*arrows* in Figs. 3b, 3c).

Axoplasmatic bulges. As in group III fibers, the terminal portion of group IV fibers consists of thick (bulged) and thin segments, which periodically alternate along the whole branches of the terminal tree. The bulges are 3 to 8 μm long; they are separated by thin segments of various length. The diameters range from about 0.2 μm at the thin segments to about 1.2 μm at the bulges.

All bulges of the nerve fibers richly contain mitochondria, glycogen particles, and frequently vesicles (Fig. 3b). The number of organelles and vesicles within the terminal portion considerably varies between different group IV fibers. *A striking difference to group III fibers is the complete absence of a neurofilament core.*

Final thickening. The terminal branches of group IV fibers end either as final thickenings, or they taper off more or less inconspicuously. Final thickenings show the same size and content of organelles as do the other bulges (Figs. 3b,

3c). Usually as in all other bulges, a fine membrane-related electron dense axoplasm is observed which strongly resembles the "receptor matrix" (Andres and von Düring, 1973). Unconspicuous end-branches are not thickened and include only few organelles, but frequently some clear vesicles.

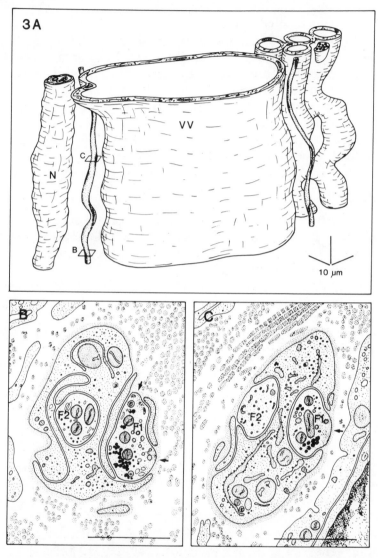

Figs. 3. Group IV fiber endings at a venous vessel in the medial collateral ligament.
A. Reconstruction based on serial semi- and ultrathin sections. Besides a small peripheral nerve (N) which contains one myelinated group III and several group IV fibers, 2 terminal nerve fiber bundles without perineurial sheath that enclose 2 group IV fibers (left) and one single fiber (right) run along the large venous vessel (VV). **B and C.** Cross sections of the nerve fiber bundle with two group IV fibers as marked in Fig. 3A at high magnification. Drawings from original electron micrographs; scale bar: 1 μm. **B.** Cross section showing a bulge of one of the nerve fibers (F1) with bare areas of the axolemma (arrows), mitochondria, glycogen particles, and vesicles. The other nerve fiber (F2) is completely ensheathed by the Schwann cell. **C.** Cross section 40 μm distal from that of Fig. 3B showing the final thickening of the nerve fiber F1 with bare area of axolemma (arrow). In single sections, the final thickenings show the same ultrastructure as do all the bulges and can only be identified by spatial reconstruction. The second nerve fiber (F2) continues.

In a few cases we have found group IV nerve fibers that simply end inside of the peripheral nerve, i.e., still surrounded by perineurium. These endings show the same ultrastructure as those outside the perineurium. The existance of such endings has already been shown in peripheral nerves of the Archilles tendon (Andres et al., 1985b).

DISCUSSION

The literature on sensory innervation of the connective tissue provides detailed accounts on the light and electron microscopical structure of corpuscular sensors, such as Golgi-Manzoni, Ruffini and Pacinian corpuscles (Polacek, 1966; Freeman and Wyke, 1967; Halata, 1975, 1977; Chouchkov, 1978; O'Connor and Gonzales, 1979; Halata and Munger, 1980; Halata et al., 1984, 1985; Strasmann et al., 1987; Munger et al., 1988; Ide et al., 1988). Close to nothing, however, is known on the ultrastructure - notably the three-dimensional ultrastructure - of fine afferent nerve fibers and their non-corpuscular or "free" nerve endings in connective tissue. Apart from the study of Andres et al. (1985b, Achilles tendon) only a few older light microscopical studies (Gardner, 1944; Polacek, 1966) have dealt with that topic for the articular capsule of the knee joint. Hence our reconstructions shed some new light on several aspects of this topic.

Nerve fibers with perineurium

All peripheral nerves are completely surrounded by a perineurial sheath, but for a few rather special exceptions (Stockinger, 1965: dental pulp; Ross and Burkel, 1970: acoustic nerve; Montes et al., 1984: Fig. 20, mouse skin). This perineurium constitutes a diffusion barrier which provides an enclosed environment for the nerve fibers within the endoneurial space, separated from the biochemical milieu of the general interstitial fluid of the body (for reviews see Shantha and Bourne, 1968; Low, 1976; Montes et al., 1984).

The endings of nerve fibers show characteristic differences in their perineurial wrapping that are directly related to their specific function. Thus in corpuscular endings of afferent nerve fibers, e.g., Pacinian or Ruffinian corpuscles, the perineurium continues around the nerve ending forming the capsule (Halata, 1975, 1977; Low, 1976; Chouchkov, 1978; Halata et al., 1985). In efferent nerve fibers the perineurium terminates at the motor end plates as an open-ended sleeve some distance before the end (Burkel, 1967; Saito and Zacks, 1969).

Nerve fibers without perineurium - The sensory terminal tree

According to our findings (see also Andres et al., 1985b), the lack of perineurium over *long* distances of fiber length is an important feature of "free nerve endings" in connective tissue. Afferent group III and IV fibers that lead to non-corpuscular endings, leave the perineurial sheath and penetrate the connective tissue or run in close contact to the blood vessel walls for several hundred microns. This *distal* or *terminal portion* of the nerve fibers lacks the blood-nerve-barrier provided by the perineurium, has furthermore no myelin sheath both in group III and IV fibers, and is thus directly exposed to any mechanical or chemical stimulus that arises in the connective tissue. The *proximal portion* of the nerve fibers, on the contrary, is much less accessible to chemical stimuli; it is always ensheathed by perineurium and - in case of group III fibers - myelinated.

In group III fibers obviously the proximal portion serves for impulse propagation only, there are no morphological structures which might be likely candidates for any sensory function of that part of the fiber. It is only the unmyelinated intermediate portion (a short transition zone of the fiber) inside and the long distal terminal portion of group III fibers outside of the perineurium that both show classical morphological indicators of receptive sites, namely swellings of the axoplasm ("bulges") stuffed with mitochondria, glycogen particles and some vesicles, as well as areas of "bare" axolemma that is not covered by Schwann cell processes (Andres, 1969; Andres and von Düring, 1973).

With regard to these well established criteria for receptive sites it is a new and most remarkable finding that the long rows of axoplasmatic bulges within the length of the distal portion show almost exactly the same ultrastructure as does the final thickening at the tip of each fiber branch; for a difference only the neurofilament core of the cytoskeleton is lacking in the latter. Hence the sensory function of the distal portion can impossibly be limited to the tip of the fiber exclusively, but must take place all along its branches, and this is why we have termed that part of the primary afferent neuron the *sensory terminal portion* or *sensory terminal tree* ("*sensible Endstrecke*").

To *group IV fibers* most facts and really all arguments apply that have already been discussed for group III. Group IV fibers also leave the perineurium, and the branches of the distal portion likewise consist of long rows of spindle-shaped axoplasmatic bulges that have areas of "bare" axolemma and are filled with mitochondria, glycogen particles, and vesicles. In fact, due to the lack of a neurofilament core in group IV fibers it is impossible to tell the final thickening from a bulge of the fiber without reconstruction. Hence we have also termed the distal portion of group IV fibers *sensory terminal portion* or *sensory terminal tree*.

Definition of the sensory terminal tree

We have defined as sensory terminal tree (or portion) of "free nerve endings" that distal part of the afferent nerve fiber which is *not surrounded by perineurium*. We have chosen this definition as it is the only one that can equally be applied both to group III and IV nerve fiber endings.

This is a positive, but not an all-including definition: All fibers that have fulfilled it in our material, will have a sensory function, as far as this can ever be judged from morphological parameters, but fibers not included by it may nevertheless carry receptive sites, and there are clear examples for this case. In group III fibers the short intermediate portion with perineurial but without myelin sheath, has an ultrastructure almost identical to that of the distal portion; and there exist group IV nerve fibers that simply end within the perineurium (Andres et al., 1985b; own observation).

Size of the sensory terminal tree

The sensory terminal portion of fine afferent nerve fibers branches several times and forms a mainly two-dimensional tree. While this sensory terminal tree of group III fibers innervates an area of up to 200 μm x 300 μm, the branches of the sensory terminal portion of group IV fibers extend over distances of more than 300 μm, thus innervating a considerably larger area than in group III. In this respect our ultrastructural data completely agree with findings of Stacey (1969), who examined the "free nerve endings" of silver-stained muscle afferents of the cat by light microscopy. He reported that the nerve endings ("axon terminals") of non-myelinated nerve fibers "are more diffuse in comparison to those of myelinated fibres, and spread out over a far greater area of tissue ... approximately 1000 μm x 300 μm" (Stacey, 1969, p. 243). - Interestingly, Stacey also described "neurofibrillary expansions of the axoplasm ... at regular intervals along the terminal" that "... occur with increasing frequency towards the tip of the terminal" (Stacey, 1969, p. 239-241). This description strongly resembles that of the axoplasmatic bulges evenly distributed over the whole terminal portion of fine afferent nerve fibers.

Discrimination of group III and group IV fiber endings

As group III nerve fibers have no more a myelin sheath in their intermediate and distal portion, the discrimination of group III and IV fibers in their terminal portion was impossible at the onset of our study. Therefore we traced all nerve fibers back up to the endpoint of perineurium, or perineurium and myelin sheath respectively. These long reconstructions with full identification of the fiber type revealed 3 morphological parameters, in which the sensory terminal portions of group III and group IV nerve fibers clearly differ. These 3 parameters allow for the first time the safe recognition of group III and group IV "free nerve endings" using short reconstructions (10-20 μm long) or even single sections only.

The first parameter for discrimination of group III and IV is the difference in *mean fiber diameter* of the terminal portion. Although the fiber size varies extremely between bulges and waist-like segments, the mean diameter of each fiber can be estimated rather exactly from the reconstruction of a 10-20 μm long segment of the nerve fiber. In this study the diameters ranged from 0.30 to 0.55 μm in group IV and from 1.00 to 1.50 μm in group III fiber endings. There is almost no overlap of the ranges, and this makes the mean fiber diameter a good criterion for discrimination.

As second parameter, the sensory terminal portion of group III nerve fibers contains a higher *amount of mitochondria* than in group IV. Single cross-sections of group III fibers frequently contain more than 10 profiles of mitochondria, but we never observed cross-sections of group IV fibers with more than 7 mitochondrial profiles.

As third and most important parameter, the sensory terminal portion only of group III fibers has a *neurofilament core*, i.e., a conspicuous bundle of parallel microfilaments sometimes intermingled with a few microtubules but free of mitochondria, glycogen particles and vesicles. This neurofilament core has been found in all cross-sections of group III fibers but for those of the final thickenings. It traverses the entire length of the sensory terminal tree, ending short of the fiber tips. The functional significance of this remarkable structure is unknown thusfar. Possibly the neurofilament core serves as a cytoskeletal "backbone" of the group III fibers.

REFERENCES

Andres, K. H., 1969, Zur Ultrastruktur verschiedener Mechanorezeptoren von höheren Wirbeltieren, *Anat. Anz.*, 124:551-565

Andres, K. H., and von Düring, M., 1973, Morphology of cutaneous receptors in: "Handbook of Sensory Physiology Vol. II," A. Iggo, ed., Springer Berlin, Heidelberg, New York: 3-28

Andres, K. H., and von Düring, M., 1981, General methods for characterization of brain regions in: "Techniques in Neuroanatomical Research," Ch. Heym and W. G. Forssmann, eds., Springer Berlin, Heidelberg, New York: 100-108

Andres, K. H., von Düring, M., Jänig, W., and Schmidt, R. F., 1985a, Degeneration pattern of postganglionic fibers following sympathectomy, *Anat. Embryol.*, 172:133-143

Andres, K. H., von Düring, M., and Schmidt, R. F., 1985b, Sensory innervation of the Achilles tendon by group III and IV afferent fibres, *Anat. Embryol.*, 172:145-156

Bauer, R., 1987, High-resolution imaging of thick biological specimens with an imaging Electron Energy Loss Spectrometer, *Zeiss Information MEM*, 5:10-15

Breathnach, A. S., 1971, "An Atlas of the Ultrastructure of Human Skin." J. & A. Churchill, London

Burkel, W. E., 1967, The histological fine structure of perineurium, *Anat. Rec.*, 158:177-190

Chouchkov, C., 1978, Cutaneous receptors, *Adv. Anat. Embryol. Cell Biol.*, 54:1-61

Freeman, M. A. R., and Wyke, B., 1967, The innervation of the knee joint: An anatomical and histological study in the cat, *J. Anat.*, 101:505-532

Gardner, E., 1944, The distribution and termination of nerves in the knee joint of the cat, *J. Comp. Neurol.*, 80:11-32

Grigg, P., Schaible, H.-G., and Schmidt, R. F., 1986, Mechanical sensitivity of Group III and IV afferents from posterior articular nerve in normal and inflamed cat knee, *J. Neurophysiol.*, 55:635-643

Halata, Z., 1975, The mechanoreceptors of the mammalian skin, ultrastructure and morphological classification, *Adv. Anat. Embryol. Cell Biol.*, 50-5:1-77

Halata, Z., 1977, The ultrastructure of the sensory nerve endings in the articular capsule of the knee joint of the domestic cat (Ruffini corpuscles and Pacinian corpuscles), *J. Anat.*, 124:717-729

Halata, Z., and Munger, B. L., 1980, The ultrastructure of Ruffini and Herbst corpuscles in the articular capsule of domestic pigeon, *Anat. Rec.*,, 198:681-692

Halata, Z., Badalamente, M. A., Dee, R., and Propper, M., 1984, Ultrastructure of sensory nerve endings in monkey (*Macaca fascicularis*) knee joint capsule, *J. Orthopaed. Res.*, 2:169-176

Halata, Z., Rettig, T., and Schulze, W., 1985, The ultrastructure of sensory nerve endings in the human knee joint capsule, *Anat. Embryol.*, 172:265-275

Heppelmann, B., Heuss, C., and Schmidt, R. F., 1988, Fiber size distribution of myelinated and unmyelinated axons in the medial and posterior articular nerves of the cat's knee joint, *Somatosens. Res.*, 5:273-281

Heppelmann, B., Meßlinger, K., and Schmidt, R. F. in press, Three-dimensional reconstruction of fine afferent nerve fibres in the knee joint capsule, *Europ. J. Cell Biol. (Proceedings) Suppl.*

Ide, C., Yoshida, Y., Hayashi, S., Takashio, M., and Munger, B. L., 1988, A re-evaluation of the cytology of cat Pacinian corpuscles. II. The extreme tip of the axon, *Cell Tissue Res.*, 253:95-103

Low, F. N.,, 1976, The perineurium and connective tissue of peripheral nerve in: "The Peripheral Nerve," D. N. Landon, ed., John Wiley and Sons, New York: 159-187

Montes, G. S., Cotta-Pereira, G., and Junqueira, L. C. U.,, 1984, The connective tissue matrix of the vertebrate peripheral nervous system in: "Advances in Cellular Neurobiology," S. Fedoroff, ed., Academic Press, Orlando: 177-218

Munger, B. L., Yoshida, Y., Hayashi, S., Osawa, T., and Ide, C., 1988, A re-evaluation of the cytology of cat Pacinian corpuscles. I. The inner core and clefts, *Cell Tissue Res.*, 253:83-93

O'Connor, B., and Gonzales, J., 1979, Mechanoreceptors of the medial collateral ligament of the cat knee joint, *J. Anat.*, 129:719-729

Polacek, P., 1966, Receptors of the joints. Their structure, variability and classification, *Acta Fac. Med. Univ. Brunensis*, 23:1-107

Ross, M. D., and Burkel, W. E., 1970, Electron microscopic observations of the nucleus, glial dome and meninges of the rat acoustic nerve, *Am. J. Anat.*, 130:73-92

Saito, A., and Zacks, S. I., 1969, Ultrastructure of Schwann and perineurial sheaths at the mouse neuromuscular junctions, *Anat. Rec.*, 164:379-390

Schaible, H.-G., and Schmidt, R. F., 1983a, Activation of Group III and IV sensory units in medial articular nerve by local mechanical stimulation of knee joint, *J. Neurophysiol.*, 49:35-44

Schaible, H.-G., and Schmidt, R. F., 1983b, Responses of fine medial articular nerve afferents to passive movements of knee joint, *J. Neurophysiol.*, 49:1118-1126

Schaible, H.-G., and Schmidt, R. F., 1985, Effects of an experimental arthritis on the sensory properties of fine articular afferent units, *J. Neurophysiol.*, 54:1109-1122

Shantha, T. R., and Bourne, G. H., 1968, The perineurial epithelium - a new concept *in*: "The Structure and Function of Nervous Tissue," G. Bourne, ed., Academic Press, New York: 379-459

Stacey, M. J., 1969, Free nerve endings in skeletal muscle of the cat, *J. Anat.*, 105:231-254

Stockinger, L., 1965, Nervenabschnitte ohne Perineurium, *Acta Anat.*, 60:244-252

Strasmann, T., Halata, Z., and Loo, S. K., 1987, Topography and ultrastructure of sensory nerve endings in the joint capsule of the Kowari (*Dasyuroides byrnei*), an Australian marsupial, *Anat. Embryol.*, 176:1-12

8

CARBONIC ANHYDRASE ACTIVITY OF PRIMARY AFFERENT NEURONS IN RAT: ATTEMPT AT MARKING FUNCTIONALLY RELATED SUBPOPULATIONS

M. Szabolcs, M. Kopp, and G. Schaden

Anatomisches Institut der Universität Wien, 3. Lehrkanzel, Wien, Austria

Earlier studies have described strong or moderate activities of carbonic anhydrase (CA; E.C.4.2.1.1) in primary afferent neurons with axons of intermediate or large size.[1-5] Enzyme activity of the individual neuron, however, remains constant throughout the entire length of its axis cylinder,[6] which greatly facilitates tracing of neurons with different carbonic anhydrase activities. This phenomenon led to the observation[4,6-8] that the spinal projection of the central processes of highly CA-positive dorsal root ganglion (DRG) cells corresponds well to the projection of type I muscle afferents. The assumption that the highly CA-reactive neurons are proprioceptive is corroborated by the fact that anulospiral nerve terminals enwrapping intrafusal muscle fiber are heavily stained.[8] The present study is an attempt at assigning such CA-reactive afferent nerve fibers to a functionally related subpopulation.[1,4,5]

We took advantage of the fact that the innervation of the rat trapezius muscle divides into an afferent and an efferent part. The *R. trapezius* of the cervical plexus contains only proprioceptive nerve fibers.[9,10] A branch of the spinal accessory nerve *(N. trapezius)* is responsible for the motor innervation.

We compared the distribution of CA-activity in muscle afferents of the *R. trapezius* with the activity in exteroceptive nerve fibers of the *N. auricularis magnus*, and in efferent nerve fibers of the *N. trapezius*. For this purpose cross sections of *N. auricularis magnus* and *N. trapezius* were analyzed proximally and distally to the union with the *R. trapezius* (Fig. 1). The CA-activity was demonstrated histochemically using Hansson's method.[11,12,13] The specifity of the reaction was verified by inhibiting the enzyme with acetoazolamide, or by omitting the substrate.[4,6,7,12]

By means of this technique numerous highly CA-reactive axons were observed only in nerve sections which contain proprioceptive nerve fibers (Fig. 2, 5, 6). The diameter of heavily stained nerve fibers ranged from 8 to 12 μm. Within this caliber range all proprioceptive nerve fibers displayed high staining

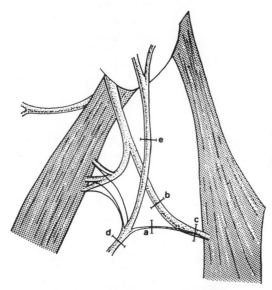

Fig. 1. R. trapezius of the Plexus cervicalis (**A**), proximal (**B**), and distal (**C**) cross section of the N. trapezius, N. auricularis magnus proximalis (**D**) and distalis (**E**).

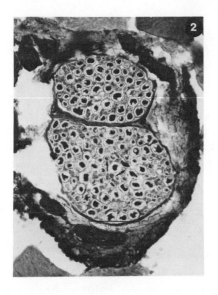

Fig. 2. Cross section of the R. trapezius of the Plexus cervicalis (proprioceptive nerve fibers). There is strong CA-activity in all large nerve fibers (8-12 µm). The thin CA-positive nerve fibers (arrow) have been identified as axons emerging from secondary nerve endings of muscle spindles, since the same nerve fibers can be found in capsules of muscle spindles.

intensity (Fig. 3). A minor part of exteroceptive nerve fibers (4% on average), however, were also highly CA-reactive (Fig. 7). None of the efferent nerve fibers between 7 and 15µm were ever CA-positive.

The strong deposition of reaction product in the respective nerve terminals[16,17] in muscle spindles (Fig. 8) and Golgi tendon organs (Fig. 9) confirms the diameter-related assignment of the highly CA-reactive, proprioceptive nerve fibers to types IA or IB.[14,15]

Nerve fibers of moderate CA-activity could be detected in all cross sections studied. Hence it may be assumed that this type of axons fulfills various functions. Among the axons displaying moderate reaction to carbonic anhydrase are, e.g., fusimotor nerve fibers (Fig. 4) and axons emerging from secondary endings of muscle spindles.

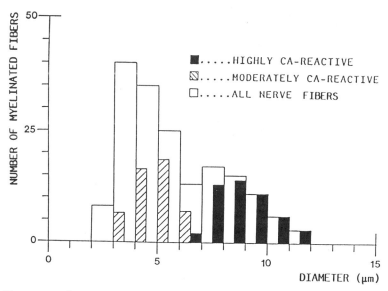

Fig. 3. Histogram of the R. trapezius indicates a strongly (8-12 μm) and a moderately (4-7 μm) CA-positive subpopulation. All thick nerve fibers stained strongly, as demonstrated by the equal height of full and hollow columns.

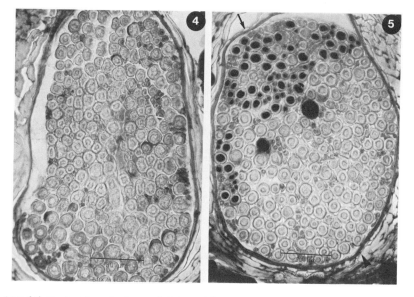

Figs. 4 and 5. Carbonic anhydrase staining of the proximal (**Fig. 4**) and distal (**Fig. 5**) cross sections of the N. trapezius. In the proximal cross section (purely efferent) all CA-positive nerve fibers are confined to the small- caliber classes. The large efferent fibers remain unstained. In the distal cross section a clear distinction can be made between the afferent (arrow) and efferent (arrow head) parts since only the large-diameter muscle afferents contain high amounts of reaction product.

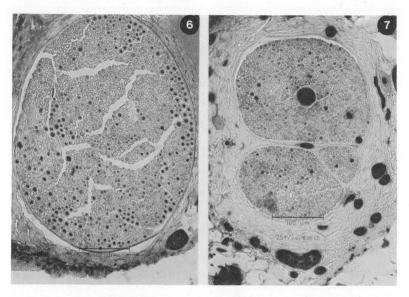

Figs. 6 and 7. Cross sections of the N. auricularis magnus proximalis (**Fig. 6**) and distalis (**Fig. 7**) stained for carbonic anhydrase. Highly CA-reactive axons can be detected only in the proximal cross section, but are absent in the distal cross section (exteroceptive nerve fibers).

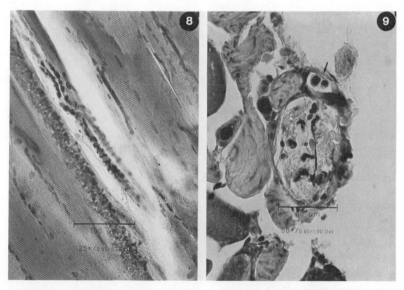

Fig. 8. Longitudinal section of a muscle spindle. Both the primary nerve endings and the appertaining IA fiber are highly CA-reactive.

Fig. 9. Cross section of a Golgi tendon organ. The IB fiber (arrow) and its endings are heavily stained.

REFERENCES

1   W. Crammer, R. Sacchi, and V. Sapirstein, Immunocytochemical localization of carbonic anhydrase in the spinal cord of normal and mutant (shiverer) adult mice with comparison among fixation methods, *J. Histochem. Cytochem.* 33:45 (1985).

2   B. Droz, and J. Kazimierczak, Carbonic anhydrase in primary-sensory neurons of dorsal root ganglia, *Comp. Biochem. Physiol.* 88B:713 (1987).

3   J. Kazimierczak, E. W. Sommer, E. Philippe, and B. Droz, Carbonic anhydrase activity in primary sensory neurons - I. Requirements for the cytochemical localization in the dorsal root ganglion of chicken and mouse by light and electron microscopy, *Cell Tissue Res.* 245:487 (1986).

4   D. A. Riley, S. Ellis, and J. Bain, Carbonic anhydrase activity in skeletal muscle fibre types, axons, spindles, and capillaries of the rat soleus and extensor digitorum longus muscles, *J. Histochem. Cytochem.* 30:1275 (1982).

5   V. Wong, C. P. Barrett, E. J. Donati, and L. Guth, Distribution of carbonic anhydrase activity in neurons of the rat, *J. Comp. Neurol.* 257:122 (1987).

6   V. Wong, C. P. Barrett, E. J. Donati, L. F. Eng, and L. Guth, Carbonic anhydrase activity in the first order sensory neuron of the rat, *J. Histochem. Cytochem.*, 31:293 (1983).

7   J. M. Peyronnard, J. P. Messier, L. Charron, J. Lavoie, F. X. Bergouignan, and M. Dubreuil, Carbonic anhydrase activity in the normal and injured peripheral nervous system of the rat, *Exp. Neurol.* 93:481 (1986).

8   D. A. Riley, S. Ellis, and J. L. W. Bain, Ultrastructural cytochemical localization of carbonic anhydrase activity in rat peripheral sensory and motor nerves, dorsal root ganglia and column nuclei, *Neuroscience* 13:189 (1984).

9   J. Gottschall, W. Zenker, W. Neuhuber, A. Mysicka, and M. Müntener, The sternomastoid muscle of rat and its innervation. Muscle fiber composition; perikarya and axons of efferent and afferent neurons, *Anat. Embryol.* 160:285 (1980).

10   E. B. Krammer, M. F. Lischka, T. P. Egger, M. Riedl, and H. Gruber, The motoneuronal organisation of the spinal accessory nuclear complex, *Adv. Anat. Embryol.* 103:1 (1987).

11   H. P. J. Hansson, Histochemical demonstration of carbonic anhydrase activity, *Histochemie* 11:112 (1967).

12   G. Lönnerholm, Carbonic anhydrase in rat liver, rabbit skeletal muscle; further evidence for the specifity of the histochemical cobalt-phosphate method of Hansson, *J. Histochem. Cytochem.* 28:427 (1980).

13   S. Rosen, and G. L. Musser, Observations on the specifity of newer histochemical methods for the demonstration of carbonic anhydrase activity, *J. Histochem. Cytochem.* 20:951 (1972).

14   C. C. Hunt, Relation of function to diameter in afferent fibres of muscle nerves, *J. Gen. Physiol.* 38:117 (1954).

15   D. P. C. Lloyd, Conduction and synaptic transmission of reflex responses to stretch in spinal cats, *J. Neurophysiol.* 6:317 (1943).

16   R. E. Burke, Firing patterns of gastrocnemius motor units in the decerebrate cat, *J. Physiol. (Lond.)* 196:631 (1968).

17   T. W. Schoultz, and J. E. Swett, The fine structure of the Golgi tendon organ, *J. Neurocytol.* 1:1 (1972).

9

AFFERENT INNERVATION OF THE ESOPHAGUS IN CAT AND RAT

Winfried L. Neuhuber[1] and Nadine Clerc[2]

Anatomisches Institut der Universität Zürich-Irchel
Zürich, Switzerland[1], and C.N.R.S., Laboratoire de
Neurobiologie, Marseille, France[2]

INTRODUCTION

Afferent neurons connecting the gastrointestinal tract with the brainstem and spinal cord are important for regulation and coordination of the various motor, absorptive and secretory functions of this organ system. Thus, afferent signals and feedback are as much important for, e.g., an orderly progression of swallowing (Falempin et al., 1986) as they are for defecation and continence (for reviews see Christensen, 1987; Gonella et al., 1987). Other gastrointestinal functions are also more or less dependent on, or influenced by, afferent information (for reviews see Mei, 1983, 1985; Roman and Gonella, 1987). Visceroafferent neurons not only transmit information from the periphery to the central nervous system, but seem to be involved also in local "effector" actions in various organs (for reviews see Dockray and Sharkey, 1986; Holzer et al., 1987; Maggi and Meli, 1988).

It appears natural to suspect the existence of specific neural structures as a basis for the various specific visceroafferent, e.g., muscular tension receptor and chemoreceptor, as well as for local effector functions. In addition, differences in the "molecular morphology" of sensory endings, e.g., equipment with specific membrane receptor molecules, may also account for specific functional properties. A functional specialization with respect to sensory modality is more obvious in vagal (Mei, 1983, 1985; Andrews, 1986) than in spinal (Jänig and Morrison, 1986) afferents. Thus, one might expect structural diversification, particularly of peripheral visceroafferent terminals, to occur in vagal rather than in spinal afferents.

Anterograde tracing techniques utilizing tritiated amino acids (Bower et al., 1978) or horseradish peroxidase (HRP; Marfurt and Turner, 1983) and its wheat germ agglutinin conjugate (WGA-HRP; Robertson and Aldskogius, 1982; Aldskogius et al., 1986) have opened new avenues for structural studies on peripheral ramifications of visceral afferents. Although correlation of anatomical data thus obtained with physiological results still relies on inference, comparisons with other sensory receptors and speculations, such studies certainly provide an important supplement to classical (for review see Kadanoff, 1966) and recent immunohistochemical (e.g., Gibbins et al., 1985) investigations. They also can help to clarify uncertainties as to the identification of peripheral fibers as visceroefferent or afferent (Brettschneider, 1966; see discussion in Stach, 1976).

This chapter deals with the afferent innervation of the esophagus in cat and rat, and focuses on recent data on distribution and structure of vagal afferent fibers within this organ. Electrophysiological studies on esophageal afferents, mainly in the cat, are reviewed, and attempts are made to reconcile anatomical and physiological results.

For methodological details, the reader is referred to the original papers.

TRACING STUDIES

Retrograde and anterograde tracing of esophageal afferents in cats

In the cat, tracing of afferent pathways has been performed using retrograde transport of HRP from the lower esophageal sphincter (LES) to nodose and dorsal root ganglia. It has been shown that the vagus and splanchnic nerves are not the only pathways involved in the afferent innervation of the LES since, after sectioning of the splanchnic nerves, labeled cell bodies have been still observed in thoracic dorsal root ganglia (Clerc, 1983, 1984). Complementary nerve section experiments have shown that sensory fibers originating from the LES run also in the sympathetic cardiac branch emerging from the stellate ganglion, and in the thoracic sympathetic branches. The involvement of these different pathways leads to a wide distribution of the cell bodies of LES sensory neurons over all thoracic and the first two lumbar spinal ganglia, whereas most of the dorsal root ganglion cells projecting through the major splanchnic nerve are concentrated at levels from T5 to T11 (Kuo and DeGroat, 1985). These different pathways may also exist, as corroborated by electrophysiological studies (Clerc and Mei, 1983), for the midesophagus and distal esophagus (identical with the thoracic esophagus below the aortic arch).

Anterograde tracing from the nodose ganglion with WGA-HRP has been used to label the peripheral processes of vagal sensory neurons innervating the esophagus, and in particular the LES (Clerc and Condamin, 1987; Clerc and Neuhuber, unpublished results). One can reasonably assert that this labeling is selective for vagal sensory fibers since, in the dorsal motor nucleus of the vagus, labeled cell bodies were either absent or sparse. In the cat, this labeling selectivity is favoured by the anatomical separation of sensory and motor components of the vagus nerve at the level of the nodose ganglion.

Two populations of WGA-HRP-labeled vagal sensory fibers might exist at least at the level of the LES (Clerc and Condamin, 1987): 1) fibers running from the serosa to the area located between longitudinal and circular muscle layers, called henceforth the "myenteric area" (Fig. 1); 2) fibers directly crossing the

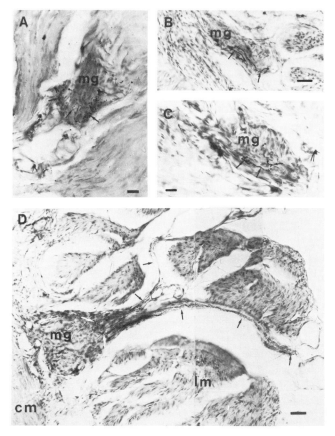

Fig. 1. Cross sections of the cat LES showing HRP-labeled vagal sensory fibers in myenteric ganglia after injection of 2 µl of 4% WGA-HRP into the nodose ganglion. In **A**, **B** and **C**, WGA-HRP-labeled vagal fibers (arrows) having sinuous pathways are shown within AChE-positive myenteric ganglia (mg). In **A** and **C**, vagal fibers have a varicose-like appearance (arrows). TMB/DAB/CoCl2 procedure combined with thiocholine histochemistry for AChE. In **D**, two bundles of labeled vagal sensory fibers (black arrows) are seen crossing the longitudinal muscle layer (lm), and entering a myenteric ganglion (mg). cm - circular muscle layer. TMB reaction, neutral red counterstain. Bars: 20 µm in A and C, and 40 µm in B and D.

longitudinal and circular muscle layers and passing to the mucosa. Fibers of both types enter the wall of the LES progressing between the longitudinal muscle fiber bundles without entering them and without showing any conspicuous ramification. Fibers belonging to the first type run in the myenteric area following either a circular or a longitudinal pathway. They might either terminate in the myenteric area, or, after a short course between both muscle layers, enter the circular muscle layer to reach the mucosa (Fig. 2). In the myenteric area, ramifications may be present but they were not visible because of the very sinuous pathway of the fibers. Fibers of the second type have never been found to ramify in the circular musculature. Instead, they run perpendicularly to the muscle fibers. At mucosal level, they display sinuous pathways and terminal arborizations which penetrate deeply into the epithelium. The absence of fiber ramifications in the circular muscle layer might indicate that these fibers do not terminate there. Seemingly they only cross this layer to reach the mucosa.

More recent observations (Clerc and Hamida, unpublished results) have shown that, in the myenteric area, the labeled fibers run either outside or inside myenteric ganglia. When they have an intraganglionic location, they may be sinuous (Fig. 1) or not (Fig. 2). Sometimes, they display a varicose-like appearance suggesting terminations of sensory fibers close to enteric neurons (Fig. 1a,c). The same pattern has been observed all along the esophagus from its cervical part to the LES (Clerc and Neuhuber, unpublished results). But at LES only, labeled fibers merely passing through the ganglion and following their pathway to the mucosa have been observed (Clerc and Hamida, unpublished results; Fig. 2). It is not possible to evaluate how frequently such a pathway occurs since it is only observable if the fibers are entirely within one sectional plane.

One can point out that relationships between vagal sensory fibers and myenteric ganglia do exist for fibers terminating in the myenteric area as well as for fibers innervating the mucosa. This latter feature might be related to the absence of any submucous plexus all along the length of the cat esophagus (Christensen and Robison, 1982).

Retrograde, transganglionic and anterograde tracing studies in rats

Retrograde tracing using HRP, WGA-HRP, cholera toxin-HRP and nuclear yellow has shown neurons in glossopharyngeal-vagal and cervicothoracic spinal ganglia which innervate the esophagus (Fryscak et al., 1984; Altschuler et al.,

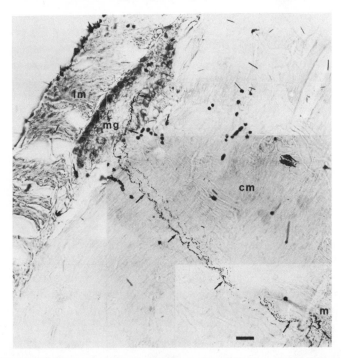

Fig. 2. Pathway of a bundle of WGA-HRP-labeled vagal sensory fibers innervating the mucosa in a cross section of the cat LES. Labeled fibers (arrows) are seen crossing a myenteric ganglion (mg), passing through the circular muscle layer (cm) perpendicularly to the muscle fibers and entering the mucosa (m) where their pathway becomes sinuous. TMB reaction, neutral red counterstain. Bar: 40 μm.

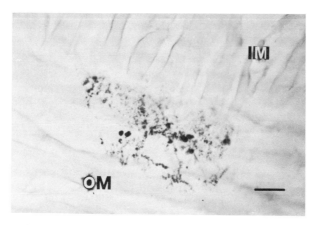

Fig. 3. WGA-HRP-labeled sensory vagal fibers arborizing between outer and inner muscle layers (OM, IM) of the rat cervical esophagus, giving rise to dense fine granular labeling and some bouton-like swellings. TMB reaction, no counterstain. Bar: 20 μm.

1989; Neuhuber, unpublished results). Transganglionic labeling of vagal afferent terminals has been detected in distinct subnuclei of the solitary complex. Most of the terminals can be localized in an area between the solitary tract and the lateral part of the dorsal motor nucleus of the vagus. This area has been included in the medial solitary subnucleus by Fryscak et al. (1984), and has been designated the central subnucleus of the solitary complex by Altschuler et al. (1989). The central subnucleus harbours neurons projecting to the esophagomotor compact formation of the nucleus ambiguus, thus forming part of a disynaptic reflex loop for the control of esophageal peristalsis (for references see Altschuler et al., 1989). Data on terminal fields of spinal afferents from the esophagus are not available.

Anterograde tracing has been performed as yet only in vagal afferents. Injections of WGA-HRP and H-3-leucine (Neuhuber, 1987), as well as free HRP (Neuhuber, unpublished results) have demonstrated labeled sensory fiber bundles entering the wall of the esophagus and reaching all layers of its cervical, thoracic and abdominal parts. However, there is an uneven distribution of labeling. Mucosal fibers have been frequently observed at the level of the upper sphincter but are rather sparse aboral to it. In the submucosa, afferent fibers forming complex loops ("Schlingenterritorien", Stöhr, 1957) occur along the entire length of the organ.

The most prominent afferent labeling has been found sandwiched between outer and inner muscle layers, in the myenteric area. Labeled afferent fibers form dense networks with bouton-like swellings interspersed (Fig. 3) which have been confirmed to be located almost exclusively around myenteric ganglia (Neuhuber, 1987). About 50% of esophageal myenteric ganglia are supplied by afferent fiber "baskets". These myenteric periganglionar sensory baskets are often interconnected by single fibers or bundles of fibers. There are no labeled fibers ramifying in the outer or inner muscle layers proper. This particular distribution of labeled afferent vagal fibers has been revealed with all three tracers. Myenteric fiber baskets have been found also in experiments in which supranodose vagotomy preceded tracer injections into nodose ganglia, thus corroborating their sensory nature.

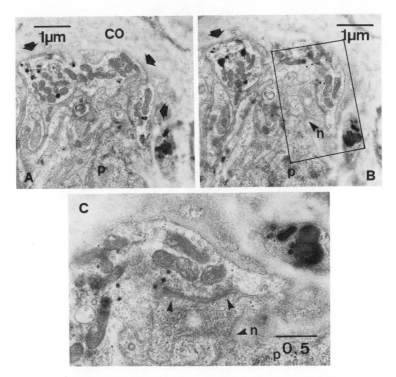

Fig. 4. WGA-HRP-labeled vagal sensory terminal (arrows, **A** and **B** represent two consecutive ultrathin sections) on the surface of a myenteric ganglion in the abdominal esophagus of the rat. Note mitochondria and vesicles of various sizes within the terminal, and the intimate relationship of the terminal to the periganglionar collagen (co). In **C** (inset from B), specialized, presumably synaptic contacts between the vagal sensory terminal and an unlabeled profile are indicated by arrowheads. Note accumulation of small clear vesicles on the "vagal" side of the contacts. The unlabeled "postsynaptic" profile appears to be contiguous through a tiny neck (n) with the perikaryon (p) of an enteric neuron.

Electron microscopy of labeled vagal sensory baskets has confirmed their location around and within myenteric ganglia (Fig. 4; Neuhuber, 1987). Enlargements measuring up to several microns are located predominantly on the surface of the ganglia, partly covered by glial cell processes, but more often facing the collagen fibrils, separated from them only by a basal lamina. Sometimes even a basal lamina seems to be absent. Thickening of the axolemma facing the collagen fibrils is present in many terminals. The often bizarre outlines of these profiles can be traced through several serial ultrathin sections leading to the impression of large laminar structures interdigitating with their finger- or spine-like processes with the periganglionar collagen on the one hand, and with the neuropil of the ganglion on the other hand. The terminals contain numerous mitochondria, neurotubules and scattered clear and large granular vesicles of various sizes embedded in a receptor matrix (cf., Andres and von Düring, this volume). This receptor matrix, however, appears to be less prominent than in other mechanoreceptors (von Düring et al., 1974; Andres et al., 1985). Typically, vesicles are clustered at specialized membrane contacts with unlabeled profiles interpreted as dendrites of enteric neurons (Fig. 4c). Thus, vagal sensory terminals seem to be presynaptic to intrinsic neurons. Many of these intrinsic profiles contain large granular vesicles.

ELECTROPHYSIOLOGICAL STUDIES

Studies in cats

Muscular receptors: A few electrophysiological studies have been devoted to the functional characteristics of the muscular receptors innervating the LES and the other parts of the esophagus of the cat. These mechanoreceptors are sensitive either to balloon distension or to muscle contraction, and their responses adapt slowly. They are connected to unmyelinated or small myelinated fibers in case of the esophagus (Mei, 1970; Harding and Titchen, 1975), or mainly to unmyelinated fibers in case of the LES, at least for the vagal muscular receptors (Mei et al., 1973; Clerc and Mei, 1983b). Some information has been provided on muscular receptors connected with neurons having two different origins: 1) vagal sensory neurons (Mei, 1970; Mei et al., 1973; Harding and Titchen, 1975; Clerc and Mei, 1983b); 2) dorsal root ganglion neurons (Clerc and Mei, 1983a). In a given location, the muscular receptors display the same functional properties whatever the origin of the sensory neurons to which they are connected, whereas they have a different sensitivity towards distension and contraction at esophageal and LES level. The esophageal receptors are more easily activated by distension than the LES ones, while the LES receptors are more sensitive to contraction than the esophageal ones (Clerc, unpublished results). These differences may reflect electrical, mechanical and functional properties of the muscle of the esophagus on the one hand, and of the LES on the other hand (Miolan and Niel, 1988). Such differences have been observed also in the different parts of the ferret stomach (Andrews et al., 1980).

Mucosal and epithelial receptors: Vagal rapidly adapting mechanoreceptors with "on-off" responses to distension have been described in the esophagus (Mei, 1970; Harding and Titchen, 1975). Some of them respond by increasing their discharge frequency in the presence of HCl (50 and 100 mM), hypertonic NaCl (more than 345 mM) and NaOH (50 mM). Since their responses are not correlated with a rise in the endoluminal pressure, these receptors are assumed to be located in the mucosa. They are connected to small myelinated fibers.

Esophageal vagal receptors insensitive to mechanical or chemical stimuli but sensitive to thermal stimulation have been recorded by El Ouazzani and Mei (1982). Three types of thermoreceptors have been identified: 1) warm receptors discharging between 39 and 50°C; 2) cold receptors responding between 10 and 35°C; 3) mixed receptors activated by both the warm and cold temperature scales. Their responses adapt slowly. From their short activation latency, it has been deduced that they are located in the vicinity of the epithelium. They are connected to unmyelinated fibers.

At LES level, Clerc and Mei (1983a,b) have looked for mucosal mechanoreceptors connected either with vagal or dorsal root ganglion neurons. The latter type has never been encountered. The LES vagal mucosal receptors are not sensitive to moderate balloon distension or muscle contraction. Occasionally they can be activated by strong distensions. The most efficient stimuli are digital compressions (accompanied by the rubbing of two areas of the mucosa onto each other, Fig. 5) and, more physiologically, mucosal stroking. The responses to these mechanical stimuli adapt rapidly. However, the same receptors can

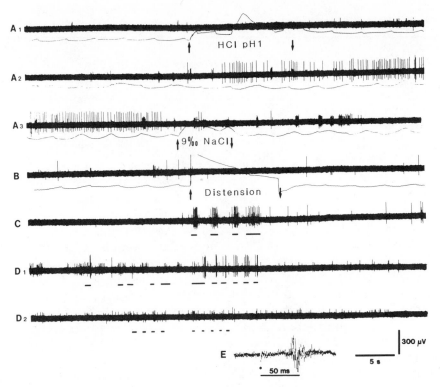

Fig. 5. Vagal mucosal receptor of the cat LES exhibiting properties of both rapidly adapting mechanoreceptors and acido-sensitive chemoreceptors. **A1, A2** and **A3**. The traces are continuous. Upper trace: Action potentials. Lower trace: LES endoluminal pressure. The receptor is activated by perfusion of the LES with chlorhydric acid solution (pH 1) after a 50 sec. latency. After rinsing with 0.9% NaCl, the unit corresponding to the acido-sensitive chemoreceptor remains silent. **B.** Upper trace: Action potentials. Lower trace: LES endoluminal pressure. LES distension does not activate the unit. **C.** LES compressions (bars under the action potential recording) induce concomittant bursts of action potentials. **D1** and **D2**. LES compressions are performed 30 sec. (in D1) and 90 sec. (in D2) after perfusion of the LES with a solution of 2% xylocaine. The response persists in D1 but is abolished in D2. **E.** Multiunit evoked potential including the unit (arrow) corresponding to the mucosal receptor studied.

display slowly-adapting responses to acid solutions: HCl pH 2.5-1 (Fig. 5), acetic acid pH 2 and 1.5, citric acid pH 1 (Clerc, 1984, and unpublished results). These receptors are activated neither by other chemical nor by thermal stimuli. Local anesthesia of the mucosa abolishes responses to mechanical and acid stimuli, which indicates that the receptors have an intramucosal location. All of them are related to unmyelinated fibers.

Serosal receptors: Serosal mechanoreceptors connected to dorsal root ganglion neurons have been evidenced in the thoracic esophagus (Clerc and Mei, 1983a). Serosal mechanoreceptors connected to both vagal and dorsal root ganglion neurons have been found at LES level (Clerc and Mei, 1983a,b). All the serosal receptors identified display the same modalities of activation and their responses adapt rapidly. They are activated by touching, stroking or stretching the serosa. They are affected neither by muscle contraction or distension, nor by stroking the mucosa. One of them has been stimulated by hot air at 50°C blown onto the serous membrane. Local anesthesia of the serosa prevents responses to the potent stimuli.

Studies in rats

In the rat, the *tunica muscularis* is composed of striated muscle fibers all along the esophagus (Gruber, 1968). In the cat, striated muscle fibers are present only in the cervical esophagus, whereas the thoracic esophagus and the LES have a smooth muscle wall. These structural differences must be reckoned with because of their consequences on esophageal motor function, and hence on the activation modalities of mechanoreceptors.

Receptors in the esophagus of the rat connected to vagal fibers have been studied by Andrew (1956, 1957) and Clarke and Davison (1975). Both, slowly adapting muscular tension receptors and mucosal receptors, have been described. Mucosal receptors are concentrated at the level of the upper sphincter (Andrew, 1957), which concurs well with the dense afferent fiber labeling as seen with anterograde tracing. The properties of these receptors do not differ markedly from those of similar receptors in other species (Iggo, 1957; Leek, 1969; Falempin et al., 1978; Andrews and Lang, 1982; see also preceding section).

DISCUSSION

One of the basic problems in studying afferent neurons is how the response characteristics to a particular stimulus can be correlated with the morphology of the receptory terminal. This morpho-functional correlation has been advanced to the limits of fine structural resolution and interpretability in "classical" sense organs as the eye and ear, and also in cutaneous receptors (for review see Iggo and Andres, 1982), although many problems remain unresolved (e.g., Munger and Ide, 1987). However, in visceral afferents, the uncertainties concern even more general questions, e.g., how to distinguish afferent and efferent terminals in the periphery, or as to the location of a mechanoreceptor within the different layers of a hollow organ. Electrophysiological studies have provided some information as to whether a mechanoreceptor might be situated in the muscle layers or in the mucosa (see above), but a correlation of the various neural structures as described by conventional techniques (Kadanoff, 1966; Rodrigo et al., 1975) with a particular afferent function has proved to be difficult.

Anterograde tracing of vagal sensory fibers innervating the esophagus demonstrates in a paradigmatic way how neural structures in the wall of the digestive tract can be identified as afferent and be described in terms of localization and ultrastructure. In both, cat and rat, vagal afferent fibers distribute primarily to the myenteric area (between outer and inner layers of the muscularis), and to the mucosa. Presumably, these two locations of terminal distribution reflect the existence of at least two different sets of afferent fibers, i.e., muscular and mucosal receptors as shown in electrophysiological studies. In the cat LES, afferent fibers are abundant in the mucosa, while at a comparable location only a few fibers can be demonstrated in the rat. Conversely, afferent vagal terminals in myenteric ganglia are profusely labeled in rats, whereas there is a comparably sparse varicose terminal-like labeling in myenteric ganglia of cats. These differences might reflect different transport conditions for WGA-HRP in different fiber types of the two species, and hence a different suitability of anterograde WGA-HRP tracing for the demonstration of mucosal versus myenteric fibers in cat and rat. Another possible reason for these discrepancies might be a different binding of WGA-HRP to, and uptake by subpopulations of nodose ganglion neurons giving rise to fibers innervating mucosa and myenteric ganglia, respectively, in the two species. Similar species-related technical differences might be the reason also for the demonstrability by WGA tracing of a retinopetal pathway in turtles but not in pigeons (Schnyder and Künzle, 1983). However, incubations of cat nodose ganglion sections with WGA-HRP results in labeling of the plasma membrane of most of the neurons (Clerc and Condamin, unpublished results), while other lectins such as *Ulex europaeus* agglutinin I may bind specifically to subsets of vagal sensory neurons in rabbits (Mori, 1987) and cats (Clerc, unpublished observation). Thus, different densities of labeled fibers in mucosa and myenteric area as found with WGA-HRP tracing in cats and rats are more likely to be due to species differences in the number of afferent fibers in the respective layers.

Anterograde tracing - a step further towards a morpho-functional correlation

Vagal afferent fibers which can be traced to the mucosa and even into the epithelium (Clerc and Condamin, 1987; Neuhuber, 1987; see also Robles-Chillida et al., 1981) obviously represent a substrate for mucosal mechano-, thermo- and chemoreceptors as described above.

As suggested by results obtained in both cat and rat, two distinct populations of sensory fibers distribute in the myenteric area and might have relationships with myenteric ganglia. The first one corresponds to fibers crossing the myenteric ganglia to the mucosa and possibly give rise to intraganglionic collaterals. The second one corresponds to arborized terminals in myenteric ganglia. These two different fiber populations may be related to different functions. The fibers crossing the myenteric ganglia to the mucosa and their putative collaterals could be involved in an axon reflex arrangement (for critical reviews see Holzer, 1988; Maggi and Meli, 1988). Such a reflex, mediated by sensory fiber collaterals via intramural cholinergic neurons, has been suggested to be responsible for the gastric contractions induced by mucosal nociceptive stimulation (Delbro, 1985). The idea of an axon reflex setting has also been adopted by other authors interpreting the results of their immunohistochemical (e.g., Green and Dockray, 1988) and tracing (Sato and Koyano, 1987) studies. The arborized terminals in myenteric ganglia, on the other hand, are candidates, *per exclusionem*, for muscular tension receptors, since fiber ramifications within the outer and inner muscle layers proper unexpectedly could be demonstrated neither in cats nor in rats.

The idea that vagal myenteric terminals subserve the function of muscular tension receptors is further supported by a look on their ultrastructure (Neuhuber, 1987, and this chapter). They form large endings lying on the surface of the ganglion in close association with periganglionar connective tissue elements. On the other hand, they are interwoven with the neuropil of the ganglion. The intimate contact of the terminals with collagen fibrils without intervening glial processes bears resemblance to other mechanoreceptors (e.g., pulmonary stretch receptors: von Düring et al., 1974; Krauhs, 1984; Golgi tendon organs: Schoultz and Swett, 1974; Ruffini corpuscules: Chambers et al., 1972; Halata, 1988; arterial baroreceptors: Böck and Gorgas, 1976; Krauhs, 1979). This strongly suggests, by analogy from criteria for sensory transduction areas in other afferents (Andres et al., 1985; see also chapters 1 and 7, this volume), that vagal terminals are receptive structures. Their particular location predestines myenteric terminals to sensors for sliding movements of the two muscle layers against each other, since shearing forces generated during these movements may be transmitted by collagen fibrils to the surface of myenteric ganglia, resulting in distortion primarily of the vagal terminals. Myenteric ganglia may also be compressed and flattened by distension of the esophagus (cf., Gabella and Trigg, 1984), which may likewise stimulate the vagal terminals. Both, tangential distortion and compression of enteric ganglia, occur during peristalsis. It is noteworthy that tension receptors are readily excited by balloon distension and contraction (see above), and by tangential lengthening of the muscle layers (Leek, 1972, and personal communication) which all produce shearing forces in the myenteric area. Moreover, mechanoreceptive units with response characteristics similar to those of muscular receptors have been recorded from myenteric ganglia (Wood, 1987). These physiological and anatomical results point to a priviledged role of myenteric ganglia in mechanoreception, and apparently support ideas expressed already by Nonidez (1946) and Gabella (1972) on a possible afferent nature of terminals located on the surface of myenteric ganglia.

Myenteric vagal afferent terminals as demonstrated in the rat with anterograde WGA-HRP tracing most likely correspond to intraganglionic laminar endings (IGLEs) as described with conventional staining methods in various species (dog: Nonidez, 1946; cat and rhesus monkey: Rodrigo et al., 1975, 1982; opossum: Christensen et al., 1987). Although vagal afferent fibers and terminals have been shown by anterograde tracing also in cat myenteric ganglia, they do not resemble IGLEs. On the other hand, IGLEs in cats degenerate upon vagal deafferentation (Rodrigo et al., 1982) which suggests their sensory nature. Thus, it seems that WGA-HRP tracing is less well suited for demonstration of IGLEs in cats than in rats. Extension of these tracing experiments to other species would be of interest.

Myenteric vagal afferent terminals contain mitochondria, often accumulated in enlargements, and clear and granular vesicles of various sizes embedded in a receptor matrix. Most of these features are common to receptory terminals in various locations (Andres, 1966; Hamilton, 1968; von Düring and Andres, 1969; Schoultz and Swett, 1974; McDonald and Mitchell, 1975; Krauhs, 1979; Ide et al., 1988). Vesicles have been assumed to play a role in the process of sensory transduction (for review see Schoultz and Swett, 1974). However, vesicles in myenteric vagal afferent terminals are frequently clustered at specialized membrane contacts with neuronal processes of myenteric ganglion cells. Thus, vagal afferent terminals appear to be, in addition to their likely mechanoreceptive function, presynaptic to enteric neurons. This interpretation of the ultrastructural features implies a local effector function of vagal afferent neurons similar to that proposed for capsaicin-sensitive peptide-containing primary afferents (for reviews see Holzer, 1988; Maggi and Meli, 1988), and recently also

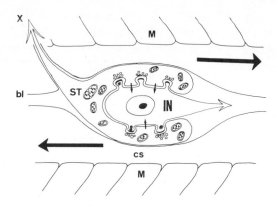

Fig. 6. A hypothesis suggesting a dual sensory - local effector role for vagal sensory terminals (ST) in a myenteric ganglion. During peristalsis, tangential forces (large black arrows) produced by sliding of the two muscle layers (M) against each other result in deformation of the myenteric ganglion. This deformation is registered by the sensory terminal and triggers both, an action potential transmitted by the vagus nerve (X) to the brainstem, and simultaneously also synaptic transmitter release (small arrows) upon an intrinsic neuron (IN). Thus, the same mechanoreceptor may be integrated in extrinsic (e.g. vago - vagal) as well as intrinsic reflex circuits. cs - collagen space.

for vestibular afferent endings on type I hair cells (Scarfone et al., 1988). A presynaptic position of vagal mechanoreceptive terminals, although shown here only for the abdominal esophagus of the rat, may be considered as the morphological substrate for the synaptic action of myenteric mechanosensitive "units", which may well be identical with peripheral vagal sensory terminals, upon other enteric neurons in the small intestine of the guinea-pig (Wood, 1987). This hypothesis is summarized in Fig. 6. A vagal sensory terminal located on the surface of a myenteric ganglion forms extensive contacts with dendrites of an enteric neuron. Stimulation by tangential forces during peristalsis not only might generate an action potential transmitted via the vagus nerve to the brainstem, but also local release of a neuroactive substance onto the enteric neuron. Conversely, substances released from large granular vesicles in dendrites of enteric neurons may influence the vagal afferent terminal in a reciprocal way. Although no reciprocal synapses similar to those seen in the carotid body (McDonald and Mitchell, 1975) have been found, a non-synaptic release of the content of large granular vesicles (e.g., Benedeczky and Halasy, 1988) seems possible. Thus, one and the same mechanoreceptor appears to be part of extrinsic as well as intrinsic reflex circuits. Moreover, the vagal terminal-enteric neuron complex could represent a device for processing of mechanosensory information *in loco*. It can be assumed that these proposed functions of vagal sensory myenteric terminals in the esophagus are important for the coordination of swallowing.

It is obvious that this hypothesis needs to be tested, and that the investigations performed on the esophagus should be extended to other parts of the digestive tract. Preliminary results (Neuhuber, unpublished) indicate that myenteric vagal afferent terminals are not confined to the esophagus and cardia, but can also be found in other parts of the stomach and in the duodenum. They are certainly not the only structure which may be devoted to tension perception in the digestive tract. Anterograde tracing from nodose and sacral spinal ganglia have shown dense afferent innervation of the circular muscle layers of the pyloric and internal anal sphincters in the rat (Neuhuber, to be published). Thus, different regions of the digestive tract may be supplied with different types of tension receptors.

Although the attempt of a morpho-functional correlation as delineated above relies more on inference than direct evidence, and it still is speculative, the concept based on the present results may provide a guideline for further, more detailed, investigations.

Acknowledgements: Part of these studies were supported by Swiss National Foundation grant 3.555-0.86 (to W.L.N.). The excellent technical assistance of Mrs S. Richter is greately acknowledged.

REFERENCES

Aldskogius, H., Elfvin, L.-G., and Andersson Forsman, C., 1986, Primary sensory afferents in the inferior mesenteric ganglion and related nerves of the guinea pig, *J. Autonom. Nerv. Syst.*, 15:179-190

Altschuler, S. M., Bao, X., Bieger, D., Hopkins, D. A., and Miselis, R. R., 1989, Viscerotopic representation of the upper alimentary tract in the rat: Sensory ganglia and nuclei of the solitary and spinal trigeminal tracts, *J. Comp. Neurol.*, 283:248-268

Andres, K. H., 1966, Ueber die Feinstruktur der Rezeptoren an Sinushaaren, *Z. Zellforsch.*, 75:339-365

Andres, K. H., Düring, M. von, and Schmidt, R. F., 1985, Sensory innervation of the Achilles tendon by group III and IV afferent fibers, *Anat. Embryol.*, 172:145-156

Andrew, B. L., 1956, The nervous control of the cervical oesophagus of the rat during swallowing, *J. Physiol. (Lond.)*, 134:729-740

Andrew, B. L., 1957, Activity in afferent nerve fibers from the cervical oesophagus, *J. Physiol. (Lond.)*, 135:54-55P

Andrews, P. L. R., 1986, Vagal afferent innervation of the gastrointestinal tract, *in:* "Visceral Sensation, Progress in Brain Research, Vol. 67," F. Cervero and J. F. B. Morrison, eds., Elsevier, Amsterdam: 65-86

Andrews, P. L. R., Grundy, D., and Scratcherd, T., 1980, Vagal afferent discharge from mechanoreceptors in different regions of the ferret stomach, *J. Physiol. (Lond.)*, 298:513-524

Andrews, P. L. R., and Lang, K. M., 1982, Vagal afferent discharge from mechanoreceptors in the lower oesophagus of the ferret, *J. Physiol. (Lond.)*, 332:29P

Benedeczky, I., and Halasy, K., 1988, Visualization of non-synaptic release sites in the myenteric plexus of the snail *Helix pomatia*, *Neuroscience*, 25:163-170

Bower, A.J., and Parry, K., 1978, A demonstration of the use of autoradiography as a method of exploring afferent autonomic innervation of the respiratory tract, *J. Physiol. (Lond.)*, 281:3-4P

Böck, P., and Gorgas, K., 1976, Fine structure of baroreceptor terminals in the carotid sinus of guinea pigs and mice, *Cell. Tiss. Res.*, 170:95-112

Brettschneider, H., 1966, Ultrastruktur der visceralen Rezeptoren und afferenten Nerven, *Acta neuroveget.*, 28:37-102

Chambers, M.R., Andres, K. H., Düring, M. von, and Iggo, A., 1972, Structure and function of the slowly adapting type II mechanoreceptor in hairy skin, *Quart. J. Exp. Physiol.*, 57:417-445

Christensen, J., 1987, Motility of the colon, *in:* "Physiology of the Gastrointestinal Tract, 2nd Ed.," L. R. Johnson, ed., Raven Press, New York: 665-693

Christensen, J., Rick, G. A., and Soll, D. J., 1987, Intramural nerves and interstitial cells revealed by the Champy-Maillet stain in the opossum esophagus, *J. Autonom. Nerv. Syst.*, 19:137-151

Christensen, J., and Robison, G. A., 1985, Nerve cell density in submucous plexus throughout the gut of the cat and opossum, *Gastroenterology*, 89:1064-1069

Clarke, G. D., and Davison, J. S., Tension receptors in the oesophagus and stomach of the rat, *J. Physiol. (Lond.)*, 244:41-42P

Clerc, N., 1983, Afferent innervation of the lower oesophageal sphincter of the cat. An HRP study, *J. Autonom. Nerv. Syst.*, 9:623-636

Clerc, N., 1984, Afferent innervation of the lower oesophageal sphincter of the cat. Pathways and functional characteristics, *J. Autonom. Nerv. Syst.*, 10:213-217

Clerc, N., and Condamin, M., 1987, Selective labeling of vagal sensory nerve fibers in the lower esophageal sphincter with anterogradely transported WGA-HRP, *Brain Res.*, 424:216-224

Clerc, N., and Mei, N., 1983 a, Thoracic esophageal mechanoreceptors connected with fibers following sympathetic pathways, *Brain Res. Bull.*, 10:1-7

Clerc, N., and Mei, N., 1983 b, Vagal mechanoreceptors located in the lower oesophageal sphincter of the cat, *J. Physiol. (Lond.)*, 336:487-498

Delbro, D., 1985, The role of substance P in the control of gut motility, in: "Tachykinin Antagonists," R. Hakanson and F. Sundler, eds., Elsevier, Amsterdam: 223-230

Dockray, G. J., and Sharkey, K. A., 1986, Neurochemistry of visceral afferent neurones, in: "Visceral Sensation, Progress in Brain Research, Vol. 67," F. Cervero and J. F. B. Morrison, eds., Elsevier, Amsterdam: 133-148

Düring, M. von, and Andres, K. H., 1969, Zur Feinstruktur der Muskelspindel von Mammalia, *Anat. Anz.*, 124:566-573

Düring, M. von, Andres, K. H., and Iravani, J., 1974, The fine structure of the pulmonary stretch receptor in the rat, *Z. Anat. Entwickl.-Gesch.*, 143:215-222

El Ouazzani, T., and Mei, N., 1982, Electrophysiologic properties and role of the vagal thermoreceptors of lower esophagus and stomach in the cat, *Gastroenterology*, 83:995-1001

Falempin, M., Madhloum, A., and Rousseau, J. P., 1986, Effects of vagal deafferentation on oesophageal motility and transit in the sheep, *J. Physiol. (Lond.)*, 372:425-436

Falempin, M., Mei, N., and Rousseau, J. P., 1978, Vagal mechanoreceptors of the inferior thoracic oesophagus, the lower oesophageal sphincter and the stomach in the sheep, *Pflügers Arch.*, 373:25-30

Fryscak, T., Zenker, W., and Kantner, D., 1984, Afferent and efferent innervation of the rat esophagus. A tracing study with horseradish peroxidase and nuclear yellow, *Anat. Embryol.*, 170:63-70

Gabella, G., 1972, Fine structure of the myenteric plexus in the guinea pig ileum, *J. Anat.*, 111:69-97

Gabella, G., and Trigg, P., 1984, Size of neurons and glial cells in the enteric ganglia of mice, guinea-pigs, rabbits and sheep, *J. Neurocytol.*, 13:49-71

Gibbins, I. L., Furness, J. B., Costa, M., MacIntyre, I., Hillyard, C. J., and Girgis, S., 1985, Co-localization of calcitonin gene-related peptide-like immunoreactivity with substance P in cutaneous, vascular and visceral sensory neurons of guinea pigs, *Neurosci. Lett.*, 57:125-130

Gonella, J., Bouvier, M., and Blanquet, F., Extrinsic nervous control of motility of small and large intestines and related sphincters, *Physiol. Rev.*, 67:902-962

Green, T., and Dockray, G. J., 1988, Characterization of the peptidergic afferent innervation of the stomach in the rat, mouse and guinea-pig, *Neuroscience*, 25:181-193

Gruber, H., 1968, Ueber Struktur und Innervation der quergestreiften Muskulatur des Oesophagus der Ratte, *Z. Zellforsch.*, 91:236-247

Halata, Z., 1988, Ruffini corpuscule - a stretch receptor in the connective tissue of the skin and locomotion apparatus, in: "Transduction and Cellular Mechanisms in Sensory Receptors, Progress in Brain Research, Vol. 74," W. Hamann and A. Iggo, eds., Elsevier, Amsterdam: 221-229

Hamilton, D. W., 1968, The calyceal synapse of type I vestibular hair cells, *J. Ultrastruct. Res.*, 23:98-114

Harding, R., and Titchen, D. A., 1975, Chemosensitive vagal endings in the esophagus of the cat, *J. Physiol. (Lond.)*, 247:52-53P

Holzer, P., 1988, Local effector functions of capsaicin-sensitive sensory nerve endings: involvement of tachykinins, calcitonin gene-related peptide and other neuropeptides, *Neuroscience*, 24:739-768

Holzer, P., Schluet, W., Lippe, Th., and Sametz, W., 1987, Involvement of capsaicin-sensitive sensory neurons in gastrointestinal function, *Acta Physiol. Hung.*, 69:403-411

Ide, C., Yoshida, Y., Hayashi, S., Takashio, M., and Munger, B. L., 1988, A re-evaluation of the cytology of cat Pacinian corpuscles. II. The extreme tip of the axon, *Cell. Tiss. Res.*, 253:95-103

Iggo, A., 1957, Gastro-intestinal tension receptors with unmyelinated afferent fibres in the vagus of the cat, *Quart. J. Exp. Physiol.*, 42:130-141

Iggo, A., and Andres, K. H., 1982, Morphology of cutaneous receptors, *Ann. Rev. Neurosci.*, 5:1-31

Jänig, W., and Morrison, J. F. B., 1986, Functional properties of spinal visceral afferents supplying abdominal and pelvic organs, with special emphasis on visceral nociception, in: "Visceral Sensation, Progress in Brain Research, Vol. 67," F. Cervero and J. F. B. Morrison, eds., Elsevier, Amsterdam: 87-114

Kadanoff, D., 1966, Histologie visceraler Rezeptoren und visceral-afferenter Nerven, *Acta neuroveget.*, 28:4-36

Krauhs, J. M., 1979, Structure of rat baroreceptors and their relationship to connective tissue, *J. Neurocytol.*, 8:401-414

Krauhs, J. M., 1984, Morphology of presumptive slowly adapting receptors in dog trachea, *Anat. Rec.*, 210:73-85

Kuo, D. C., and deGroat, W. C., 1985, Primary afferent projections of the major splanchnic nerve to the spinal cord and gracile nucleus of the cat, *J. Comp. Neurol.*, 231:421-434

Leek, B. F., 1969, Reticulo-ruminal mechanoreceptors in sheep, *J. Physiol. (Lond.)*, 202:585-609

Leek, B. F., 1972, Abdominal visceral receptors, *in:* "Handbook of Sensory Physiology, Vol. 3," E. Neil, ed., Springer, Berlin: 113-160

Maggi, C. A., and Meli, A., 1988, The sensory-efferent function of capsaicin-sensitive sensory neurons, *Gen. Pharmac.*, 19:1-43

Marfurt, C. F., and Turner, D. F., 1983, Sensory nerve endings in the rat oro-facial region labeled by the anterograde and transganglionic transport of horseradish peroxidase: a new method for tracing peripheral nerve fibers, *Brain Res.*, 261:1-12

McDonald, D. M., and Mitchell, R. A., 1975, The innervation of glomus cells, ganglion cells and blood vessels in the rat carotid body: a quantitative ultrastructural analysis, *J. Neurocytol.*, 4:177-230

Mei, N., 1970, Mécanorécepteurs vagaux digestifs chez le chat, Exp. Brain Res., 11:502-514

Mei, N., 1983, Sensory structures in the viscera, *in:* "Progress in Sensory Physiology, Vol. 4," D. Ottoson, ed., Springer, Berlin: 1-42

Mei, N., 1985, Intestinal chemosensitivity, *Physiol. Rev.*, 65:211-237

Mei, N., Aubert, M., Crousillat, J., and Ranieri, F., 1973, Sensory innervation of the lower oesophagus of the cat. Comparison with other parts of the digestive system, *Proc. IVth Internat. Symp. Gastrointest. Mot.*, Banff, Canada

Miolan, J. P., and Niel, J. P., 1988, Involvement of cholinergic neurones in the permanent depolarization of the muscular cells of the lower oesophageal sphincter in the cat *in vitro*, *J. Physiol. (Lond.)*, 406:138P

Mori, K., 1987, Specific carbohydrate expression by small-diameter subclasses of rabbit trigeminal, glossopharyngeal, and vagal afferent fibers studied with the lectin *Ulex europaeus* agglutinin I, *Neurosci. Res.*, 4:291-303

Munger, B. L., and Ide, C., 1987, The enigma of sensitivity in Pacinian corpuscules: a critical review and hypothesis of mechano-electric transduction, *Neurosci. Res.*, 5:1-15

Neuhuber, W. L., 1987, Sensory vagal innervation of the rat esophagus and cardia: a light and electron microscopic anterograde tracing study, *J. Autonom. Nerv. Syst.*, 20:243-255

Nonidez, J. F., 1946, Afferent nerve endings in the ganglia of the intramural plexus of the dog's oesophagus, *J. Comp. Neurol.*, 85:177-189

Robertson, B., and Aldskogius, H., 1982, The use of anterogradely transported wheat germ agglutinin-horseradish peroxidase conjugate to visualize cutaneous sensory nerve endings, *Brain Res.*, 240:327-330

Robles-Chillida, E. M., Rodrigo, J., Mayo, I., Arnedo, A., and Gomez, A., 1981, Ultrastructure of free-ending nerve fibers in oesophageal epithelium, *J. Anat.*, 133:227-233

Rodrigo, J., deFelipe, J., Robles-Chillida, E. M., Pérez Anton, J. A., Mayo, I., and Gomez, A., Sensory vagal nature and anatomical access paths to esophagus laminar nerve endings in myenteric ganglia. Determination by surgical degeneration methods, *Acta anat.*, 112:47-57

Rodrigo, J., Hernandez, C. J., Vidal, M. A., and Pedrosa, J. A., 1975, Vegetative innervation of the esophagus. II. Intraganglionar laminar endings, *Acta anat.*, 92:79-100

Roman, C., and Gonella, J., 1987, Extrinsic control of digestive tract motility, *in:* "Physiology of the Gastrointestinal Tract, 2nd Ed.," L. R. Johnson, ed., Raven Press, New York: 507-553

Sato, M., and Koyano, H., 1987, Autoradiographic study on the distribution of vagal afferent nerve fibers in the gastroduodenal wall of the rabbit, *Brain Res.*, 400:101-109

Scarfone, E., Demêmes, D., Jahn, R., De Camilli, P., and Sans, A., 1988, Secretory function of the vestibular nerve calyx suggested by presence of vesicles, synapsin I, and synaptophysin, *J. Neurosci.*, 8:4640-4645

Schnyder, H., and Künzle, H., 1983, Differential labeling in neuronal tracing with wheat germ agglutinin, *Neurosci. Lett.*, 35:115-120

Schoultz, T. W., and Swett, J. E., 1974, Ultrastructural organization of the sensory fibers innervating the Golgi tendon organ, *Anat. Rec.*, 179:147-162

Stach, W., 1976, Afferente Nervenendigungen im Dünndarm und Magen. Licht- und elektronenmikroskopische Untersuchungen, *Z. mikrosk.-anat. Forsch.*, 90:790-800

Stöhr jr., Ph., 1957, Mikroskopische Anatomie des vegetativen Nervensystems, *in:* "Handbuch der mikroskopischen Anatomie des Menschen, Vol. IV/5," W. Bargmann, ed., Springer, Berlin

Wood, J. D., 1987, Physiology of the enteric nervous system, *in:* "Physiology of the Gastrointestinal Tract, 2nd Ed.," L. R. Johnson, ed., Raven Press, New York: 67-109

10

THE DEVELOPMENT OF PRIMARY SENSORY NEURONS

Alun M. Davies

Department of Anatomy, St. George's Hospital
Medical School, London, England

INTRODUCTION

Primary sensory neurons originate from progenitor cells of ectodermal origin. These progenitors migrate to the sites of developing sensory ganglia where they differentiate. Sensory neurons are initially bipolar and extend two axonal processes in opposite directions towards their peripheral and central target fields. The innervation of these target fields is associated with a period of neuronal death during which superfluous neurons are eliminated followed by a period of modification and refinement of connections.

The accessibility of sensory ganglia from the earliest stages of their development is the main reason why much of our understanding of the cellular and molecular basis of neuronal development is founded on sensory neurons. The target fields of certain sensory ganglia are also well-defined and accessible for experimental studies, and this has permitted direct investigation of the regulatory influence of the target field on neuronal development.

SENSORY NEURON PROGENITOR CELLS

Cell tracing studies, largely carried out in avial embryos, have shown that sensory neurons differentiate from progenitor cells that originate from thickenings of the embryonic ectoderm, the neural crest and neurogenic placodes (Noden, 1978; Narayanan and Narayanan, 1978; Le Douarin, 1982; Le Douarin et al., 1986; D'Amico-Martel and Nodon, 1983). The neurons of dorsal root ganglia and certain cranial sensory ganglia originate from the neural crest, whereas the neurons of the remaining cranial sensory ganglia originate from neurogenic placodes (Table 1). Placode-derived neurons start differentiating before neural crest-derived neurons and are initially larger than these neurons. The satellite cells and Schwann cells associated with all sensory ganglia arise exclusively from the neural crest.

TABLE 1   Ectodermal derivation of primary sensory neurons

Neural crest-derived	Placode-derived
Dorsomedial part of the trigeminal ganglion	Ventrolateral part of the trigeminal ganglion
Trigeminal mesencephalic nucleus	Geniculate ganglion
	Vestibular ganglion
Root ganglion of facial nerve (embedded in the vestibular ganglion)	Acoustic ganglion
	Petrosal ganglion
Root ganglion of the glossopharyngeal nerve	Nodose ganglion
Jugular ganglion	
Dorsal root ganglia	

The neural crest gives rise to a wide variety of cell types in addition to those of the peripheral nervous system. Cell lineage studies have shown that neural crest cells are not committed to single fate prior to migration (Bronner-Fraser and Fraser, 1988). Cells labelled at this stage give rise to progeny in the majority of crest-derived tissues. As cells migrate from the crest to the periphery their potential to give rise to several different tissues or cell types becomes progressively restricted (Baroffio et al., 1988). In particular, separate progenitor cells for sensory neurons and autonomic neurons become apparent (Le Douarin, 1984; Ziller et al., 1987).

Back-transplantation experiments in which sensory and autonomic ganglia of quail embryos are inserted into the neural crest migration pathway of developmentally younger chick embryos indicate that sensory and autonomic progenitor cells are not completely segregated during migration (Le Douarin, 1986). Differences in the environment that the progenitors end up in govern which kind will differentiate and survive (Le Douarin, 1984). For example, the differentiation of sensory neuron progenitors in early dorsal root ganglia appears to depend on a diffusible factor emanating from the neural tube. Sensory neurons fail to develop when an impermeable barrier is placed between the neural tube and the site of dorsal root ganglion formation (Kalcheim and Le Douarin, 1986). At least some of these cells survive when the membrane is impregnated with an extract of neural tube, and a similar result is also brought about by the combination of two purified proteins adsorbed onto the membrane: brain-derived neurotrophic factor and laminin (Kalcheim et al., 1987).

AXONAL OUTGROWTH

The distance sensory axons have to grow to reach their targets varies fom one ganglion to another. Differences in target distance are particularly marked between cranial sensory ganglia, for example, nodose ganglion axons grow over ten times further to reach their targets than vestibular ganglion axons. A comparative study of the rates at which the axons of these and other cranial sensory ganglia grow to their targets *in vivo* has revealed a direct relationship between growth rate and target distance: the further the targets the faster the axons grow. These differences in growth rate appear to be intrinsic features of early neurons because single isolated neurons from each ganglion grow at comparable rates *in vitro* (Davies, 1989). It remains to be established whether this relationship between growth rate and target distance is a common feature of other classes of neurons in the developing nervous system.

AXONAL GUIDANCE

As elsewhere in the developing vertebrate nervous system, sensory axons grow with considerable precision to reach their target fields. For example, cutaneous sensory axons from DRG grow along a characteristic set of pathways to establish their dermatome at a precise location from the outset (Scott, 1982).

Sensory axons may be guided in two main ways: by the tissues lying on route to their targets and by target-derived chemotropic factors. Evidence for the former class of mechanisms comes principally from studies of the innervation of limb buds, evidence for the latter comes from work on the trigeminal system. These mechanisms will be discussed in turn.

Both sensory and motor axons grow within the confines of a distinct pattern of pathways that constitute the nerve plexus at the base of developing limb buds. When sensory and motoneurons are experimentally displaced before their axons grow into the region of mesenchyme in which the plexus forms, they trace out a normal plexus even though they may innervate inappropriate targets or reach their normal targets by a variety of novel routes within the plexus (Lance-Jones and Landmesser, 1981; Honig et al., 1986). The form of the plexus is governed in part by the presence in the limb bud of regions that are non-permissive for axonal growth such as the precartilaginous pelvic girdle anlage which constrain the growth of early axons (Tosney and Landmesser, 1984). The distribution of both neural cell adhesion molecules (N-CAM) immunoreactivity (Tosney et al., 1986) and regions of mesenchymal cell death (Schroeter et al., 1988) within the presumptive plexus region roughly corresponds to the form of the early plexus. How these distinct patterns are established and the extent to which they influence the form of the plexus by providing permissive or preferred pathways for growing axons remains unclear.

There is evidence that motor axons provide some sensory axons with guidance cues. When the ventral neural tube is ablated prior to motor axon outgrowth muscle nerve branches fail to form and cutaneous branches are enlarged, suggesting that the growth of proprioceptive axons to muscle is dependent on the presence of motor axons (Swanson and Lewis, 1986; Landmesser and Honig, 1986). Cutaneous sensory nerves, however, do not require guidance cues provided by motor axons since normal cutaneous innervation patterns are established in the absence of motor axons (Scott, 1988).

Although the above findings suggest that cutaneous sensory neurons are specified with respect to their targets, they can be respecified. When the anterior-posterior order of short stretch of the dorsal root ganglion (DRG) is reversed by the early rotation of the dorsal half of the neural tube (incorporating the neural crest), the cutaneous innervation patterns established by DRG are consonant with their new position. When the entire section of neural tube is rotated, the DRG tend to establish innervation patterns according to their original position (Scott, 1986). This suggests that the ventral cord can influence the specification of DRG neurons to project along a particular pathway. The molecular basis of the specific guidance cues that direct the axons of DRG neurons to their targets is unknown.

The homing behaviour of the axons of displaced neurons (Lance-Jones and Landmesser, 1980; Ferguson, 1983; Harris, 1986) indicates that specific cues are not restricted to a defined set of pathways but are more widely distributed. One mechanism that can account for homing behaviour is the chemotropic response of growth cones to a specific attractant diffusing from the target field. Evidence for specific chemotropism has come from co-culture studies of the mouse trigeminal ganglion and its cutaneous target field (Lumsden and Davies, 1983, 1986, 1987). When co-cultured at the developmentally appropriate stage neurites grow directly and exclusively to their own target tissue under the influence of an attractant produced by the epithelium. The factor, which is distinct from nerve growth factor (NGF) and laminin, is produced during the period of normal outgrowth, declines after the axons reach their targets and is not produced by adjoining cutaneous tissue. Similar studies of other sensory ganglia suggest that specific chemotropism is not a general guidance mechanism for sensory axons.

TARGET FIELD INNERVATION AND THE ROLE OF
NEUROTROPHIC FACTORS

The number of neurons in sensory ganglia is adjusted to the requirements of their target fields by a phase of cell death that begins shortly after the neurons innervated their targets. Manipulation of target field size (Oppenheim, 1981) and disruption of target field innervation (Yip and Johnson, 1984) have shown that both the peripheral and central target fields of sensory neurons play a role in regulating neuronal number.

Work on NGF has substantiated the hypothesis that developing target fields regulate the number of innervating neurons by the production of a limited quantity of a neurotrophic factor which the neurons require for their survival (Levi-Montalcini and Angeletti, 1968; Thoenen and Barde, 1980; Davies 1988). The most important evidence is that developing neurons whose survival is promoted by NGF *in vitro*, namely sympathetic neurons and a subset of sensory neurons, are also dependent on NGF *in vivo*. Anti-NGF antibodies eliminate these neurons during the phase of target field innervation, whereas exogenous NGF rescues neurons that would otherwise die.

Neurotrophic factors from the periphery

Sensory neurons obtain NGF from their peripheral target fields. NGF mRNA is present in appreciable amounts in skin (Davies et al., 1987a) but only low levels are present in the spinal cord (Shelton and Reichhardt, 1986). Accordingly, NGF is present in developing skin (Davies et al., 1987a) and in the peripheral branches of sensory neurons but is undetectable in spinal cord and in the central branches of these neurons (Korsching and Thoenen, 1985).

Elucidation of the spatial and temporal aspects of NGF synthesis and NGF receptor expression during cutaneous innervation (Davies et al., 1987a) has clarified the role of NGF in development and has provided some understanding of how NGF regulates the numbers and distribution of nerve terminals. Studies of the most densely innervated cutaneous target field of the mouse embryo (the maxillary process) have shown that NGF synthesis commences with the arrival of the earliest sensory axons. The expression of NGF receptors on the innervating trigeminal neurons also begins when their axons reach the target field. Prior to target field innervation the survival and growth of sensory neurons is independent of NGF and other target-derived neurotrophic factors (Davies and Lumsden, 1984; Ernsberger and Rohrer, 1988). The demonstration that NGF is not present in the target field prior to its innervation and that sensory axons lack NGF receptors when they are growing to their targets discounts the view that target-derived NGF plays a role in long-range chemotropic guidance of early sensory axons (Menisini-Chen et al., 1978; Gundersen and Barrett, 1979; Levi-Montalcini, 1982).

Assay of NGF mRNA in isolated cutaneous epithelium and mesenchyme together with localization by *in situ* hybridization has shown that its concentration correlates with innervation density. It is highest in epithelium (presumptive epidermis), lower in subjacent mesenchyme (presumptive dermis) and lowest in deep mesenchyme (presumptive subcutaneous tissue). The concentration of NGF mRNA in the epithelium of different target fields is also related to their innervation density (S. Harper and A. M. Davies, unpublished observations). These findings suggest that the local availability of NGF regulates innervation density. They also discount the view, based on immunohistochemical studies of NGF in the denervated iris, that NGF is synthesized exclusively by Schwann cells (Rush, 1984).

Although studies of NGF mRNA localization in developing cutaneous target fields indicate the kinds of cells that synthesize NGF, they provide no direct information on whether newly synthesized NGF is diffusible in the target field or whether its distribution and availability is restricted to specific sites. This information is important for understanding the nature of the competition between neurons for NGF. The finding that the NGF receptor gene is expressed in developing cutaneous target fields (S. Wyatt and A. M. Davies, unpublished observations) raises the possibility that low-affinity receptors may immobilize NGF to the surfaces of specific cells in the target field. This would provide a mechanism for selectively supporting the survival of neurons in accordance with the distribution of their axons in the target field since only neurons whose axons contact these specific target cells would be able to obtain a supply of NGF.

NGF is not the only neurotrophic factor which promotes the survival of sensory neurons. Extracts of a variety of peripheral tissues contain factors that are distinct from NGF which promote the survival of embryonic sensory neurons in culture. This raises the issue of the kinds of sensory neurons that are supported by NGF and other peripheral target-derived factors. Several *in vivo* and *in vitro* studies of the effects of NGF have given rise to the view that NGF-dependence is directly related to sensory neuron ontogeny in that all neural crest-derived sensory neurons require NGF for survival whereas placode-derived neurons do not (Pearson et al., 1983; Davies and Lindsay, 1985; Lindsay and Rohrer, 1985). Recent work, however, has clearly shown that NGF-dependence is not a feature of all neural crest-derived sensory neurons. Trigeminal mesencephalic nucleus (TMN) neurons, a population of neural crest-derived proprioceptive neurons, are not supported by NGF in culture (Davies et al., 1987b). Likewise, several experimental findings are consistent with the idea that the neural crest-derived proprioceptive neurons of DRG are also independent of NGF for survival (Davies, 1987). Although TMN neurons do not survive in the presence of NGF, they are supported by a factor present in their peripheral target tissue, skeletal muscle (Davies, 1986). In contrast to NGF, this muscle-derived factor has little or no effect on the survival of cutaneous sensory neurons. These findings suggest that the neurotrophic factor requirements of developing sensory neurons are related to the structures they innervate. For a detailed discussion of neurotrophic factor specifity in the sensory nervous system see Davies (1987).

Neurotrophic factors from the central nervous system (CNS)

A candidate for a neurotrophic factor that mediates the trophic support of the CNS on developing sensory neurons is brain-derived neurotrophic factor (BDNF; Barde et al., 1982). This protein promotes the survival of embryonic DRG neurons in culture (Lindsay et al., 1985) and rescues DRG neurons if administered to embryos during the period of natural neuronal death (Hofer and Barde, 1988). Although BDNF-dependent neurons are present in all populations of sensory neurons, their proportion varies from 10 to 80% or more (Davies et al., 1986a). In DRG, the majority of neurons respond to both BDNF and NGF *in vitro* in the early stages of their development (Ernsberger and Rohrer, 1988). As development procedes, BDNF-dependent and NGF-dependent neurons become largely distinct (Lindsay et al., 1985). This suggests that the neurotrophic factor requirements of at least some sensory neurons become more restricted as they mature.

Co-operation of neurotrophic factors

TMN neurons have been useful in resolving the issue of whether the survival of sensory neurons is regulated by the same or by different factors from the periphery and CNS. During the stage of target field innervation, the survival of the great majority of these neurons in culture is promoted by either of two neurotrophic factors: BDNF and a distinct factor present in skeletal muscle (Davies et al., 1986b). There is no additional survival in the presence of saturating levels of both factors, indicating that each neuron responds to both factors. The combined effect of both factors is additive at concentrations that promote half-maximal survival alone and is greater than additive at very low concentrations. This suggests that peripheral and central neurotrophic factors may potentiate each other at low concentrations. The finding that the responsiveness of TMN neurons to each factor is maximal during the period of natural neuronal death is further evidence that both factors cooperate in regulating sensory neuron survival during development.

TARGET RECOGNITION AND THE SPECIFICATION OF CONNECTIVITY

Neurons that innervate the same kind of sensory receptors in the periphery synapse with specific second order neurons in particular laminae of the spinal cord or equivalent nuclei in the brainstem. Several findings raise the possibility that certain oligosaccharides associated with cell-surface glycoproteins and glycolipids play a role in target cell recognition. In the developing rat, subsets of DRG neurons that project to particular laminae in the spinal cord are labelled by antibodies that recognise either lactosyl or globosyl oligosaccharides (Dodd and Jessell, 1985, 1986). Furthermore, DRG neurons that express lactosyl oligosaccharides make and release the complementary lactosyl-binding lectins (Regan et al., 1986). Although there is as yet no direct evidence that these oligosaccharides are involved in target cell recognition in the developing sensory nervous system, it has been clearly demonstrated that lactosyl oligosaccharides play an important role in cell-cell recognition and adhesion in early embryonic development (Rutishauser and Jessell, 1988).

An important question is whether the specific peripheral and central connections that primary sensory neurons make is determined prior to target field innervation or whether connectivity is governed by the particular targets these neurons encounter. The finding that sensory axons of the same modality tend to be clustered together in cutaneous nerves (Roberts and Elardo, 1986) suggests that connectivity is specified prior to innervation rather than induced by various target cells encountered by chance. On the other hand it is clear that the environment of developing sensory neurons can modify their peripheral and central connections. When thoracic DRG (predominantly cutaneous sensory) are transplanted to the brachial region in tadpoles, a proportion of the neurons in these ganglia innervate stretch receptors in muscle and make appropriate terminations in the spinal cord (Smith and Frank, 1987). The sequential timing of peripheral target field innervation and synaptogenesis in the spinal cord raises the possibility that the peripheral target may play a role in specifying central terminations (Smith and Frank, 1988; Davies et al., 1989). The periphery can also affect the phenotype of sensory neurons. Cross-anastomosis of cutaneous and muscle afferent nerves in adult rats results in changes in the neuropeptide content of sensory neurons that accords with their new targets (McMahon and Gibson, 1987).

Acknowledgements: The author's research is supported by the Cancer Research Campaign, Medical Research Council and Wellcome Trust.

REFERENCES

Barde, Y. A., Edgar, D., and Thoenen, H., 1982, Purification of a new neurotrophic factor from mammalian brain, *Embo J.*, 1:549-553

Baroffio, A., Dupin, E., and Le Douarin, N. M., 1988, Clone-forming ability and differentiation potential of migratory crest cells, *Proc. Natl. Acad. Sci. USA*, 85:5325-5329

Bronner-Fraser, M., and Fraser, S. E., 1988, Cell lineage analysis reveals multipotency of some avian neural crest cells, *Nature*, 335:161-164

D'Amico-Martel, A., and Noden, D. M., 1983, Contributions of placodal and neural crest cells to avian cranial peripheral ganglia, *Am. J. Anat.*, 166:445-468

Davies, A. M., 1986, The survival and growth of embryonic proprioceptive neurons is promoted by a factor present in skeletal muscle, *Dev. Biol.*, 115:56-67

Davies, A. M., 1987, Molecular and cellular aspects of patterning sensory neurone connections in the vertebrate nervous system, *Development*, 101:185-208

Davies, A. M., 1988, Role of neurotrophic factors in development, *Trends Genetics*, 4:139-143

Davies, A. M., 1989, Intrinsic differences in the growth rate of early nerve fibres related to target distance, *Nature*, 337:553-555

Davies, A. M., Bandtlow, C., Heuman, R., Korsching, S., Rohrer, H., and Thoenen, H., 1987, Timing and site of nerve growth factor synthesis in developing skin in relation to innervation and expression of the receptor, *Nature*, 326:353-358

Davies, A. M., and Lindsay, R. M., 1985, The avian cranial sensory ganglia in culture, *Devl. Biol.*, 111:62-72

Davies, A. M., and Lumsden, A. G. S., 1984, Relation of target encounter and neuronal death to nerve growth factor responsiveness in the developing mouse trigeminal ganglion, *J. Comp. Neurol.*, 253:13-24

Davies, A. M., Lumsden, A. G. S., and Rohrer, H., 1987, Neural crest-derived proprioceptive neurons express NGF receptors but are not supported by NGF in culture, *Neuroscience*, 20:37-46

Davies, A. M., Thoenen, H., and Barde, Y. A., 1986, Different factors from the central nervous system and periphery regulate the survival of sensory neurones, *Nature*, 319:497-499

Davies, A. M., Thoenen, H., and Barde, Y. A., 1986, The response of chick sensory neurons to brain-derived neurotrophic factor, *J. Neurosci.*, 6:1897-1904

Davis, B. M., Frank, E., Johnson, F. A., and Scott, S. A., 1989, Development of central projections of lumbosacral sensory neurons in the chick, *J. Comp. Neurol.*, (in press)

Dodd, J., and Jessell, T. M., 1985, Lactoseries carbohydrates specify subsets of sensory dorsal root ganglion neurons projecting to the superficial dorsal horn of the rat spinal cord, *J. Neurosci.*, 5:3278-3294

Dodd, J., and Jessell, T. M., 1986, Cell surface glycoconjugates and carbohydrate binding proteins: possible recognition signals in sensory neurone development, *J. Exp. Biol.*, 124:225-238

Ernsberger, U., and Rohrer, H., 1988, Neuronal precursor cells in chick dorsal root ganglia: Differentiation and survival in vitro, *Dev. Biol.*, 126:420-432

Ferguson, B. A., 1983, Development of motor innervation of the chick following dorso-ventral limb bud rotations, *J. Neurosci.*, 3:1760-1772

Gundersen, R., and Barrett, J. N., 1979, Neuronal chemotaxis: chick dorsal root axons turn towards high concentrations of nerve growth factor, *Science*, 206:1079-1080

Harris, W. A., 1986, Homing behaviour of axons in the embryonic vertebrate brain, *Nature*, 320:266-269

Hofer, M. M., and Barde, Y. A., 1988, Brain-derived neurotrophic factor prevents neuronal death in vivo, *Nature*, 331:261-262

Honig, M. G., Lance-Jones, C., and Landmesser, L., 1986, The development of sensory projection patterns in embryonic chick hind limb under experimental conditions, *Dev. Biol.*, 118:532-548

Kalcheim, C., Barde, Y. A., Thoenen, H., and Le Douarin, N. M., 1987, In vivo effect of brain-derived neurotrophic factor on the survival of developing dorsal root ganglion cells, *Embo J.*, 6:2871-2873

Kalcheim, C., and Le Douarin, N. M., 1986, Requirement of a neural tube signal for the differentiation of neural crest cells into dorsal root ganglia, *Dev. Biol.*, 116:451-466

Korsching, S., and Thoenen, H., 1985, Nerve growth factor supply for sensory neurons: site of origin and competition with the sympathetic nervous system, *Neurosci. Lett.*, 39:1-4

Lance-Jones, C., and Landmesser, L., 1980, Motoneuron projection patterns in the chick hind limb following early partial spinal cord reversals, *J. Physiol. (Lond.)*, 302:581-602

Landmesser, L., 1984, The development of specific motor pathways in the chick embryo, *Trends Neurosci.*, 7:336-339

Landmesser, L., and Honig, M. G., 1986, Altered sensory projections in the chick hind limb following early removal of motoneurons, *Dev. Biol.*, 118:511-531

Le Douarin, N. M., 1982, "The Neural Crest," Cambridge University Press, London, New York.

Le Douarin, N. M., 1984, A model for cell line divergence in the ontogeny of the peripheral nervous system, *in:* "Cellular and Molecular Biology of Neuronal Development," I.B. Black, ed., Plenum, New York: 3-28

Le Douarin, N. M., 1986, Cell line segregation during peripheral nervous system ontogeny, *Science*, 231:1515-1522

Le Douarin, N. M., Fontaine-Perus, J., and Couly, G., 1986, Cephalic ectodermal placodes and neurogenesis, *Trends Neurosci.*, 9:175-180

Levi-Montalcini, R., 1982, Developmental neurobiology and the natural history of nerve growth factor, *A. Rev. Neurosci.*, 5:341-362

Levi-Montalcini, R., and Angeletti, P. U., 1968, Nerve growth factor, *Physiol. Rev.*, 48:534-569

Lindsay, R. M., and Rohrer, H., 1985, Placodal sensory neurons in culture: nodose ganglion neurons are unresponsive to NGF, lack NGF receptors but are supported by a liver-derived neurotrophic factor, *Dev. Biol.*, 112:30-48

Lindsay, R. M., Thoenen, H., and Barde, Y. A., 1985, Placode and neural crest-derived sensory neurons are responsive at early developmental stages to brain-derived neurotrophic factor, *Dev. Biol.*, 112:319-328

Lumsden, A. G. S., and Davies, A. M., 1983, Earliest sensory nerve fibres are guided to peripheral targets by attractants other than nerve growth factor, *Nature*, 306:786-788

Lumsden, A. G. S., and Davies, A. M., 1986, Chemotropic effect of specific target epithelium in development of the mammalian nervous system, *Nature*, 323:538-539

Lumsden, A. G. S., and Davies, A. M., 1987, Chemotropic influence of specific target epithelium on the growth of embryonic sensory neurites, in: "Epithelial-mesenchymal interactions in neural development," J.R. Wolfe and J. Sievers, eds., NATO-ASI Series, H5:323-340

McMahon, S. B., and Gibson, S., 1987, Peptide expression is altered when afferent nerves reinnervate inappropriate tissue, *Neurosci. Lett.*, 73:9-15

Menesini-Chen, M. G., Chen, J. S., and Levi-Montalcini, R., 1978, Sympathetic nerve fiber ingrowth in the central nervous system of neonatal rodents upon intracerebral NGF injections, *Arch. Ital. Biol.*, 116:53-84

Narayanan, C. H., and Narayanan, Y., 1978, Determination of the embryonic origin of the mesencephalic nucleus of the trigeminal nerve in birds, *J. Embryol. Exp. Morph.*, 43:85-105

Noden, D. M., 1978, The control of avian cephalic neural crest cytodifferentiation. II. Neural tissues, *Dev. Biol.*, 67:313-329

Oppenheim, R. W., 1981, Neuronal death and some related phenomena during neurogenesis: A selected historical review and progress report, in: "Studies in Developmental Biology," W.M. Cowan, ed., Oxford University Press: 74-133

Pearson, J., Johnson, E. M., and Brandeis, L., 1983, Effects of antibodies to nerve growth factor on intra uterine development of derivates of cranial neural crest and placode in the guinea pig, *Dev. Biol.*, 96:32-36

Regan, L., Dodd, J., Barondes, S., and Jessell, T. M., 1986, Selective expression of endogenous lactose-binding and lactoseries glycoconjugates in subsets of rat sensory neurons, *Proc. Natl. Acad. Sci. USA*, 83:2248-2252

Roberts, W. J., and Elardo, S. M., 1986, Clustering of primary afferent fibers in peripheral nerve fascicles by sensory modality, *Brain Res.*, 370:149-152

Rush, R. A., 1984, Immunohistochemical localization of endogenous nerve growth factor, *Nature*, 312:364-367

Rutishauser, U., and Jessell, T. M., 1988, Cell adhesion molecules in vertebrate neural development, *Physiol. Rev.*, 68:819-857

Schroeter, S., Pkrzywinski, J. A., and Tosney, K. W., 1988, Mesenchymal cell death delineates axon pathways in the hind limb and does so independently of neural interactions, *Soc. Neurosci. Abst.*, 14:870

Scott, S. A., 1982, The development of the segmental pattern of skin sensory innervation in embryonic chick hind limb, *J. Physiol. (Lond.)*, 330:203-230

Scott, S. A., 1986, Skin sensory innervation patterns in embryonic chick hind limb following dorsal root ganglion reversals, *J. Neurobiol.*, 17:649-668

Scott, S. A., 1988, Skin sensory innervation patterns in embryonic chick hind limbs deprived of motoneurons, *Dev. Biol.*, 126:362-374

Smith, C. L., and Frank, E., 1987, Peripheral specification of sensory neurons transplanted to novel locations along the neuraxis, *J. Neurosci.*, 7:1537-1549

Smith, C. L., and Frank, E., 1988, Specifity of sensory projections to the spinal cord during development in bullfrogs, *J. Comp. Neurol.*, 269:96-108

Swanson, G. J., and Lewis, J. H., 1986, Sensory nerve routes in chick wing buds deprived of motor innervation, *J. Embryol. Exp. Morph.*, 95:37-52

Thoenen, H., and Barde, Y. A., 1980, Physiology of nerve growth factor, *Ann. Rev. Neurosci.*, 60:284-335

Tosney, K. W., and Landmesser, L., 1984, Pattern and specifity of axonal outgrowth following varying degrees chick limb bud ablation, *J. Neurosci.*, 4:2518-2527

Tosney, K. W., Watanabe, M., Landmesser, L., and Rutishauser, U., 1986, The distribution of N-CAM in the chick hind limb during axon outgrowth and synaptogenesis, *Dev. Biol.*, 114:437-452

Yip, H. K., and Johnson, E. M., 1984, Developing dorsal root ganglion neurons require trophic support from their central processes: Evidence for a role of retrogradely transported nerve growth factor from the central nervous system to the periphery, *Proc. Natl. Acad. Sci. USA*, 81:6245-6249

Ziller, C., Fauquet, M., Kalcheim, C., Smith, J., and Le Douarin, N. M., 1987, Cell lineages in peripheral nervous system ontogeny: medium-induced modulation of neuronal phenotypic expression in neural crest cell cultures, *Dev. Biol.*, 120:101-111

11

REMODELING OF NEURONAL SUBPOPULATIONS IN DORSAL ROOT GAN-
GLION: ROLE OF CHEMICAL FACTORS AND INTERCELLULAR CONTACTS

B. Droz, I. Barakat, J. Kazimierczak,
E. Philippe and A. Rochat

Institut d'Histologie et d'Embryologie, Université de
Lausanne, Lausanne, Switzerland

INTRODUCTION

The primary sensory neurons enclosed in the dorsal root ganglion (DRG) offer neurobiologists the possibility to test the influence exerted by environmental conditions on the phenotypes expressed by DRG cells. Although the population of the primary sensory neurons originates from a common source, the neural crest, DRG cells differentiate into various neuronal subpopulations which may be distinguished by their peripheral and central projections, electrophysiological properties, enzymatic equipment, pharmacological receptors, neurotransmitter content, cell surface antigens as well as cytochemical and ultrastructural characteristics. The proportion of these neuronal subpopulations shows great fluctuations with the age (Droz and Kazimierczak, 1987), position of the DRG along the neuraxis (Masurich et al., 1986a), or alteration of the innervated targets (Carr and Simpson, 1978; Philippe et al., 1988). It is therefore important to specify the cellular mechanisms which are involved in the remodeling of the neuronal population in DRG.

Theoretically, the remodeling of the population of primary sensory neurons could result from three main processes. The first one consists of the loss of a particular subpopulation of neurons. After removal of peripheral targets at a defined stage of development, subsets of DRG cells disappear by cell death (Carr and Simpson, 1978). In this chapter, we will show that a defined subpopulation of neurons is able to migrate out of a ganglion; consequently the DRG is enriched in the persisting neuronal subclasses. The second possibility is the generation of new neurons. If these newly-formed neurons belong to a defined subpopulation, the population of the DRG will be modified by the addition of these newly-formed neurons. While mitosis have been suspected to account for the genesis of new neurons (Scott, 1977; Devor and Govrin-Lippmann, 1985), evidence is presented that quiescent postmitotic neuroblasts could differentiate later and give rise to a defined subset of sensory neurons. The third mechanism is a change in the expression of the phenotype which characterizes a given subpopulation of DRG cells. In this case, the hypothesis is that environmental factors could influence the expression of the neuronal phenotypes. Mudge (1981) has clearly demonstrated that sensory neurons respond to a factor from non-neuronal cells that influences the biosynthesis of somatostatin and substance P in cultured DRG cells. Barakat et al. (1986) have shown that factors present in the horse serum promote the expression and maintenance of carbonic anhydrasein subsets of cultured ganglion cells. In the present survey of the contribution of our group, three distinct cellular mechanisms accounting for the role exerted by innervated targets on the phenotype of the innervating sensory neurons will be discussed.

I. MIGRATION OF A DEFINED SUBPOPULATION OF SENSORY NEURONS OUT OF DRG EXPLANTS: ROLE OF THE EXTRACELLULAR MATRIX

The DRG contain two main classes of ganglion cells: the large A neurons and the small B neurons which are also subdivided into various subclasses (Andres, 1961; Rambourg et al., 1983). In the chicken, the small B neurons (Omlin et al., 1985) and postmitotic neuroblasts committed to differentiate into small B ganglion cells (Philippe et al., 1986) accumulate in their Golgi apparatus a membrane-bound antigen which is recognized by antisera directed to myelin-associated glycoprotein (MAG). According to Milner and Sutclife (1988), MAG and the neural protein IB 236 possess identical sequences (Arquint et al., 1987) and belong to the Ig superfamily related to the neural cell-adhesion molecules or N-CAM. Thus, MAG-immunoreactivity was used as a cell marker of the subpopulation of small B neurons to investigate the influence of the extracellular matrix on chick DRG explants at the 10th embryonic day (E10). When DRG explants were cultured on collagen substrate, the explant core contained MAG-immunoreactive B ganglion cells intermixed with non-immunoreactive A ganglion cells. With time, MAG-immunoreactive ganglion cells, identified by the neuronal cell marker NSE (neuron-specific enolase), were found to undergo migration gradually out of the explant core. The results obtained by Rochat et al. (1988) indicate that small B neurons or postmitotic neuroblasts committed into B ganglion cells keep on the migratory aptitude characterizing neural crest cells and still express this mobile capacity on a collagen substrate at a relatively late stage of development. In contrast, DRG explants grown on a polyornithine substrate did not reveal any migration of B ganglion cells. Furthermore, the strong MAG-immunoreactivity present at the cell surface of ganglion cells in DRG grown on polyornithine, but not on collagen, suggest that these neurons are immobilized by interaction of the cell surface component with the positive charges of the substrate (Rochat et al., 1988).

Back-transplantation experiments of DRG at E10 have shown that some DRG cells are able to migrate out of the grafted ganglion in vivo (Schweizer et al., 1983). Furthermore, transplantation of DRG onto the chorioallantoic membrane of a host embryo reveals that the migratory aptitute of B ganglion cells persists at E8-E10, but declined with the age of the grafted ganglion (Philippe, unpublished). Thus the congenital lack of small B ganglion cells could result not only from a defect in the genesis of these neurons but also from an abnormal migration of B neurons which are not immobilized in the DRG.

II. DELAYED DIFFERENTIATION OF POSTMITOTIC NEUROBLASTS INTO SUBSTANCE P-IMMUNOREACTIVE NEURONS

In the chick embryo, the production of neurons by mitotic division of neuroblasts is stopped by E7.5 (Carr and Simpson, 1978). Yet, mixed cell cultures containing both neurons and non-neuronal cells from DRG at E10 show a net increase in the neuron number (120-130%) after 6 days of culture (Barakat and Droz, 1987). This augmentation of the neuronal population could not result from a proliferation of undifferentiated progenitor cells, which are no longer detected after E7 (Rohrer et al., 1985), or of neuroblasts since only 0.06% of the neurons were labeled after a 24 h-incubation with ^{3}H-thymidine. By using sequential microphotographs of selected fields taken every hour 3 days after plating, Barakat and Droz (1987) observed that flat dark cells of apparently non-neuronal nature started to elongate, then to display an ovoid and refringent cell body from which neurites were growing. In the electron microscope, the ultrastructural organization of these evolving cells reproduced the transitional changes taking place in primitive neuroblasts maturing into intermediate neuroblasts (Pannese, 1974). During their maturation, these newly-formed ganglion cells acquired gradually neuronal cell markers such as a positive immunostaining for neuron-specific enolase, N-CAM or 150 and 200 kD neurofilament subunits. The most interesting feature is that all the newly-produced neurons express the same phenotypic characteristics: small size, immunoreactivity to MAG and substance P antisera (Barakat and Droz, 1987; Ernsberger and Rohrer, 1988). Thus, these lately differentiated neurons reinforce the subpopulation of preexisting small B substance P-containing neurons. Since the differentiation of quiescent neuroblasts into nociceptive neurons is dependent on NGF (Barakat and Droz, 1987) and stimulated by high K$^+$ concentration (Chalazonitis and Fischbach, 1980), it would be essential to reappraise the importance of this phenomenon in vivo and to determine by which factors it is controlled.

III. INFLUENCE OF TARGETS ON THE EXPRESSION OF PARTICULAR NEURONAL PHENOTYPES

The concept that innervated targets exert a control on the phenotype of the innervating sensory neurons is highly attractive but difficult to prove. In their extensive review, Hughes and Carr (1978) concluded that the inductive effects produced by peripheral targets on the differentiation of sensory neurons could be rated as minimal. However, recently Marusich et al. (1986b) have shown that the expression of the SN1 epitope by a subpopulation of sensory neurons is regulated by interaction with skin. The major obstacle to prove or disprove the role of a peripheral target upon the expression of a defined phenotype is due to the discrimination between a selective survival of the subpopulation expressing this phenotype or an inductive effect on the gene responsible for the expression of this phenotype. To illustrate how this difficulty may be overcome, we will

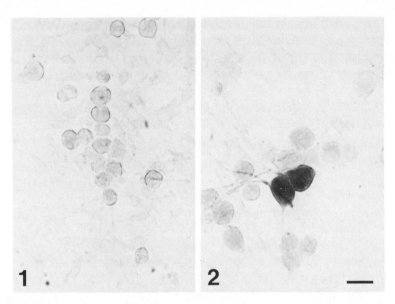

Figs. 1 and 2. Expression of calbindin D-28K immunoreactivity by DRG cells grown from E6 chick embryos for 5 days in F14 medium supplemented with NGF and 10% heat-inactivated horse serum. Bar: 25 µm. In control culture (**Fig. 1**), all the DRG cells are devoid of any immunostaining. After addition of soluble muscle extract (130 µg of protein per ml) from 2 week-old chicken (**Fig. 2**), two nerve cell bodies and neurites exhibit a strong immunostaining while several DRG cells are free of immunoreaction.

examine the role played by skeletal muscles on the subpopulations of sensory neurons expressing calbindin D-28K. Calbindin D-28K which participates in the control of intracellular Ca^{++} variations may be specifically immunodetected (Celio and Norman, 1985) in subsets of cytologically characterized DRG cells (Philippe and Droz, 1988) which project to skeletal muscles (Philippe and Droz, 1989).

In vivo experiments were performed to analyse the effects of target deprivation on the appearance of calbindin-immunoreactive DRG cells (Philippe et al., 1988). In the chick embryo, the first calbindin-immunostained DRG cells appear by E10, that is at a time coinciding with the establishment of specific afferent muscular connections (Tello, 1922). DRG cells were transplanted from chick embryos at E8 (prior to formation of neuromuscular spindles) and transplanted onto the chorioallantoic membrane of a host embryo. After 4 days, all the grafted DRG were free of calbindin-immunoreactive DRG cells. This lack of calbindin-immunoreactive neurons contrasts with the 20% of calbindin-immunoreactive neurons disclosed in control DRG at E12. When DRG at E8 were co-transplanted with musculature cells, about 14% of the DRG cells counted in the transplanted ganglion were calbindin-immunoreactive. As compared with control DRG from intact embryos at E12, a great part of the calbindin-immunoreactive subpopulation of neurons is restored. In such an experiment, the effect exerted by skeletal muscle cells may be due to a neurotrophic factor (Davies, 1986) which would prevent the death of sensory neurons covertly committed to express calbindin as well as to an inducing factor which would initiate the genic expression of calbindin by responsive subsets of sensory neurons.

In vitro DRG cell cultures were able to bring an appropriate answer. DRG cells grown from chick embryos at E6 are devoid of any calbindin-immunostained neurons (Fig. 1) even after 7 or 10 days in culture (Barakat and Droz, 1989). Yet when soluble extracts prepared from skeletal muscles are added to the culture medium, subpopulations of DRG cells grown at E6 become calbindin-immunoreactive (Fig. 2) within 3 days and the percentage of calbindin-positive neurons follows a dose-response curve, up to a maximum. Since the muscular extract could improve adhesion and survival of seeded neuroblasts committed to express calbindin, the muscular extract was added 3 days after plating: the proportion of calbindin-immunoreactive neurons was unchanged. Hence the emergence of calbindin-positive neurons does not correspond to a selective survival of the neuronal subpopulation. On the opposite, an inductive effect of the muscular factor on calbindin expression by sensory neurons is supported by the dose-dependent response which reaches a plateau; this result indicates that only the neuronal population which is responsive to the muscular factor can be recruited. Furthermore muscular extracts would contain a factor which acts directly on the responsive neurons and not through a non-neuronal cell mediation. Co-cultures of DRG cells at E6 with myotubes or cultures in a conditioned medium confirm the role of a muscular factor rather than of cell-to-cell contacts (Bossart et al., 1968). This muscular factor is a protein distinct from NGF but requiring cooperation with NGF to trigger off the gene expression of calbindin D-28K by sensory neurons (Barakat and Droz, 1989). To sum up, the appearance of subpopulations of calbindin-expressing sensory neurons, which in vivo project to skeletal muscles, is regulated by interaction with musculature cells. This effect is reproduced by a soluble protein factor from muscles. This muscular factor promotes the induction of the gene expression of the calbindin phenotype.

IV. CONCLUSIONS

The development of particular phenotypes in the population of the primary sensory neurons is controled by the genetic programme and by epigenetic interactions with the environment. As illustrated in the present review, the population of DRG may by changed by the selective loss of a defined subpopulation, due to either neuronal death or substrate-dependent migration out of the ganglion. Furthermore the emergence of NGF-dependent newly-formed neurons, which possess the same phenotypic characters as a preexisting subpopulation, will reinforce this subpopulation of substance P-containing neurons. Finally changes in the genic expression of a particular phenotype may be regulated by peripheral targets. Besides intercellular contacts, factors provided by innervated tissues act on responsive neuronal subsets to initiate the expression of a phenotype such as calbindin. As a result of the different processes of remodeling the DRG cell population possesses prominent capacities of flexibility and adaptability.

Acknowledgements: Work supported by a grant of the Swiss National Foundation No. 3.397.86.

REFERENCES

Andres, K. H., 1961, Untersuchungen über den Feinbau von Spinalganglien, *Z. Zellforsch.*, 55:1-48

Arquint, M., Roder, J., Loo-Sar, C., Down, J., Wilkinson, D.,Baley, H., Braun, P. and Dunn, R., 1987, Molecular cloning and primary structure of myelin-associated glycoprotein, *Proc. Nat. Acad. Sci.* (USA), 84:600-604

Barakat, I., and Droz, B., 1987, Differentiation of postmitotic neuroblasts into substance P-immunoreactive sensory neurons in dissociated cultures of chick dorsal root ganglion, *Devl. Biol.*, 122:274-286

Barakat, I., and Droz, B., 1989, Inducing effect of skeletal muscle extracts on the appearance of calbindin-immunoreactive dorsal root ganglion cells in culture, *Neuroscience*, 28:39-47

Barakat, I., Kazimierczak, J., and Droz, B., 1986, Carbonic anhydradrase activity in primary sensory neurons. II. Influence of environmental factors on the phenotypic expression of the enzyme in dissociated cultures of chicken dorsal root ganglion cells, *Cell Tissue Res.*, 245:497-505

Bossart, E., Barakat, I., and Droz, B., 1988, Expression of calbindin immunoreactivity by subpopulations of primary sensory neurons in chick embryo dorsal root ganglion cells grown in coculture or conditioned medium, *Dev. Neurosci.*, 10:81-90

Carr, V. McM., and Simpson, S. B., 1978, Proliferation and degenerative events in the early development of chick dorsal root ganglia: responses to altered peripheral fields, *J. Comp. Neurol.*, 182:741-756

Celio, M. R., and Norman, A. W., 1985, Nucleus basalis Meynert neurons contain the vitamin D-induced calcium binding protein (calbindin D-28K), *Anat. Embryol.*, 173:143-148

Chalazonitis, A., and Fischbach, G. D., 1980, Elevated potassium induces morphological differentiation of dorsal root ganglionic neurons in dissociated cell culture, *Dev. Biol.*, 78:173-183

Davies, A. M., 1986, The survival and growth of embryonic proprioceptive neurons is promoted by a factor present in skeletal muscle, *Devl. Biol.*, 115:56-67

Devor, M., and Govrin-Lippmann, R., 1985, Neurogenesis in adult rat dorsal root ganglia, *Neurosci. Lett.*, 61:189-194

Droz, B., and Kazimierczak, J., 1987, Carbonic anhydrase in primary sensory neurons, *Comp. Biochem. Physiol.*, 88B: 713-717

Ernsberger, U., and Rohrer, H., 1988, Neuronal precursor cells in chick dorsal root ganglia: Differentiation and survival in vitro, *Devl. Biol.*, 126:420-432

Marusich, M. F., Pourmehr, K., and Weston, J. A., 1986a, A monoclonal antibody (SN1) identifies a subpopulation of avian sensory neurons whose distribution is correlated with axial level, *Devl. Biol.*, 118:494-504

Marusich, M. F., Pourmehr, K., and Weston, J. A., 1986b, The development of an identified subpopulation of avian sensory neurons is regulated by interaction with the periphery, *Devl. Biol.*, 118:505-510

Milner, R. J., and Sutcliffe, J. G., 1988, Molecular neurobiological strategies applied to the nervous system, *Disc. in Neurosci.*, Vol. 5, 2:11-63

Mudge, A. W., 1981, Effect of chemical environment on levels of substance P and somatostatin in cultured sensory neurons, *Nature*, 292:764-767

Omlin, F. X, Matthieu, J. M., Philippe, E., Roch, J. M., and Droz, B., 1985, Expression of myelin-associated glycoprotein by small neurons of the dorsal root ganglion in chickens, *Science*, 227:1359-1360

Pannese, E., 1974, The histogenesis of the spinal ganglia, *Adv. Anat. Embryol. Cell Biol.*, 47:3-97

Philippe, E., and Droz, B., 1988, Calbindin D-28K immunoreactive neurons in chick dorsal root ganglion: ontogenesis and cytological characteristics of the immunoreactive sensory neurons, *Neuroscience*, 26:215-224

Philippe, E., and Droz, B., 1989, Calbindin immunoreactive sensory neurons of dorsal root ganglion project to skeletal muscle in the chick, *J. Comp. Neurol.*, (in press).

Philippe, E., Garosi, M., and Droz, B., 1988, Influence of peripheral and central targets on subpopulations of sensory neurons expressing calbindin immunoreactivity in the dorsal root ganglion of the chick embryo, *Neuroscience*, 26: 225-232

Philippe, E., Omlin, F. X., and Droz, B., 1986, Myelin-associated glycoprotein immunoreactive material: an early neuronal marker of dorsal root ganglion cells during chick development, *Dev. Brain Res.*, 27:275-277

Rambourg, A., Clermont, Y., and Beaudet, A., 1983, Ultrastructural features of six types of neurons in rat dorsal root ganglia, *J. Neurocytol.*, 12:47-66

Rochat, A., Omlin, F. X., and Droz, B., 1988, Substrate-dependent migration of myelin-associated glycoprotein immunoreactive cells in cultured explants of dorsal root ganglia from chick embryos,*Dev. Neurosci.*, 10:236-244

Rohrer, H., Henke-Fahle, S., El-Shakawy, T., Lux, H. D., and Thoenen, H., 1985, Progenitor cells from embryonic chick dorsal root ganglia differentiate in vitro to neurons: biochemical and electrophysiological evidence, *EMBO J.*, 4:1709-1714

Scott, B. S., 1977, The effect of elevated potassium on the time course of neuron survival in cultures of dissociated dorsal root ganglia, *J. Cell Physiol.*, 91:305-316

Schweizer, G., Ayer-Le Lièvre, C., and Le Douarin, N. M., 1983, Restriction capacities in the dorsal root ganglia during the course of development, *Cell. Diff.*, 13:191-200

Tello, J. F., 1922, Die Entstehung der motorischen und sensiblen Nervenendigungen. I. In dem lokomotorischen System der höheren Wirbeltiere: Muskuläre Histogenese, *Z. Anat. Entw. Gesch.*, 64:348-440

12

NEUROPEPTIDES IN PRIMARY AFFERENT NEURONS

Eberhard Weihe

Anatomisches Institut der Johannes Gutenberg-
Universität Mainz, Mainz, F.R.G.

INTRODUCTION AND HISTORICAL BACKGROUND

Traditionally, the primary sensory neurons, having their perikarya in the spinal or cranial sensory ganglia with processes directed towards the periphery and the central nervous system, have been regarded to function as receptive and afferent systems which reflexly activate central effector systems.[1] This, however, does not apply to the small diameter primary afferents as suggested by the observation made about a century ago that antidromic stimulation of transected dorsal roots or sensory nerves caused vasodilatation and inflammatory signs in the skin.[2,3] The novel concept which has been confirmed by many investigators ascribes to small diameter (particularly unmyelinated C) primary sensory neurons dual central afferent as well as peripheral local effector and axon-reflex functions.[3,4]

The question arose what the mediator substance of the peripheral effects would be. Dale[5] argued that the identification of the peripheral messenger would automatically mean that the central messenger(s) had been found. He assumed that the small primary sensory neuron uses the same messenger substance(s) at the peripheral and central ramifications. A strong messenger candidate for such dual functions turned out to be the undecapeptide substance P (SP), first discovered by von Euler and Gaddum[6] as an unidentified depressor substance and then chemically characterized by Chang and Leeman.[7] Based on his observations that extracts of dorsal roots were ten times as potent in a guinea-pig ileum bioassay for SP than extracts of ventral roots, Lembeck proposed a role for this peptide in primary afferent neurotransmission and in the axon-reflex.[8] This postulate became a concept when Jancsó[9] introduced as a further milestone for the understanding of primary afferent functions the neurotoxin capsaicin, the pungent ingredient in a variety of red peppers of the genus Capsicum. Initially, capsaicin has been thought to quite selectively affect small diameter (C) primary afferents containing SP.[3,10,11] In neonatal animals, capsaicin irreversibly destroyed the majority of small diameter fibers and, consequently, depleted SP[3,11] In adult animals, capsaicin caused no or only partial degeneration of small primary afferents but, nevertheless, depleted them from SP. Capsaicin-treated rats not only lost responsiveness to chemical noxious stimuli but also the ability to develop neurogenic inflammation, thus, their afferent and local effector functions.[9] These observations, in conjunction with the demonstration of peripheral and central (spinal) release of SP, strongly supported the view that primary sensory SP is an important messenger candidate of nociceptive afferent transmission and a mediator of neurogenic inflammation.[3,12,13,14,15]That the tachykinin SP is not the only candidate became evident when the list of neuropeptides detected in sensory ganglia began to grow[3,14] In the light of recent progress in molecular genetics,[16,17,18] it was no surprise that other mammalian tachykinins (TKs) had to be added to the list and have been shown to co-exist with SP. More unexpected was the fact that numerous peptides not related to the tachykinis, e.g., calcitonin gene-related peptide (CGRP), have been found to co-exist with TKs in varying combinations. This points to the principle of multiple peptide co-messenger functions[19].

For the better understanding of peptide functions in primary sensory neurons three main questions have to be answered: firstly, which peptides respectively which peptide families coexist with each other and/or with a possible classical transmitter; secondly, what are the exact target relations of the various peptide containing sensory fibers in the central and peripheral system; thirdly, under which physiological and/or pathophysiological conditions are they being released and what are their exact actions.

In the present review I try to address these questions by summing up the current knowledge based on the literature and own, mainly immunohistochemical observations. Concerning neuroanatomical, neurophysiological and neuropharmacological research on primary sensory neurons and peptides of the past and the presence, the vast number of studies makes it unavoidable to select references and to often refer to excellent reviews extensively covering this multidisciplinary field.[1,3,4,11,12,13,14,15,19,20,21,22,23,24] Further, I try to outline the exciting evidence for a functional significance of primary sensory peptides and receptors in animal and human pathophysiology with particular respect to inflammation, peripheral and central nociception/antinociception, allergy and immune disorders and their therapy.

TECHNICAL CONSIDERATIONS

Immunohistochemistry and radioimmunoassay (RIA) are the major analytical methods on which investigations of peptides rely. More recently, molecular biology has been added as a third powerful tool to localize and measure peptide mRNA. In order to interprete the results correctly, some principle difficulties concerning the specificity of these methods have to be recalled.

The problems are particularly critical in immunohistochemistry. In spite of extensive specificity controls, the molecular specificity of antisera can not be guaranteed. Principally, antisera can recognize C- or N-terminal extensions of the homologous antigene. Therefore, it is often impossible to say, whether the immunostaining represents the authentic posttranslationally processed peptide or the unprocessed peptide still incorporated in the precursor. Another intriguing possibility is that peptide metabolites contain the immunogenic epitope or parts of it and, therefore, give positive immunoreactions and falsely indicate the presence of the authentic biologically active peptide. Further, antisera directed against closely related peptides bearing a great extent of sequence homology are bound to mutually crossreact. This can be the case with C-terminally directed antisera against mammalian and amphibian tachykinins which have a great similarity in their C-terminal sequences.[23] For example, antisera against C-terminal SP also recognize other mammalian tachykinins, namely neurokinin A (NKA) and neurokinin B (NKB).[16,17,18] This possibility has not been considered before the presence of NKA in primary sensory neurons was known. Retrospectively, therefore, past studies employing C-terminal anti-SPs have not selectively mapped the localization of SP but also of NKA (and in the CNS also of NKB).[16]

The situation is similarly complex with the opioids comprising peptides derived from the three precursors pro-enkephalin (PRO-ENK), pro-dynorphin (PRO-DYN) and pro-opiomelanocortin (POMC).[25,26,27] Principally, these considerations apply to all peptide families with complex coding.

The most serious problem is unexpected crossreactivity of antisera with unrelated peptides. There are two examples in the recent literature for such an annoying situation: firstly, antisera against corticotropin-releasing factor (CRF) crossreact with SP,[28] therefore, CRF-immunoreactivity reported to be present in primary sensory neurons may reflect SP or even pan-tachykinin immunoreactivity; secondly, cholecystokinin (CCK) immunoreactivity in sensory neurons, at least of the rat, may be due to crossreactivity of some anti-CCKs with CGRP.[29]

Qualitative and quantitative differences in peptide staining (number of cells, intensity of staining) which have been reported to exist between species[3,4,30] could be due to minimal (so far unknown) inter-species aberrations from peptide homology. Then the same antiserum applied in different species would not be equally sensitive in all species to be compared. Consequently, a wrong impression of species differences in peptide expression could be created. However, a negative result of peptide staining in one species contrasting to positive immunostaining in others may be simply due to major aberration from peptide homology. Before definitely concluding absence of a peptide in a particular species, the species-specific peptide sequence must be known. RIA of peptides gives much less reason to be concerned about specificity, at least when RIA is combined with prior separation of different molecular forms of peptides by high performance liquid chromatography (HPLC).

Quantitative estimations and results obtained with double-staining techniques, intended to reveal coexistence patterns, have to be interpreted with outmost caution. Differences reported could be due to inter-laboratory variations in sensitivity and specificity of antisera.[31] Many other variable factors have to be considered, such as fixation technique, sensitivity of the detection systems (e.g., streptavidin-biotin-peroxidase visualization is much more sensitive than immune fluorescence), use of colchicine to enhance cellular immunostaining, or the use of explanted dorsal root ganglia.

It is particularly problematic to compare results obtained with colchicine treated and/or explanted sensory ganglia.[3] Explantation per se could change the level of transmitters including peptides. Colchicine-induced alterations of cell size[31] or peptide-phenotype of sensory neurons cannot be ruled out. Certainly, peptide phenotypes of colchicine-treated or explanted sensory neurons do not reflect the physiological in vivo situation of sensory neurons as integrated functional units of the organism.

Further, peptide phenotypes may be differentially affected in vivo by different anaesthetics commonly used before fixation, a possibility which has not yet been carefully examined.

Quantification and localization of peptide mRNA by hybridization assay and in situ hybridization, respectively, are complementary to the immunohistochemical and biochemical methods. However, the presence of the mRNA per se does not automatically answer the important question whether biologically active peptides are posttranslationally processed. Conversely, the presence of peptides in sensory neurons does not necessarily imply that they are synthesised there.[16,17,18] Peptides (or immunogenic peptide metabolites) could be taken up from the blood stream.[27,30]

In conclusion, the combined application of the three different methods, though each of them relatively powerful per se, reach a higher degree of validity when applied in combination. Sensu strictu, it is only justified to talk of a peptidic sensory neuron when it has been unequivocally demonstrated that the mRNA encoding the respective peptide(s) and at least one biologically active molecular form of peptide are contained in it.

MULTIPLICITY OF PEPTIDES IN SPINAL AND TRIGEMINAL SENSORY GANGLIA

The number of peptides and peptide families found in sensory ganglia by immunohistochemistry and by biochemistry (bioassay/RIA either in combination with HPLC or not) is growing.[1,3,4,15,16,19,23,30,32] Based on such data in spinal and trigeminal ganglia, an alphabetical list of the current peptide mediator candidates in primary sensory neurons is compiled (Table 1) and details of peptide families and individual peptide species are outlined below. The special situation of the vagal and glossopharyngeal sensory ganglia, which has been less extensively studied, will be commented separately.

The quantity of the various peptide-coding sensory neurons varies segmentally.[3,33,34,35,36,37] Their distribution per individual ganglion is often non-uniform. Classifications of peptide-containing neuronal cell bodies as small, intermediate and large are in contrast to the classical view of the presence of small dark and large pale cell bodies.[34,37] However, it is not possible to assume strict correlation of fiber diameter and cell size.[38,39] Some large or intermediate sensory neurons may give rise to unmyelinated C fibers.

For some of the peptides, hybridisation assay and hybridisation histochemistry have confirmed or disproved the conclusions drawn from the immunohistochemical and biochemical results. The evidence that multiple tachykinins, CGRP and somatostatin can be expressed in sensory neurons has been corroborated by hybridisation assays and hybridisation cytochemistry showing that mRNAs for these peptides are encoded by sensory neurons.[16,17,40,41,42]

Table 1

PEPTIDES IN SPINAL SENSORY NEURONS

		References
ANF	Atrial natriuretic factor n.i.	74,75
ANG II	Angiotensin II n.i.	1,3
AVP	Arginine vasopressin*	36
BOMB/GRP	Bombesin/Gastrin releasing peptide*	13,66,67
CCK	Cholecystokinin*, ***	31,35,37,57
a-/ß-CGRP	Calcitonin gene-related peptide*	19,31,44,45, 46,47,50,51, 52,57,85,90
CRF	Corticotropin releasing factor*,***	1,3,28
GAL	Galanin*	19,31,43,82
GRF	Growth hormone releasing factor n.i.	76
	Met-Tyr-Lys Methionyl-tyrosyl-lysine n.i.	1,3,65
OP	Opioid peptides	
- DYN A	- Dynorphin A*	27,30,32,45, 54,56,57,58,64
- DYN B	- Dynorphin B n.i.	27,30,32,33, 45,54,55
- LE	- Leu-enkephalin*	27,30,32,35, 45,54,55
- ME	- Met-enkephalin n.i.	35,64
- NEO	- a/ß-neoendorphin n.i.	27,30,32,45, 54,55,64
OXY	Oxytocin n.i.	36,73
SOM	Somatostatin*	31,34,35,37,51,68
TKs	Tachykinins*	19,51,27,30,45, 51,55
- NKA	- Neurokinin A	30,43,44,47,50, 84,88
- NPK	- Neuropeptide K	3
- SP	- Substance P	19,24,34,36, 37,43, 44,46,47,50,57,66, 68,84,85,90
VIP/PHI/PHM	Vasoactive intestinal polypeptide*/ Peptide histidine isoleucine/ methionine n.i.	3,31,35,45,68,82

*   depleted by capsaicin

**   not depleted by capsaicin

***   due to unexpectedly cross-reacting antibodies, CCK-immunoreactivity possibly represents CGRP and is absent in rat [29]; CRF-immunoreactivity is likely to represent SP-immunoreactivity [28]

n.i. capsaicin-sensitivity not investigated

Cholecystokinin mRNA has been found in guinea-pig but not in rat sensory neurons, pointing to major species differences in peptide gene expression.[40]

The Tachykinin Family (TK)

Tachykinins have a common C-terminal sequence Phe-X-Gly-Leu-Met-NH$_2$ (X is the variable amino acid). In contrast, the N-terminus of tachykinins is much more heterogeneous.[16,17,18,23] The family includes the mammalian tachykinin peptides and a variety of non-mammalian peptides, variously isolated from frog skin (e.g., physalaemin, kassinin) and octopus salivary glands (eledoisin).

The mammalian TKs include the undecapeptide substance P (SP), the decapeptides neurokinin A (NKA) and neurokinin B (NKB), alternatively referred to as neuromedin K, and the neuropeptide K (NPK) which consists of 36 amino acids and has NKA at its C-terminus.[16,17,18] Thus, NPK is the N-terminally extended form of NKA. There are two preprotachykinin (PPT) genes, PPT-A and PPT-B. PPT-A is the gene encoding substance P.[16,17,18] It comprises seven exons and, by alternative RNA splicing mechanisms, it gives rise to three distinct mRNAs, and hence three distinct precursors, namely a-PPT-A, ß-PPT-A and gamma-PPT-A. SP is contained in all three precursors. NKA is present in ß- and gamma-PPT-A. NPK is only yielded by ß-PPT-A. PPT-B yields NKB.[16,17,18]

In primary sensory afferents, only PPT-A and its posttranslational processing products SP, NKA and NPK are present. In contrast to some regions in the CNS and in the intestine, PPT-B and, consequently, its final product NKB are absent from primary afferents.[16,17,18] Previously reported staining for eledoisin-like peptides[3] in primary afferents is most likely artifactual.

In accordance with the characteristic features of the gene expressions of the mammalian tachykinin precursors, SP, NKA and NPK have been isolated from sensory ganglia and characterized by HPLC and RIA. N-terminal antisera differentiating between SP and NKA revealed what could have been expected, namely that subpopulations of TK primary afferents contain only SP or SP + NKA.[3,4,43,44] This is the case in spinal, trigeminal, vagal and glossopharyngeal sensory ganglia of a variety of species including humans.[3,4,30,43,44,45,46,47,48,49,50,51] The existence of a further predictable subpopulation coding for SP +NKA + NPK has not yet been demonstrated but can be anticipated.[3] Earlier reports, often using C-terminally directed anti-SPs, could not pay attention to the possible cross-reaction of anti-SP with NKA and NPK in sensory neurons.

Small dark (B type) sensory cells staining for SP and/or NKA outnumber SP and/or NKA-ir intermediate or large neuronal cell bodies. Contrary to the often presented view that TK immunoreactivity is only present in a very few large cell bodies[3,4] I regularly see TK immunoreactivity in a substantial proportion of large and intermediate cell bodies. However, in large and intermediate cells, the TK immunoreactive material has a more granular appearance and is less densely packed than in small cells. It is noteworthy that the intensity of immunostaining generally varies greatly between the different cells from very weak to very strong. TK immunostained cells, particularly the small cells, often occur in groups of several cells which are closely packed together. More frequently, however, they occur individually.

The Calcitonin Gene-Related Peptide (CGRP) Family

From the analysis of calcitonin gene structures, the existence of CGRP has been predicted. In rat and humans, two separate CGRP genes have been recognized encoding a-CGRP and ß-CGRP (or I and II), which differ by three and one amino acids, respectively.[52] Chromatographical and radioimmunological characterization revealed that a-CGRP is the predominant form in rat sensory neurons although ß-CGRP is also present. In accordance, in situ hybridization detects a-CGRP and ß-CGRP mRNA in sensory neurons. In contrast, only ß-CGRP mRNA (and ß-CGRP) is present in rat intestinal neurons.[42,52] Because of the minimal differences in sequence homology of the two rat and human CGRPs, all respective antisera so far available mutually crossreact. Therefore, a- and ß-CGRP cannot yet be differentiated by immunohistochemistry. The term CGRP immunoreactivity implies that differential expression of the two CGRPs or the proportions of a-CGRP:ß-CGRP cannot be visualized by immunocytochemistry.

CGRP-immunoreactivity is present in most of the small, in many intermediate and in a substantial number of large cell bodies of spinal and trigeminal ganglia. In the rat, and in many other species the CGRP-staining population of afferents is usually larger than that staining for TKs.[3,4] In the guinea-pig, however, CGRP and SP stained primary sensory neurons are equally frequent and fully coincide.[30]

The Opioid Family

There are three opioid precursors, proenkephalin (PRO-ENK), prodynorphin (PRO-DYN) and proopiomelanocortin (POMC).[21,25,26,27,53] Immunohistochemical studies indicate that several PRO-DYN-derived opioid peptides (leu-enkephalin, dynorphin A, dynorphin B and a-/ß-neoendorphin) are present in spinal and trigeminal ganglia of adult guinea-pigs and in neonatal mouse and RIA shows presence of dynorphins in spinal ganglia of rat and rabbits.[54,55,56,57,58,59] In contrast, immunohistochemistry indicates that the PRO-ENK-derivatives octapeptide, heptapeptide, amidorphin, BAM 12 P and peptide F are absent from guinea-pig spinal and trigeminal ganglia.[54,55] Dynorphin B occurs in cat dorsal root ganglia, particularly at the sacral level, as determined and characterized by HPLC and RIA.[33] Immunohistochemistry did not reveal opioid-ir neuronal cell bodies in adult rat sensory ganglia[30,60,61] although RIA data indicate the presence of DYN in extracts of rat spinal ganglia.[58] There is immunohistochemical and radioimmunological evidence that DYN occurs in trigeminal sensory nerves of humans.[62] The PRO-DYN-derivative leumorphin is absent from monkey spinal ganglia but present in spinal cord.[63] Low concentrations of the PRO-DYN- (DYN A and NEO) and the PRO-ENK-derivative met-enkephalin have been measured in spinal ganglia and dorsal roots of humans.[64]

Thus, it is unclear whether only PRO-DYN- or also PRO-ENK- opioid peptides are contained in sensory neurons. Further chromatographic characterization of the opioid-ir material is necessary to clarify these questions. Their synthesis in sensory neurons must be confirmed by molecular genetics techniques. There are certainly species variations in the sensory opioid expression. The opioid population of spinal and trigeminal sensory neurons of guinea-pigs is smaller than that coding for TKs and/or CGRP (OP < TKs < CGRP).[27,30,45,54,55,57]

Staining for the non-opioid POMC-derivative a-MSH is localized in some intermediate and all large ganglionic cell bodies of spinal and trigeminal ganglia. This distribution pattern resembles that of neurofilament, and it is possible that anti-MSH crossreacts with neurofilament which may contain a non-POMC-derived a-MSH-like peptide.[27]

Methionyl-Tyrosyl-Lysine (Met-Tyr-Lys)

This tripeptide has been characterized in extracts of the dorsal horn and dorsal root ganglia. However, it has not yet been immunohistochemically localized.[65]

Cholecystokinin (CCK)

The past view that CCK is synthetized in rat sensory neurons[3,4] has been challenged by the fact that CCK mRNA is absent from rat sensory neurons.[40] However, CCK mRNA is present in sensory neurons of guinea-pig.[40] The contradictory results obtained in rat can be explained by the fact that some antisera against CCK recognize CGRP.[29,31] In the light of the large number of small, intermediate and large cells staining for CGRP in virtually all mammalian species, a general over-estimation of CCK immunostained sensory neurons in previous reports is not unlikely. Also in species where CCK mRNA is present, previous immunohistochemical data have to be re-evaluated.

Bombesin/Gastrin-Releasing-Peptide (BOMB/GRP)

BOMB/GRP-like immunoreactivity has been found in small ganglionic cells of spinal ganglia of the rat.[66,67] It has been characterized by gel chromatography, HPLC and RIA.[66] Absence of ir-BOMB from sensory neurons of dorsal root ganglia has been reported in the cat.[68]

Somatostatin (SOM)

The tetradecapeptide SOM is present in a minor population of small diameter sensory neurons of spinal and trigeminal ganglia. In contrast to SP, it appears to be absent from intermediate cells.[31,37]

Galanin (GAL)

The 29 amino acid peptide GAL, initially isolated from porcine intestine took its name from its N-terminal glycine and C-terminal amidated alanine residues.[69,70] It has been detected by immunohistochemistry and RIA in sensory ganglia of several mammalian species.[29,31,69,70] There appears to be a single GAL encoding gene. The expression of rat GAL is tissue-specific. Its amino acid sequence is 90% similar to porcine GAL.[69,70] Immunostaining was usually of low intensity and often confined to the perinuclear area. It prevailed in small ganglionic cells and appears to be strikingly intensified by peripheral axotomy.[3]

Vasoactive Intestinal Polypeptide (VIP)

VIP and peptide histidine isoleucine (PHI) are contained in the same precursor. Peptide histidine methionine (PHM) is the human variant of PHI.[71] Therefore, complete overlap of staining for VIP, PHI or PHM is expected. VIP/PHI/PHM immunoreactivities in spinal sensory neurons are not easily visualized, even when colchicine is used.[31,45] Some VIP neurons are also seen in guinea-pig trigeminal ganglia.[32,59] Clear staining of cell bodies and increased numbers of immunopositive cells has been induced by cutting the peripheral axons.[72] Therefore, the quantitative estimation of VIP/PHI/PHM sensory neurons is difficult.

Arginin Vasopressin (AVP) and Oxytocin (OXY)

The presence of the hypothalamic hormones AVP and OXY has been demonstrated by immunohistochemistry and the immunoreactive material has been characterized by chromatography of extracts of human spinal ganglia.[36,73] In contrast to these reports, I have not been able to immunohistochemically visualize AVP- or OXY-ir neuronal cell bodies in guinea pig spinal or trigeminal ganglia although the commercial antisera used (Immunonuclear) clearly stained nerve fibers in the adjacent spinal cord present on the same sections.

Atrial Natriuretic Factor (ANF)

Most recently, there is evidence from immunohistochemistry[74,75] and from RIA[75] that the cardiac hormone ANF is contained in spinal and trigeminal ganglia. The number of ANF containing sensory cell bodies is low.

Corticotropin-Releasing Factor (CRF)

In the light of unexpected crossreactivity of antisera against CRF with SP (and possibly other tachykinins) the presence of authentic CRF in sensory neurons has to be questioned and the view that it is a primary sensory transmitter candidate must be challenged.[28]

Growth Hormone-Releasing Factor (GRF)

GRF-like ir has been recently found in trigeminal and spinal ganglia of adult and fetal rats. GRF was present in only 1% of the sensory neurons and restricted to the perinuclear region.[76] In contrast to all the other peptides, GRF was absent from sensory fibers. No significant changes during development were seen.

Neuropeptide Y (NPY)

NPY is not present in primary sensory neurons. However, some NPY fibers supply blood vessels of dorsal root ganglia. In conjunction with previous evidence for sympathetic innervation of spinal and trigeminal ganglia derived from fluorescence histochemistry it is most likely that the NPY innervation originates from postganglionic noradrenergic sympathetic neurons, where NPY is a well known cotransmitter candidate.[19]

MULTIPLE PEPTIDE CODING OF VAGAL AND GLOSSOPHARYNGEAL SENSORY NEURONS

Immunohistochemistry demonstrated the presence of SP, NKA, CGRP, CCK, VIP/PHI and somatostatin (but not for neurotensin) in the nodose, jugular and petrosal ganglia.[43,49,50] The distribution of peptide-coded neuronal cell bodies within these ganglia is not uniform.[49] In the rat, peptidic neurons were generally more frequent in the rostral than in the caudal nodose ganglion and more numerous in petrosal and jugular ganglia than in the nodose ganglion. VIP exhibited a more caudal distribution in the nodose ganglion.[49] Weakly stained SOM is only present in very few neurons. The presence of CCK in rat vagal and glossopharyngeal afferents remains controversial because the anti-CCK used crossreacted, though not extensively, with CGRP.[29,50] However, in the nodose ganglion of cat the presence of authentic CCK-8 has been demonstrated by the use of chromatography and RIA. Further, the presence of GAL cell bodies in the nodose and jugular ganglia of the rat has been reported.[43,50] According to own observations, the situation is principally similar in the guinea-pig where, in addition, we have presented preliminary evidence that opioids are present in some neurons of the vagal sensory ganglia.[30] Angiotensin (ANG) II, present in spinal sensory neurons, appears to be absent from vagal and glossopharyngeal afferents.[3,49] The possible presence and most likely differential distribution of the other sensory peptide messenger candidates (see Table 1) in vagal and glossopharyngeal sensory ganglia remains to be evaluated and complemented by molecular genetic techniques.

CAPSAICIN-SENSITIVITY OF SENSORY NEURONS

Only the most essential points can be mentioned. The issue of capsaicin-sensitivity of primary afferents has been extensively reviewed.[3,4,11,77] It is generally thought that capsaicin has no direct excitatory or neurotoxic actions on non-primary sensory neurons. In low doses, it excites sensory nerve endings (mainly C fibers) and releases peptides. This effect is followed by rapid desensitization. In addition, capsaicin blocks nerve conduction. Low doses are used in isolated organ preparations and in local applications in order to study the local effector (sensory-efferent role) of sensory peptides (see below).

High systemic doses of capsaicin have differentially degenerating and peptide-depleting effects in neonatal and adult animals. These capsaicin actions not only depend on the dose used but also on the species investigated. In neonatal rats, 50 mg/kg of capsaicin only affects C fibers (50-90%) while higher doses not only degenerate C but also a variable number of A∂ fibers. In adult rats, no or only a partial degeneration of small diameter fibers is observed whereas, in the adult guinea-pig, capsaicin causes degeneration of unmyelinated fibers.[3,4] As it is to be anticipated, capsaicin-induced degeneration of small diameter fibers is paralleled by a reduction of peptide-staining sensory neurons. Further, there is also a marked depletion of peptides from non-degenerated primary sensory neurons of adult animals.[3,4,11]

It is now clear that systemic administration of capsaicin ($\geq$ 50 mg/kg) depletes, to some extent, most of the primary sensory peptides as summarized in Table 1.[1,3,4,11] It is noteworthy, that the depletion of peptides is not total. Some capsaicin-insensitive peptide-containing sensory neurons always remain. They are usually more numerous in animals treated as adults than in those receiving capsaicin neonatally. Thus, capsaicin is a non-peptide selective neurotoxin for small diameter primary afferents. It is not known whether capsaicin affects non-peptidic primary afferents.[3,4,11,77]

Although capsaicin seems to be ineffective in influencing non-primary sensory afferents, indirect effects on neurons connected with primary afferents are conceivable.[3,4,11] In fact, degeneration of spinothalamic tract neurons as a consequence of capsaicin treatment has been reported.[78] Further, there are some effects of capsaicin in the CNS which can only be explained by the presence of an unknown capsaicin-sensitive neuronal population or by a more widespread distribution of primary capsaicin-sensitive afferents than currently believed. In this context, a very recent study is of major significance. It shows, unexpectedly, that neurons in the supraoptic nucleus of the cat are directly excited by III-IV (A ∂/C) muscle afferents.[79] Such afferents can be expected to be capsaicin-sensitive and to contain CGRP and partly also TKs.[80]

RESPONSE OF PRIMARY SENSORY PEPTIDES TO PERIPHERAL NERVE SECTION, DORSAL RHIZOTOMY AND LOCAL CAPSAICIN APPLICATION

Peripheral nerve section has striking and differential effects on the peptide content of some primary sensory neurons.[81] The number of dorsal root ganglion cells staining for VIP/PHI or GAL is markedly increased and their immunostaining is intensified in the response to peripheral axotomy.[72,82] After axotomy, GAL is not only seen in small but also in many intermediate and large sensory perikarya.[82] Interestingly, the increase in the number of GAL-positive cells appeared to be not restricted to the side ipsilateral to the transsection but there was also a small increase in the contralateral dorsal root ganglia.[82] The change of VIP/PHI is paralleled by an increase of VIP/PHI staining in the dorsal horn, which is also seen after peripheral nerve crush.[81] In contrast, crush has no effect on the levels of SP, whereas nerve section depletes SP from the dorsal horn. Capsaicin, locally applied to peripheral nerves, depletes SP and increases the VIP/PHI content in the dorsal horn.[81] Dorsal rhizotomy is well known to deplete many of the sensory peptides from the dorsal horn.[3,4,12,13] However, this effect is never total because of substantial spinal origin of fibers staining for these peptides. There is one exception, namely CGRP, which is completely depleted from the rat lumbar dorsal horn when extensive bilateral dorsal rhizotomy is performed.[83] Unilateral dorsal rhizotomy reveals that CGRP primary sensory fibers also project to the contralateral side,[83] and there is similar evidence for tachykinins.[84] The functional significance of these changes in peptide expression is unclear. In the light of the glycogenolytic potencies of VIP, one may speculate that the increase in VIP, leading to an increased glycogen production and glucose utilization, reflects the organisms response to promote regeneration in the injured neuron.[82] It is tempting to conceive that the peptide regulatory profiles of NGF[85] and the axotomy-induced alterations of microtubule and neurofilament function also come into play.[86]

REGULATION OF SENSORY PEPTIDE GENE EXPRESSION BY NERVE GROWTH FACTOR (NGF)

NGF is essential for the survival of sensory neurons during ontogeny but not once they are mature.[85,87] Sciatic nerve section in the rat induces loss of SP from lumbar dorsal root ganglia. Since the loss of SP is reversed by NGF, one may interpret that deprivation of sensory neurons from target-derived NGF may be causative for the loss of SP. In support of this view, it has been very recently reported that NGF regulates the expression of mRNAs encoding the precursors of SP and coexisting CGRP in mature sensory neurons.[85] One may speculate that other sensory peptides may be similarly dependent to and regulated by the retrograde effector function of NGF. Macrophages play a role in the function of mRNA encoding NGF to regulate the degeneration and regeneration of the rat sciatic nerve.[88] This points to a functional coupling of the peptidic neuroimmune axis and NGF. That NGF regulatory roles of peptide levels are not specific for

Table 2

CO-EXISTENCE PATTERNS AND PATHWAY-SPECIFIC SUBSETS OF
DORSAL ROOT PRIMARY SENSORY NEURONS

Peptide combinations	Pathway	References
TKs/CGRP	ubiquitous	3,4,19,46, 50,57
TKs/CGRP/CCK*/DYN skin		57
TKs/CGRP/CCK*	microvasculature of skeletal muscle	57
TKs/CGRP/DYN	pelvic viscera, heart, respiratory tract	30,54,57
TKs/CGRP/PRO-DYN-OP	ubiquitous (except of systemic vasculature)	27,30,32,54
TKs/CGRP/SOM/GAL	?	31
TKs/CGRP/GAL	?	43,50
TKs/SOM somatic		31,34,37,51,68
TKs/SOM/VIP/CCK*	?	68
TKs/AVP/OXY	?	36
TKs/GAL	?	31
TKs/VIP	?	31,68
CGRP/SOM	?	31
CGRP/GAL	?	31
CGRP/VIP/PHI	?	31

* CCK may be overestimated or absent due to crossreactivity of anti-CCK with CGRP; further references see text

spinal afferents is indicated by the NGF-induced increase of SP in cultured rat vagal sensory neurons.[89]

SUBSETS OF PRIMARY AFFERENTS CODE FOR VARIOUS PEPTIDE COMBINATIONS

It is not surprising that peptides derived from the same precursors coexist (Table 2) as it has been outlined above for the TKs, opioids and VIP/PHI/PHM. That unrelated peptides also co-exist is more remarkable. The co-existence of multiple combinations of unrelated peptides appears to be rather the rule than the exception. So far, there has been no evidence for ANG II,[3] ANF[74] or GRF[76] to co-exist with any of the peptides listed in Table 1. For all the other peptides, various combinations have been reported in dorsal root ganglia (Table 2).[1,3,4,19,23,27,30,31,36,42,50,55,57,66,67,68,90] Similar combinatory sets have been seen in trigeminal, vagal and glossopharyngeal sensory ganglia, though the list for cranial ganglia is less complete.[27,30,49,50,51,59] Probably due to differences in the techniques (see Technical Considerations), several conflicting results have been reported. Thus, BOMB and SOM have been first described to be not co-existing with SP. Later on, partial coexistence of BOMB and SP and of SOM and SP have been found.[31,37,66,67]

For some of the subpopulations of afferents characterized by specific combinatory sets of peptides pathway-specific peripheral distributions have been reported.[57] For others, such a possibility remains to be clarified (Table 2). The coexistence patterns determined in perikarya provide useful guidance for the search of specific coexistence patterns in peripheral and central endings. However, it is also conceivable that peptide combinations may differ in central vs peripheral endings or even in individual ramifications and endings. Further, the relative proportions of peptides may vary according to the particular functional state of the sensory neuron. This may be reflected by sometimes reciprocal differences in the intensity of immunostaining for pairs of coexisting peptides. Taken together, coexistences of various peptide families, each possibly giving rise to numerous biologically active molecules provides sensory neurons with an enormous armament to fulfill their functions. This may be even greater, if one takes into account that some peptides may be further processed, e.g., SP to N- and C-terminal sequences which have different biological actions.[91] On the other hand, some peptides may be phylogenetic vestiges.[92] The principle question arises whether the coexisting peptides fulfill transmitter and/or modulator functions. SP has been regarded as the first neuropeptide transmitter candidate, although this view, at least as regards nociception, has been often challenged.[3,4,12,13,14,93,94,95,96,97]

Nonetheless, multiple peptide coexistence may point to multiple ways of peptide cofunction in central sensory (including nociceptive) processing and various peripheral efferent mechanisms. Details of their dual afferent and sensory-efferent functions and their significance in pathophysiology will be outlined in the respective sections (see below).

COEXISTENCE OF A CLASSICAL (GLUTAMATE) AND A PEPTIDE (SUBSTANCE P) TRANSMITTER CANDIDATE

The excitatory amino acid glutamate has been long regarded as a classical transmitter candidate in primary sensory neurons.[1] However, it has been only very recently that colocalization of glutamate and SP in (mainly small) primary sensory neurons has been demonstrated.98 Colchicine treatment reveals that in the rat about 90% of SP positive neurons of dorsal root ganglia are also staining for glutamate. This is compatible with the view that primary afferent endings, both in the periphery and in the CNS, may release a fast acting classical transmitter and a peptide which induces prolonged slow depolarization.[98]

THE CENTRAL CONNECTION: TARGET DISTRIBUTION

All peptides present in primary sensory neurons are also contained in the CNS and in the areas where the different spinal, trigeminal and other cranial sensory afferents have their endings.[3,4,12,13,14,22,24] This complicates the differentiation of non-primary sensory fibers and central primary sensory endings staining for the same peptides or peptide combinations. Deafferentiation provided some evidence for the distribution of SP endings.[12,13,14,22,24] However, one has to take into account that deafferentiation not only destroys primary peptidic afferents in the spinal cord but also can have secondary effects on the peptide levels of central target neurons, e.g., intensification of dynorphin B staining.[99]

Spinal Endings

At least in the spinal dorsal horn, the difficulties in determining the specific targets of peptidic primary sensory afferents and in selectively differentiating them from primary non-sensory peptide-coded fibers can be circumvented by focussing on CGRP. Extensive deafferentiation studies indicate that CGRP in the dorsal horn, at least in the lumbar segments of the rat, is exclusively of primary sensory origin.[83] Very occasionally observed supraspinal CGRP neurons projecting to the spinal cord and CGRP motoneurons appear to be no problem.[100] Since most of the peptides present in primary sensory afferents are coexisting in CGRP primary afferents,[3,4,30,31] the overall distribution pattern of CGRP in the dorsal horn can be assumed to reflect that of most peptidic afferents. By using double or even multiple labelling on identical sections with antisera against CGRP and any primary sensory peptide candidate(s), respectively, one would reasonably expect to map out the primary sensory component from the non-primary sensory fibers. The difficulty to trace the non-CGRP peptidic fibers of minor subpopulations of primary afferents, however, remains. In this case, one is still relying on the combination of retrograde tracing and immunohistochemistry which, however, does not allow to determine the whole spectrum of central ramifications and target distributions of the particular peptide-containing afferents in question.

CGRP fibers and varicosities form a dense plexus in Rexed's laminae I, IIo, V, and in the region of the central canal. Only a few fibers are present in laminae IIi, III, IV, and in laminae VI, VII, and IX.[46,101,102] In addition, CGRP-ir primary sensory fibers are present in the Lissauer tract.[101,102] Further, they are contained in the dorsolateral funiculus where they are to 95% depleted by dorsal rhizotomy, as determined by immuno-electron microscopy.[102] This indicates that small primary afferent fibers are more widely distributed in spinal white matter than it has been previously thought.[101]

A principally equivalent distribution pattern of CGRP is seen in the spinal trigeminal nucleus. The distribution of SP in spinal cord overlaps with that of CGRP, except in the ventral horn where CGRP motoneurons but virtually no CGRP fibers are present.[46] The CGRP-positive but SP-negative motoneurons are targetted by numerous SP fibers. However, this SP fiber population in the ventral horn is unaffected by capsaicin and dorsal rhizotomy, indicating that it is not primary sensory.[12,13] It is noteworthy that some CGRP fibers and SP fibers are apparently partly codistributed in intermediomedial autonomic nuclei. CGRP fibers and SP fibers target dendrites and somata of deep dorsal horn neurons which express opioid immunoreactivity.[60,61,103] This is also the case with small and large neurons of the superficial dorsal horn.[61] The functional significance of such a constellation will be discussed below in the context of inflammation and pain. In the dorsal horn, SP fibers outnumber CGRP fibers, indicating that there is a substantial non-primary sensory (intrinsic and descending) neuronal system containing SP.[12,13]

Vagal and Glossopharyngeal Afferents

The nucleus tractus solitarius (NTS) is filled up by virtually all peptides which have been detected so far in the central and peripheral nervous system.[104] In the

NTS, primary (apparently non-peptidic) afferent fibers mediating gustatory impulses and peptidic primary afferent inputs from receptors that innervate visceral organs via the vagal and glossopharyngeal nerves, converge in the rostral one third and caudal two thirds, respectively. There is a loose somato-topic organization in the termination of the afferent inputs in various subnuclei of the NTS. However, here it is not possible to differentiate the endings of the various peptide containing glossopharyngeal and vagal afferents from similarly peptide-coded NTS interneurons and CNS projections to the NTS.[104] CGRP afferents to the NTS may be an exception in as much as they are mainly of primary sensory origin.

AFFERENT NEUROTRANSMITTER OR NEUROMODULATOR FUNCTION AND MODALITY-SPECIFICITY OF PRIMARY SENSORY PEPTIDES?

A Specific Role of Primary Sensory Peptides in Nociception/Antinociception?

The initially presented view that SP, present in primary sensory neurons, may specifically be involved in the transmission of nociceptive impulses is no longer accepted. To the other extreme, a function of SP in nociception at all has been questioned by a most recent series of papers.[94,95,96,97] However, there is still plenty of evidence that SP is released in the spinal cord upon noxious stimulation, elegantly demonstrated by using a novel technique of antibody microprobes.[105] In support of these data noxious stimulation has been recently shown by quantitative in situ hybridization to increase the mRNA levels for TKs in spinal sensory ganglia.[106] Further, a similar increase in SP- and CGRP-mRNA content of spinal ganglia occurs as a response to experimentally induced hyperalgesia and inflammation in the rat.[41] In the line with these observations, an increased spinal release of SP has been reported in hyperalgesia and inflammation.[107] However, only marginal changes (intensification) of TK/CGRP immunstaining of dorsal root ganglionic cells but no increase in TK/CGRP cell numbers could be detected by immunohistochemistry.[60] The combinatory evidence of these studies is in favour of a role of SP, other TKs and CGRP in nociceptive spinal transmission. Whether this is in a true transmitter mode or in a neuromodulatory way cannot yet be answered.[12,13]

Glutamate, shown to coexist in primary sensory SP afferents,[98] could fulfill the role of a rapid nociceptive transmitter and SP, CGRP, and many other peptides (see section of coexistence) could be involved as co-transmitter in the more prolonged processing of nociceptive impulses, particularly under the condition of acute and chronic inflammatory pain. It is noteworthy that GAL could be inhibitory in this presumed multipeptide concert of nociceptive transmission.[108]

The main candidates for inhibitory actions on primary sensory afferents are, of course, the endogenous opioids.[21] Multiple opioids released from intrinsic spinal neurons are thought to presynaptically inhibit the release of nociceptive transmitters.[12,13,21] In addition, at least in some species, endogeneous primary sensory opioid peptides are autoinhibitory transmitter candidates in spinal and trigeminal nociceptive afferents.[30]

Under conditions of inflammatory pain in the rat, however, there is no evidence yet for an activation of primary sensory opioids.[60,61] Most interestingly, primary sensory CGRP fibers and SP fibers (probably of primary sensory origin) project to and form contacts with superficial and deep dorsal horn neurons which selectively increase their staining for PRO-DYN- and/or PRO-ENK-opioids in the response to chronic inflammatory hyperalgesia of the rat.[60,61,103,109,110] Further, the mRNA levels of PRO-ENK and PRO-DYN and the levels of the respective opioid peptides are elevated, particularly as regards PRO-DYN opioid products and mRNA.[61,111,112] This points to a functional activation of the spinal opioid system as a response to chronic inflammatory pain, depending on an increased input of primary sensory peptides like SP and CGRP.[60,61] It is debatable whether pain or inflammation are causative for this response. Nevertheless, it is tempting to suggest that there is not only an influence of primary sensory peptidic afferents on the spinal opioid neurons but also vice versa. This intriguing possibility of an antidromic regulation of primary afferents has been recently debated.[4] I postulate that such a presumed crosstalk between primary afferents and spinal second order neurons, mediated by multiple peptides with opposite actions, may be crucial in the process of chronic inflammatory pain. On the other hand, deafferentiation causes a similar increase of DYN in spinal neurons[99] which raises the possibility that the opioid response is not specifically related to inflammation or pain, but may be a trophic compensatory mechanism. On the other hand, deafferentiation may also involve pain and local inflammation.

Modality Specificity of Sensory Peptides in Nociceptive Transmission?

The controversial issue of the exact role and modality specificity of pro-nociceptice (SP, CGRP, GAL, possibly VIP) and of anti-nociceptive (opioids, GAL, CCK, SOM) primary sensory peptides cannot be debated in detail.[12,13,108,113] Antinociceptive[108] and pro-nociceptive[113] actions of GAL have been reported. There is some evidence that primary afferent peptides have modality specific nociceptive actions.[108,113,114,115] SP and CGRP elevate the spinal excitability to thermal and mechanical nociceptive stimuli.[113,114] In contrast, SOM is active in thermonociception but not in mechanonocicption.[114] VIP may be released by thermal nociceptive afferents which respond poorly to mechanonociceptive stimuli.[115] GAL appears to be involved in mechanical nociception only.[113] Anti-nociceptive actions of SOM may be due to neurotoxic effects.[116] That CGRP is pro-nociceptive is indicated by anti-nociceptive effects of intrathecally administered anti-CGRP.[117] Such strategies of peptide immunoneutralisation are promising tools in differentiating selective peptide actions in spinal nociception/antinociception. Of particular interest is the characterization of the peptide phenotype of chemonociceptive afferents which may be particularly crucial in neurogenic inflammation. Taken together, however, there appears to be no clear-cut relationship between modality and peptide phenotype.

A Role of Sensory Peptides in Non-Nociceptive Afferent Transmission?

All the peptides present in primary sensory neurons could contribute to non-nociceptive transmissions, i.e., mechanoreception, chemoreception, and thermoreception. That portion of CGRP which appears to be present in Aß-afferents can certainly be expected to be not involved in nociception.[27,30] Peptidic small diameter afferents may contribute to viscero- and somato-autonomic reflexes at the spinal level either directly or indirectly.[3,4,12,13,20] The presence of SP and CGRP in primary sensory fibers in intermediomedial spinal autonomic nuclei can be regarded as a neurochemical correlate for direct sensory autonomic circuits. In contrast, it is unlikely that SP fibers in intermediolateral nuclei are of primary sensory origin. Therefore, the excitatory effect of SP on sympathetic preganglionic neurons reflects the possible function of non-primary sensory SP fibers.

Vagal and Glossopharyngeal Afferents

The peptide-coding of vagal and glossopharyngeal afferents deserves special attention because afferents of the IXth and Xth cranial nerves mediate special autonomic regulatory reflexes (baroreception, chemoreception, gastric stretch reception, hepatic osmoreceptors, portal vein glucose receptors etc).[3,4,49,50,51] The sensory ganglia of the vagal and glossopharyngeal nerves are functionally diverse. The general somatic afferent neurons are located in the superior ganglia of the vagal (jugular ganglion) and of the glossopharyngeal nerves. The general visceral afferent neurons are located in the nodose (Xth) and petrose (IXth) ganglion.

SP and, implicitly NKA and NPK, are candidates to play a role in the transmission of chemoreceptive afferent impulses from the carotid chemoreceptors.[49] Further, it has been discussed as chemo- and baro-afferent transmitter from the aortic arch. Based on the wide spectrum of peptides, all present in the different subdivisions of the vagal/glossopharyngeal afferent system, it is suggested that they all could play a role in the various modalities transmitted by the cranial sensory system to the integratory area of the NTS.[104]

THE PERIPHERAL CONNECTION: ORGAN-, PATHWAY- AND TARGET-SPECIFICITY OF SUBSETS OF SENSORY PEPTIDIC NERVES

Intermingling of various Sensory and Autonomic Peptidic Fibers

The differentiation of peptidic sensory fibers from equally peptide-stained peripheral autonomic fibers poses some problems. In addition, it is necessary to segregate peptidic sensory fibers of the spinal and trigeminal system from those of the other cranial sensory ganglia. Further, some primary sensory peptides are variously coded in postganglionic cranial parasympathetic neurons adding to the inherent difficulties to differentiate intermingled sensory and autonomic fibers.[19,27,30,32,118,119] However, in non-cranial somatic tissues and in systemic blood vessels, both tachykinin- and CGRP-coded fibers can be regarded as staining sensory fibers specifically.[19,32,119] In contrast, TK and CGRP innervation of visceral tissues and organ-vasculature can also be delivered by vagal fibers, intrinsic neurons (especially in the gastroenteropancreatic but also in the respiratory system) and by a minor subset of neurons in local microganglia coding for TK and/or perhaps CGRP.[19,27,30,32,45,57] As a principle, an exact differentiation of the different peptidic sensory fibers and their peripheral and central projection sites and target relations can be only achieved by the combined effort of immunohistochemistry, retrograde tracing, capsaicin treatment, and selective nerve dissection or ganglionectomy. Such studies have been performed to trace the sensory innervation of the urinary tract[120] and of the thyroid gland[50] and in many other tissues.[3,4] Further details cannot be outlined here.

Taken together, the situation is extremely complex, particularly in the cervical region where sensory peptidic innervation from the glossopharyngeal, vagal, facial, trigeminal nerve, and from dorsal root ganglia of various cervico-thoracic segments converge and intermingle with peptidic fibers from sympa-thetic and cranial parasympathetic sources and from microganglia.[31,32,37,43,49,50] In this context, the even more puzzling possibility should be pointed out, namely, that SP, generally believed to be absent from sympathetic fibers, can be expressed by sympathetic postganglionic sympathetic neurons after reserpine treatment. Further, some primary sensory neurons can express catecholami-nergic (possibly dopaminergic) phenotypes.[34,119] Their putative peptide co-transmitters are not yet known.

In contrast to the ubiquituos distribution of TK and CGRP afferents in the somatic and visceral periphery,[3,4,27,30,45] the peripheral distribution of subsets of primary afferents containing opioids, CCK, SOM appears to be more pathway- and target-specific.[57] SOM may be absent from visceral but is present in somatic afferents.[51] Vagal TK/CGRP afferents contribute little to the sensory innervation of the gastroenteropancreatic system as compared to spinal TK/CGRP affe-rents.[51]

Common Target Relations

Besides non-target related distribution as 'free nerve endings', fibers staining for TK and CGRP have common targets in somatic and visceral tissues, i.e., vascular smooth muscle of all segments of the vasculature, endothelium (particularly of postcapillary venules), some epithelial cells, and paracrine cells, mast cells, immune cells, some endocrine cells and autonomic ganglion cells (Table 3).[27,30,32,119]

The presence of numerous TK/CGRP-fibers and of a comparatively lower number of opioid fibers around postcapillary venules in a wide variety of somatic (skin)[119] and visceral (urinary tract, respiratory tract)[27,30,121,122] deserves special attention because it is in this segment of the microvasculature where protein extravasation and edema formation are initiated.[3,4,30] Interestingly, mast cells are often in the vicinity of these areas.[30,119] TKs induce protein extravasation which is either mast cell mediated or not.[3,4] Endogenous primary sensory or circulating opioids may inhibit the release of TKs and therefore of protein extravasation.[3,4,27,30]

Peptidic Sensory-Neuroimmune Connection

Of particular interest is the presence of sensory TK/CGRP fibers in various lymphoid tissues. Our group has shown that peptide-coded neuroimmune interconnection is a general principle in various mammals including hu-man.[123,124,125] Sensory TK/CGRP fibers are in close contact with immunocompe-tent cells of primary (thymus) and secondary (lymph nodes) lymphoid tis-sues.[27,30,123,124,125] Sensory peptidic fibers are closely interrelated with numerous

Table 3

COMMON PERIPHERAL TARGETS OF SENSORY TK/CGRP VARICOSE
FIBERS AND TARGET-RELATED LOCAL EFFECTOR ACTIONS OF TKS
VS CGRP*

	TKs	CGRP
Vascular smooth muscle	vasodilatation	< dto
Endothelium	protein extravasation (particularly of venules)	
Non-vascular smooth muscle (gut, urinary tract, respiratory tract, reproductive system)	contraction and and/or relaxation	dto
Epithelia (intraepithelial (location; epidermis, cornea, oral mucosa, repiratory and lower urinary tract)	trophic actions? barrier regulation?	dto
Endocrine cells (adrenal cortex and medulla, thyroid and parathyroid gland, Langerhans island, pineal gland, etc.)	direct effects questionable; indirect effects via sensory-endocrine reflexes	
Mast cells SP (ubiquitously in somatic degradation and visceral tissues)	histamine release; generation of arachi- donic acid derivates;	inhibits
Immunocompetent cells in lymphoid and other tissues (leucocytes, macrophages)	immunomodulation: T-lymphocyte prolife- ration; increase in immunglobulin production; chemotaxis; phagocytosis; lysosomal enzyme release; IL-1 release;	yes dto n.i.

* For references see text

mast cells present in large numbers in the thymus and lymph nodes.[125] Similarly, various mucosa associated lymphoid tissues are the target of TK/ CGRP fibers. In addition, I have preliminary evidence for the presence of TK/ CGRP fibers in the bone marrow. Also in the spleen, TK/CGRP fibers are not restricted to blood vessels, as currently beleived, but often form contacts with immunocompetent cells (Zentel and Weihe, unpublished). Further, TK/CGRP fibers contact macrophages in various tissues, particularly in the skin.[119] As a rule, the peptidic sensory innervation of lympoid tissues appears to be denser than that by autonomic peptidic (NPY, VIP/PHI) fibers. It is noteworthy, that peptidic sensory and autonomic fibers in the immune system (like in the peripery in general) are partly intermingled, particularly around vessels but also in mast cell rich areas.[119,125] The peptidic primary sensory system provides the chemoanatomical basis for bidirectional crosstalk between the CNS and the immune system. The wide spectrum of putative neuroimmunomodulatory role of sensory peptides and their possibly crucial significance in disease are outlined below (see also Table 3).

Peptidic Sensory-Autonomic Collaterals

Further, I want to stress the attention to the possibility of a general collateral primary sensory afferent innervation of autonomic ganglia. Many collaterals of visceral primary afferents staining for TKs and CGRP (and probably also a few staining for PRO-DYN-opioids) project to prevertebral autonomic ganglia and contact dendrites of postganglionic sympathetic neurons synaptically and nonsynaptically as revealed by light and electron microsopic immunohistoche-mistry.[30,126,127] Further, SP collaterals form contacts with other nerve endings.[126] From the regular TK/CGRP innervation of non TK/CGRP postganglionic neu-rons in other prevertebral and paravertebral sympathetic ganglia as well as in the various cranial parasympathetic ganglia (and in local microganglia),[30,32,45] I deduce the general principle of sensory-collateral-autonomic intercommunica-tion with multiple peptides involved.[32]

HETEROGENEITY OF PHYSIOLOGICAL AND PATHOPHYSIOLOGICAL EFFERENT ROLES OF SENSORY PEPTIDES

From the specific target relations of peptidic sensory fibers (summarized in Table 3), multiple peripheral sensory-efferent functions can be deduced and explained in conjunction with results obtained by capsaicin treatment, antidro-mic nerve stimulation and the neuropharmacology of the sensory peptides and respective antagonists.[3,4] In the recent light of a vast heterogeneity of possible local effector functions of peptidic primary afferents[3,4] in physiological and, particularly, pathophysiological, mechanisms only a few examples can be outlined.

Tachykinins and CGRP: Major Sensory-Efferent Co-messengers

Local effector roles can be theoretically fulfilled by all peptides present in primary sensory peripheral endings. Certainly, SP and CGRP are the most important candidates because the sensory fibers encoding them are ubiquituos-ly distributed in somatic and visceral regions.[3,4]

In contrast, other peptidic fibers may be more organ- or target-specific[57] and, therefore, an ubiquitous local effector role for them cannot be postulated. To stress the principal spectrum of peptide sensory-efferent roles, one may concentrate on SP and CGRP and, in selected peripheral regions, on their possible interaction with endogenous opioids.

Vasodilatation, Protein extravasation, Smooth Muscle Spasm, the Mast Cell and Opioid Connection

Generally, CGRP is a more potent vasodilator than TKs. TKs are mediator candidates for protein extravasation whereas CGRP is inactive in this respect. However, CGRP enhances the action of SP by inhibiting its degradation.[3] Further, there are complex mutual interactions between TKs, CGRPs and degrading enzymes released from mast cells.[3,4,119] The TKs and CGRPs have differentially potent actions on various non-vascular smooth muscles and a variety of non-muscle cells. Their effects are mediated by distinct receptor subtypes, in the case of TKs by NK-1, NK-2, NK-3 and possibly NK-4 recep-tors.[3,4,18,119]

Vasodilatation and protein extravasation induced by antidromic nerve stimulation can be inhibited by opioid agonists.[3,4,30] The concept is that opioids presynaptically inhibit the release of vasoactive peptides from peripheral sensory endings. It has been postulated that endogenous opioids present in peripheral sensory endings may tonically autoinhibit the release of other sensory peptides.[27,30,54,55] On the other hand, it has been reported that some opioids, namely DYN 1-13, themselves cause protein-extravasation.[3,4] This implies that endogenous opioids could be even mediators of neurogenic extravasation. However, DYN 1-13 is basic and, therefore, bound to release histamine from mast cells, as do many basic substances, including morphine. Thus, such opioid effects may be not mediated by stereospecific opioid receptors. Similarly, it is under debate to which extent tachykinins induce histamine release due to their basic nature or via specific TK receptor subtypes.[3,4]

Cardiac Functions and CGRP

The heart receives a much denser CGRP than SP innervation.[4,19,45] In contrast to SP fibers, CGRP fibers regularly supply cardiac muscle, particularly in the atria.[45] Upon pharmacological application, SP has no effect on the contraction force of cardiac muscle, but CGRP has.[3,4] Thus, the positive inotropic action of capsaicin could be explained by the fact that it releases CGRP from nerves supplying the target structure and the absence of an effect of SP on the force of contraction can be explained by the lack of a general SP innervation of the cardiac muscle.[3,4,45] From the dense CGRP innervation of the conduction system, chronotropic effects can be anticipated. In fact, CGRP is positively chronotropic.[3,4] In addition, one could speculate that CGRP sensory innervation may influence cardiac ANF secretion.

Urinary tract

Differential sensory efferent functions of SP and CGRP and, in addition, species differences have been demonstrated in the urinary tract.[3,4,128] Species differences in the smooth muscle contraction and relaxation effects of SP and CGRP between rat and guinea-pig could find their morphological correlate in a more intense SP innervation in guinea-pig as compared to the rat.[122] Various PRO-DYN-derivatives have been shown to coexist in a subpopulation of vascular (venules) and non-vascular (smooth muscle) TK/CGRP sensory fibers of the urinary tract of guinea-pig. Thus, endogenous opioids could autoinhibit protein extravasation and urinary spasm.[122]

Airways

In the respiratory system, both TKs and CGRP contract bronchotracheal smooth muscle. TK/CGRP fibers in the lower respiratory tract originate mainly from vagal sensory fibers which are more numerous than sensory fibers projecting via sympathetic afferents. However, there are also some intrinsic TK neurons in the cat and dog but apparently not in the guinea-pig and rat.[129] TK/CGRP vagal sensory fibers in the lower airways are thought to mediate, via local effector and axon reflex mechanisms, the response to airway irritation characterized by protein extravasation from postcapillary venules (TKs), vasodilatation (CGRP > TKs), increased mucous production (TKs) and bronchoconstriction

(NKA > SP).[130] In such responses to various allergic and-non allergic irritations, e.g., cigarette smoke containing irritant chemicals (e.g., aldehydes, acrolein), not only TK/CGRP vagal and sympathetic sensory fibers but also cholinergic and various peptidergic intrinsic neurons are involved.[130] Further, interactions with mast cells, common targets of autonomic and sensory fibers, are contributing to the response. Vagal sensory fibers have opioid receptors.[3,4,30] This observation prompted the hypothesis that opiates may have peripheral as well as central antitussive actions. In fact, local actions of inhaled codeine and morphine in the lung of guinea-pigs attenuate cough and bronchoconstrictor reflexes to acid.[131] Capsaicin challenge in humans, expected to release Tks and CGRP from sensory fibers of the respiratory tract, revealed that inhaled opiates peripherally reduced the increase in respiratory resistance.[131] In contrast, the opiate inhibition of the capsaicin-induced cough reflex was a central effect.

In this context, it is of interest that our group provided evidence for the presence of opioid peptides in nerve fibers of the respiratory tract.[27,30,121] However, the exact origin of the opioid fibers in the airways is not yet fully elucidated. They may be of multiple origin, vagal afferent, vagal preganglionic, sympathetic afferent and efferent, and intrinsic and PRO-ENK as well as PRO-DYN may contribute. A fraction of opioid fibers appears to coincide with TK fibers. Taken together, this points to the possibility that the response of the airways to irritation may be controlled by endogenous opioids of heterogenous neuronal, including primary sensory, origin.

Other tissues and Sensory-Endocrine Intercommunication

Principally similar mechanisms as exemplarily lined out in the preceeding paragraphs apply to many other tissues and have been studied extensively, particularly in the eye. Further details are beyond the scope of this review and I refer to previous reviews.[3,4]

Sensory-endocrine intercommunications deserve a few more words

Capsaicin treatment attenuates many endocrine functions. This appears to be due to disturbances of reflex pathways, that is to central effects at the interface of the incoming primary afferents and various second order effector neurons.[3,4,20] However, a direct local effector role of peptidic sensory afferents to regulate endocrine secretion in the periphery cannot be ruled out.[30] On the other hand, the endocrine system producing a plethora of messengers including peptides could communicate with peptidic sensory fibers. Circulating opioids (from the adrenal medulla or the pituitary) and CRF could target sensory fibers and modulate their excitability, particularly under stress conditions. Perhaps, some of the opioid analgesia in stress is due to hormonal presynaptic inhibition of primary nociceptive neurons, well known to contain opioid receptors.[27,30] Also CRF can modulate primary sensory function. In fact, it attenuates neurogenic inflammation.[3,4]

SENSORY PEPTIDES IN INFLAMMATORY PAIN AND PEPRIPHERAL OPIOID ANTINOCICEPTION

Pain- and inflammation-induced changes in peptide levels do not only occur in ganglionic cells[60,106] and spinal endings[60,106] of small diameter primary afferents but also in the periphery. Thus, levels of SP in inflamed tissues are elevated.[3,4,132] The changes may have their morphological correlate in an apparent sprouting of CGRP/TK sensory fibers in the affected areas.[60,133] This, apparently, does also apply to humans as suggested by immunohistochemistry of inflamed skin[119] and of pancreas in chronic painful pancreatitis.[134] Further, the levels of SP are elevated in synovial fluids of arthritic patients.[3,4,20,132]

Most interestingly, inflammatory pain is attenuated by peripheral opioids while peripheral opioid antinociception is ineffective in uninflamed tissues.[135,136] All three opioid receptors (μ, kappa, ∂) appear to play a role. One possible target of peripheral opioid treatement are the primary sensory afferents.[20,27,30,60,134,135,136] Opioids could inhibit the potentially increased peripheral release of pro-inflammatory peptides from the different target endings of the peptidic sensory afferents in the inflamed tissue and, thereby, reduce protein extravasation and activation of inflammatory mediators. This view is supported by the fact that opioid receptors are present on SP containing primary afferents and that exogenous opioids and opiates peripherally attenuate protein extravasation and vasodilatation induced by antidromic nerve stimulation.[3,4,20,30,60] A possible role of endogenous primary sensory opioids in inflammation has to be also taken into account.

Bradykinin receptors are localized to small diameter nociceptive fibers and bradykinin must be continuously produced in inflamed tissue to maintain hyperalgesia.[137] Thus, alternatively, peripheral opioid analgesia 'lit up' in inflammation may work via inhibition of bradykinin production. The levels of the nonapeptide bradykinin are certainly elevated in acute and chronic inflammation as the analgesic actions of bradykinin antagonists indicate.[137] Interestingly, the bradykinin-induced peripheral release of SP may be independent of electro-physiological excitation of sensory endings.[138] Whether bradykinin chronically alters peptide gene expression and contents of primary sensory neurons may be worthwhile to investigate.

Further, it will be interesting to see, whether glutamate, colocalized with SP in dorsal root ganglionic cells,[98] is present in peripheral endings. According to Dale's principle[3,4,5,30] glutamate should not only transmit afferent nociceptive impulses but also plays a role in neurogenic inflammation. Thus, it is a putative mediator in acute and chronic inflammatory pain. Possibly, the peripheral antinociceptive opioid effects are mediated by inhibition of putative glutamate release in the periphery.

In addition, local immune cells expressing opioid receptors, opioid mRNAs and possibly opioid peptides may be involved.[60,119,139] It is conceivable that the peripheral antinociceptive effect of opioids may be indirect via an immunomo-dulatory action on the infiltrated inflammatory cells. A potential role (autocrine?) of opioid peptides in immune cells needs to be investigated in inflamed tissues.[136] In the inflamed periphery, opioids may be not only anti-nociceptive but also anti-inflammatory. In both respect, they may be the functional antipodes to the pro-inflammatory and pro-nociceptice tachykinins. The role of hormonal opioids and of other circulating neuroendocrine hormones with peripheral aninociceptive action profiles has to be taken into account.

SENSORY PEPTIDES IN NEURO-IMMUNE INTERACTIONS AND TROPHIC MECHANISMS: BASIS OF DISEASE?

Sensory neuropeptides, in particular SP, have a wide spectrum of immuno-modulatory actions.[3,4,123,124,125,140] Some neuropeptides, e.g., opioids may be even synthetized in immunocompetent cells.[139,141,142] PRO-ENK mRNA is expressed in T cells, macrophages and mast cells.[142] Further, sensory peptides are mediator candidates of various trophic actions.[143] I postulate that the putative immuno-modulatory and trophic actions of sensory peptides may be linked to regulate the organism's immune and trophic homeostasis and its response to injury and inflammation. Changes in peptide metabolism of primary afferents may have some causal relationship with diseases of immunological (and psychological) background like allergy, asthma, psoriasis, neurodermitis etc.[20,119,130] There may be even a neuroimmune connection to tumor growth[144] and to immune mechanisms in AIDS. In fact, VIP and AIDS related peptides depress lymphocyte traffic in vivo[145] The influence of TKs and CGRP on lymphocyte traffic is not known. Further, a role of sensory peptides in autoimmune disease is conceivable. I shall briefly outline some recent observations in these directions. The polymodal immunoregulatory actions of SP give the lead.[146] SP alters the function of mononuclear and polymorphonuclear leucocytes in the inflammatory response, i.e., chemotaxis, phagocytosis, lysosomal enzyme release.[146] SP stimulates cell-mediated immunity at low concentrations in that synthesis of immunoglobulins by mature T- and B-lymphocytes is increased. Functionally relevant TK receptors have been characterized on T-lymphocytes. These observations are compatible with the view that SP is a specific T cell mitogen.[146] Direct evidence for an important immunoregulative function of sensory neuropeptides in vivo is provided by recent studies. Firstly, sensory peptidic nerves have target relations with immunocompetent cells in lymph nodes, thymus, bone marrow, spleen, various, mucosa associated tissues, disseminated macrophages.[3,4,30,123,124,125] Secondly, capsaicin-induced depletion of sensory peptides alters the proliferation of lymphocytes to immune stimulation.[147,148] In inflamed tissues, sensory peptidic fibers sprout into areas where mast cells and various immune cells accumulate. In lymphoid tissues, sensory fibers target mast cells. This points to mast cell mediated immune functions of peptidic fibers.[119,123,124,125] Histamine itself is well known to be immunoactive.

SP modulates immediate hypersensitivity responses. It does not release histamine from basophils but from mucosal and cutaneous mast cells.[146] Further support for the possible relevance of sensory tachykinins to arthritic disease has been deduced from the influence of Tks on interleukin-1 (IL-1) production in a mouse cell line.[149] SP, NKA and NKB induced IL-1 secretion[149] indicating that a multiplicity of TK receptors is involved. The mitogenic effects of SP on fibroblasts[146] may be indirectly mediated by IL-1 which proliferates fibroblasts.[149] Indirect actions of SP (and possibly other sensory peptides) via IL-1 may even be a general principle, both in homeostaic and pathologic mechanisms. IL-1 is a peptide hormone with a wide spectrum of biological actions. These comprise T cell activation, secretion of prostaglandins and collagenase from fibroblasts, chondrocyte activation, bone resorption and fever.[149] SP acts in this line in as much as it stimulates prostaglandin synthesis by macrophages and activates synoviocytes to produce PGE2 and collagenase. The neuropeptide-induced release of the cytokinine IL-1 from a lymphoid cell line is compatible with the view that neurogenic stimulation of immunologic mechanisms may not only be etiopathic in inflammatory disease like arthritis but also in autoimmune disorders, e.g., multiple sclerosis. SP is certainly not the only primary sensory peptide possibly involved. For virtually all of the other primary sensory peptides, immunomodulatory actions have been reported which cannot be outlined here.

There may be not only destructive actions of sensory peptides but also trophic actions of preserving and repair character. In support of such a view, N-terminal sequences of SP have been found to enhance the regeneration of nerves in chick embryos.[150] Beneficial trophic effects of sensory peptides in mammals deserve more attention. A sprouting of SP/CGRP fibers as observed in inflammation and burning injury may reflect neurotrophic repair actions and importance in healing. The inflammation-induced activation of peptide metabolism in small diameter primary afferents is possibly related to an increase in NGF activity. This may be deduced from the recently detected effects of NGF to regulate peptide expression in adult sensory neurons.[85]

In this context it is noteworthy that the axonal transport of SP is impaired in experimental diabetes.[150] This points to the possibility that changes in primary SP and, implicitly, of other sensory peptides, are relevant to diabetic neuropathy.

That peptides in sensory afferents are involved in allergic processes is suggested by an increased release of CGRP and SOM, measured in the nasal lavage after antigen challenge of allergic patients.[151] It has been postulated that CGRP is causative for nasal congestion and SOM in the late involvement of basophils and mast cells in allergic rhinitis.[151] Similarly, sensory neuropeptides are thought to be neuroimmune mediators in asthma.[130]

SUMMARY AND CONCLUSIONS

The present review summarizes selected aspects of the increasing knowledge about the morphology and possible physiological and pathophysiological function of peptide-coded primary afferents of mammals. The picture emerges that mainly small diameter (C and A∂) sensory neurons of spinal and cranial sensory ganglia but not of special sensory organs (retinal, spiral and vestibular ganglion, olfactory and gustatory primary afferents) contain a principally similar spectrum of multiple peptides, excepted of some species, regional and quantitative differences. Immunohistochemistry, biochemistry and molecular biology reveal the presence of an increasing list of gut-brain peptides and peptide hormones. The tachykinins (TKs) substance P (SP) and neurokinin A (NKA) and the calcitonin gene-related peptide (CGRP) prevail.

Not only multiple peptides derived from the same precursors (e.g., SP and NKA from preprotachykinin) are expressed in identical ganglionic cells and peripheral as well as central ramifications, but also multiple peptides derived from unrelated precursors, e.g., TKs, CGRP, cholecystokinin (CCK) and pro-dynorphin opioids. Such multiple combinations of a plethora of peptides in subsets of primary sensory neurons are rather the rule than the exception.

Further, a sparse extrinsic innervation by fibers coding for neuropeptide Y (NPY) is delivered to blood vessels of spinal and cranial sensory ganglia. It most likely reflects sympathetic innervation of sensory ganglia.

The distribution and targets of central and peripheral ramifications of peptide-containing primary sensory neurons are difficult to differentiate from non-primary sensory fibers in the CNS target areas and from peripheral autonomic fibers, which both are often similarly peptide-coded. In the spinal dorsal horn, however, CGRP may be an almost selective marker of the C and A ∂ primary afferents.

The specific central target distribution which may be much more widespread than currently believed, and the multiple peripheral target relations are the morphological basis for the wide spectrum of dual afferent and sensory-efferent (local effector role) of primary sensory peptides. To fulfill these functions, they most likely act in multiple combinations as indicated by multiple peptide coexistence. Their net function is the balance between partly synergistic and partly opposite actions of the various peptide components. The presence of opioids in and of opioid receptors on spinal and vagal sensory fibers points to the possibility that the functional activity of primary afferents is 'auto-tuned' (inhibited) by endogenous opioids. The latter constellation may be important in both sensory-afferent and sensory-efferent, nociceptive/antinociceptive and chemo-/baroreceptive neurotransmission and/or neuromodulation. As a rule, no absolute specific afferent modality can be attributed to a single peptide or peptide receptor although some may be reatively specific. The peripheral action profiles of many sensory peptides are complex and imply vasodilatation (CGRP > TKs), protein extravasation (TKs), non-vascular smooth muscle contraction or relaxation, immunoregulation and various trophic actions. Further, somato-visceral and viscerosomatic sensory reflex pathways are influenced by peptidic sensory collaterals at different levels of the neuraxis.

The peripheral local effector role may be of crucial importance under pathophysiological conditions such as injury, acute and chronic inflammation, allergy, autoimmune disease, acute and chronic pain. This is reflected by the increase in levels of TKs and a sprouting of CGRP fibers in inflamed peripheral tissue of the rat. Also in arthritic patients, SP levels are high in synovial fluid and the density of TK/CGRP fibers is increased in chronic painful pancreatitis.

Interestingly, SP and CGRP fibers innervate somata and dendrites of opioid coding neurons in the spinal cord which selectively respond to chronic arthritic inflammation of the rat with an increase in their opioid message.

The plasticity of sensory neurons in peptide expression, apparently regulated by peripheral target-derived NGF, may be an important link in neuro-immuno-endocrine, neurotrophic and possibly neuroteratologic responses. Further progress in multidisciplinary research of primary sensory peptidic neurons may lead to the development of and treatment with peripherally acting antinociceptive, antiinflammatory and antiallergic drugs with the advantage of minor side effects as compared to systemic application.

Acknowledgements: This study was supported by the Deutsche Forschungs-gemeinschaft (DFG grant We 910/2-1). I thank A. Leibold and M. Stuhlträger for technical assistance, D. Nohr and J. Zentel for helpful discussions and U. Hulick for typing the references.

REFERENCES

1. T. E. Salt and R. G. Hill, Neurotransmitter candidates of somatosensory primary afferent fibres, *Neuroscience* 10:1083 (1983).
2. G. Gaertner, Über den Verlauf der Vasodilatoren, *Klin. Wschr.* 51:980 (1889).
3. P. Holzer, Local effector functions of capsaicin- sensitive sensory nerve endings: involvement of tachykinins, calcitonin, gene-related peptide and other neuropeptides, *Neuroscience* 24:739 (1988).
4. C. A. Maggi and A. Meli, The sensory-efferent function of capsaicin-sensitive sensory neurons, *Gen. Pharmac.* 19:1 (1988).
5. H. H. Dale, Pharmacology and nerve endings, *Proc. R. Soc. Med.* 28:319 (1935).
6. U. S. von Euler, and J. H. Gaddum, An unidentified depressor substance in certain tissue extracts, *J. Physiol. (Lond.)* 72:74 (1931).
7. M. M. Chang and S. E. Leeman, Isolation of a sialogogic peptide from bovine hypothalamic tissue and its characterisation as substance P, *J. biol. Chem.* 245:4784 (1970).
8. F. Lembeck, Zur Frage der zentralen Übertragung afferenter Impulse - III. Mitteilung. Das Vorkommen und die Bedeutung der Substanz P in den dorsalen Wurzeln des Rückenmarks, *Arch. exp. Path. Pharmak.* 219:197 (1953).
9. N. Jancsó, A. Jancsó-Gabor, and J.Szolcsanyi, Direct evidence for neurogenic inflammation and its prevention by denervation and by pretreatment with capsaicin, *Br. J. Pharmac.* 31:138 (1967).
10. F. Lembeck, Columbus, capsicum and capsaicin: past, present and future, *Acta Physiol. Hungarica* 69:265 (1987).
11. S. H. Buck and T. F. Burks, The neuropharmacology of capsaicin: review of some recent observations, *Pharmacol. Reviews* 38:179 (1986).
12. W. Zieglgänsberger, Central control of nociception, *in:* "Handbook of Physiology - The Nervous System IV," V.B. Mountcastle, F.E. Bloom, and S.R. Geiger eds. Williams & Wilkins, Baltimore (1986).
13. J. M. Besson and A. Chaouch, Peripheral and spinal mechanisms of nociception, *Physiol. Rev.* 67:67 (1987).
14. T. M. Jessell, Nociception, *in:* "Brain Peptide," D. T. Krieger, M. J. Brownstein, and J. B. Martin, eds., J. Wiley & Sons, N.Y.,Chichester, Brisbane (1983).
15. J. C. Foreman, Peptides and neurogenic inflammation, *British Med.Bull.* 43:386 (1987).
16. M. K. Warden, and W. S. Young, III, Distribution of cells containing mRNAs encoding substance P and neurokinin B in the rat central nervous system, *J. Comp. Neurol.* 272:90 (1988).
17. J. E. Krause, J. M. Chirgwin, M. S. Carter, Z. S. Xu, and A. D. Hershey, Three rat preprotachykinin mRNAs encode the neuropeptides substance P and neurokinin A, *Proc. Natl. Acad. Sci. USA* 84:881 (1987).
18. Y. Masu, H. Tamaki, Y. Yokota, and S. Nakanishi, Tachykinin precursors and receptors: molecular genetic studies, *Regulatory Peptides* 22:9 (1988).
19. T. Hökfelt, D. Millhorn, K. Seroogy, Y. Tsuruo, S. Ceccatelli, B. Lindh, B. Meister, T. Melander, M. Schalling, T. Bartfai, and L. Terenius, Coexistence of peptides with classical neurotransmitters, *Experientia* 43:768 (1987).
20. F. Lembeck, The 1988 Ulf von Euler Lecture, *Acta Physiol. Scand.* 133:435 (1988).
21. M. J. Millan, Multiple opioid systems and pain, *Pain* 27:303 (1986).
22. M. A. Ruda, G. J. Bennett, and R. Dubner, Neurochemistry and neural circuitry in the dorsal horn, *Prog. Brain Res.* 66:219 (1986).
23. A. C. Cuello, Peptides as neuromodulators in primary sensory neurons, *Neuropharmacol.* 26:971 (1987).
24. S. P. Hunt, Cytochemistry of the spinal cord, *in:* "Chemical Neuroanatomy," P. C. Emson, ed., Raven Press, N.Y. (1983).
25. M. J. Millan, and A. Herz, The endocrinology of the opioids, *Neurobiology* 26:1 (1985).
26. A. Goldstein, Biology and chemistry of the dynorphin peptides, *in:* "The peptides," S. Udenfriend and J. Meienhofer, eds., Academic Press, N.Y. (1984).
27. E. Weihe, D. Nohr, W. Hartschuh, B. Gauweiler, and T. Fink, Multiplicity of opioidergic pathways related to cardiovascular innervation: Differential contribution of all three opioid precursors, *in:* "Opioid Peptides and Blood Pressure Control," K.O. Stumpe, K. Kraft, and A.I. Faden, eds., Springer Verlag Heidelberg (1988).

28. F. Berkenbosch, J. Schipper, and F. J. H. Tilders, Corticotropin-relaeasing factor immunostaining in the rat spinal cord and medulla oblongata: an unexpected form of cross-reativity with substance P, *Brain Res.* 399:87 (1986).

29. G. Ju, T. Hökfelt, J. A. Fischer, P. Frey, J. F. Rehfeld, and G. J. Dockray, Does cholecystokinin- like immunoreactivity in rat primary sensory neurons represent calcitonin gene-related peptide? *Neurosci. Lett.* 68:305 (1986).

30. E. Weihe, D. Nohr, and W. Hartschuh, Immunohistochemical evidence for a co-transmitter role of opioid peptides in primary sensory neurons. *Progr. Brain Res.* 74:189 (1988).

31. G. Ju, T. Hökfelt, E. Brodin, J. Fahrenkrug, J. A. Fischer, P. Frey, R. P. Elde, and J. C. Brown, Primary sensory neurons of the rat showing calcitonin gene-related peptide immunoreactivity and their relation to substance P-, somatostatin-, galanin-, vasoactive intestinal polypeptide- and cholecystokinin-immunoreactive ganglion cells, *Cell Tissue Res.* 247:417 (1987).

32. E. Weihe, D. Nohr, B. Gauweiler, T. Fink, E. Nowak, S. Konrad, Immunohistochemical evidence for a diversity of opioid coding in peripheral sympathetic, parasympathetic and sensory neurons: a general principle of prejunctional opioid autoinhibition? *in*: "Regulatory Role of Opioid Peptides Symposium to the Second World Congress of Neuroscience, Budapest," P. Illes, and C. Farsang, eds., VCH Verlagsgesellschaft Weinheim, N.Y. (1988).

33. A. I. Basbaum, L. Cruz, and E. Weber, Immunoreactive dynorphin B in sacral primary afferent fibers of the cat, *J. Neurosci.* 6:127 (1986).

34. J. Price, An immunohistochemical and quantitative examination of dorsal root ganglion neuronal subpopulations, *J. Neurosci.* 5:2051 (1985).

35. M. Kawatani, J. Nagel, and W. C. de Groat, Identification of neuropeptides in pelvic and pudendal nerve afferent pathways to the sacral spinal cord of the cat, *J. Comp. Neurol.* 249:117 (1986).

36. M. A. Kai-Kai, B. H. Anderton, and P. Keen, A quantitative analysis of the interrelationships between subpopulations of rat sensory neurons containing arginine vasopressin or oxytocin and those containing substance P, fluoride-resistant acid phosphatase or neurofilament protein, *Neuroscience* 18:475 (1986).

37. M. M. Tuchscherer, and V. S. Seybold, Immunohistochemical studies of substance P, cholecystokinin-octapeptide and somatostatin in dorsal root ganglia of the rat, *Neuroscience* 14:593 (1985).

38. U. Hoheisel, and S. Mense, Observations on the morphology of axons and somata of slowly conducting dorsal root ganglion cells in the cat, *Brain Res.* 423:269 (1987).

39. U. Hoheisel, and S. Mense, Non-myelinated afferent fibres do not originate exclusively from the smallest dorsal root ganglion cells in the cat, *Neurosci. Lett.* 72:153 (1986).

40. N. Réthelyi, P. K. Lund, E. R. Perl, Peptide precursor mRNAs in the primary sensory neurons, *Neuroscience* 22, Suppl. S15 (1987).

41. M. J. Iadarola, and G. Draisci, Elevation of spinal cord dynorphin mRNA compared to dorsal root ganglion peptide mRNSAs during peripheral inflammation, *in*: "The arthritic rat as a model of clinical pain?," J. M. Besson and G. Guilbaud, Excerpta Medica, Amsterdam, New York, Oxford (1988).

42. S. J. Gibson, J. M. Polak, A. Giaid, Q. A. Hamid, S. Kar, P. M. Jones, P. Denny, S. Legon, S. G. Amara, R. K. Craig, S. R. Bloom, R. J. A. Penketh, C. Rodek, N. B. N. Ibrahim, and A. Dawson, Calcitonin gene- related peptide messenger RNA is expressed in sensory neurones of the dorsal root ganglia and also in spinal motoneurones in man and rat, *Neurosci. Lett.* 91: 283 (1988).

43. F. Grunditz, R. Håkanson, F. Sundler, and R. Uddman, Neurokinin A and galanin in the thyroid gland: neuronal localization, *Endocrinology* 121:575 (1987).

44. A. Saria, R. Gamse, J. M. Lundberg, T. Hökfelt, E. Theodorsson-Norheim, J. Petermann, and J.A. Fischer, Co-existence of tachykinins and calcitonin gene-related peptide in sensory nerves in relation to neurogenic inflammation, *in*: "Tachykinin Antagonists," R. Håkanson and F. Sundler, eds., Elsevier Science Publishers B.V. (Biomedical Division) Amsterdam, New York, Oxford (1985).

45. E. Weihe, Peripheral innervation of the heart, *in*: "Silent Ischemia," Th. von Arnim, ed., Steinkopff Verlag, Darmstadt (1987).

46. A. Franco-Cereceda, H. Henke, J. M. Lundberg, J. B. Petermann, T. Hökfelt, and J. A. Fischer, Calcitonin gene-related peptide (CGRP) in capsaicin-sensitive substance P-immunoreactive sensory neurons in animals and man: distribution and release by capsaicin, *Peptides* 8:399 (1987).

47. F. Sundler, E. Brodin, E. Ekblad, R. Håkanson, R.Uddman, Sensory nerve fibers: distribution of substance P, neurokinin A and calcitonin gene- related peptide, *in*: "Tachykinin Antagonists," R. Håkanson and F. Sundler, eds., Elsivier Science Publishers B.V. (Biomedial Division), Amsterdam, New York, Oxford (1985).

48. K. Saito, S. Greenberg, and M. A. Moskowitz, Trigeminal origin of ß-preprotachykinin products in feline pial blood vessels, *Neurosci. Lett.* 76:69 (1987).

49. C. K. Helke and K. M. Hill, Immunohistochemical study of neuropeptides in vagal and glossopharyngeal afferent neurons in the rat, *Neuroscience* 26:539 (1988).

50. T. Grunditz, R. Håkanson, F. Sundler, and R. Uddman, Neuronal pathways to the rat thyroid revealed by retrograde tracing and immunocytochemistry, *Neuroscience* 24:321 (1988).

51. T. Green and G. J. Dockray, Characterization of the peptidergic afferent innervation of the stomach in the rat, mouse and guinea-pig, *Neuroscience* 25:181 (1988).

52. P. K. Mulderry, M. A. Ghatei, R. A. Spokes, P. M. Jones, A. M. Pierson, Q. A. Hamid, S. Kanse, S. G. Amara, J. M. Burrin, S. Legon, J. M. Polak, and S. R. Bloom, Differential expression of a-CGRP by primary sensory neurons and enteric autonomic neurons of the rat, *Neuroscience* 25:195 (1988).

53. V. Höllt, P. Sanchez-Blazquez, and J. Garzon, Multiple opioid ligands and receptors in the control of nociception, *Phil. Trans R. Soc. Lond. B* 308:299 (1985).

54. E. Weihe, W. Hartschuh, E. Weber, Prodynorphin opioid peptides in small somatosensory primary afferents of guinea pig, *Neurosci. Lett.* 58:347 (1985).

55. E. Weihe, A. Leibold, D. Nohr, T. Fink, and B. Gauweiler, Co-existence of prodynorphin - opioid peptides and substance P in primary sensory afferents of guinea-pigs. Proc. Int. Narc. Res. Conf., San Francisco, *NIDA Research Monograph* 75:295 (1986).

56. P. M. Sweetnam, J. H. Neale, J. L. Barker, and A. Goldstein, Localization of immunoreactive dynorphin in neurons cultured from spinal cord and dorsal root ganglia, *Proc. Natl. Acad. Sci. USA* 79:6742 (1982).

57. I. L. Gibbins, J. B. Furness, and M. Costa, Pathway- specific patterns of the co-existence of substance P, calcitonin gene-related peptide, cholecystokinin and dynorphin in neurons of the dorsal root ganglia of the guinea-pig, *Cell Tissue Res.* 248:417 (1987).

58. L. J. Botticelli, B. M. Cox, and A. Goldstein, Immunoreactive dynorphin in mammalian spinal cord and dorsal, *Proc. Natl. Acad. Sci. USA* 78:7783 (1981).

59. W. Kummer and C. Heym, Correlation of neuronal size and peptide immunoreactivity in the guinea-pig trigeminal ganglion, *Cell Tissue Res.* 245:657 (1986).

60. E. Weihe, D. Nohr, M. J. Millan, C. Stein, S. Müller, C. Gramsch, and A. Herz, Peptide neuroanatomy of adjuvant-induced arthritic inflammation in rat, *Agents and Actions* 25:255 (1988).

61. E. Weihe, M. J. Millan, V. Höllt, D. Nohr, and A. Herz, Induction of the gene encoding prodynorphin by experimentally induced arthritis enhances staining for dynorphin in the spinal cord of rats, *Neuroscience* (in press) (1988).

62. M. A. Moskowitz, K. Saito, L. Brezina, and J. Dickson, Nerve fibers surrounding intracranial and extracranial vessels from human and other species contain dynorphin-like immunoreactivity, *Neuroscience* 23:731 (1987).

63. K. Uda, H. Okamura, H. Imura, C. Yanaihara, N. Yanaihara, and Y. Ibata, Distribution of human leumorphin-like immunoreactivity in the monkey spinal cord revealed by immunocytochemistry, *Neurosci. Lett.* 62:39 (1985).

64. R. Przewlocki, C. Gramsch, A. Pasi, and A. Herz, Characterization and localization of immunoreactive dynorphin, a-neo-endorphin, Met- enkephalin and substance P in human spinal cord, *Brain Res.* 280:95 (1983).

65. S. D. Logan, C. J. Lote, J. H. Wolstencroft, J. P. Gent, J. E. Fox, D. Hudson, and M. Szelke, Isolation, identification and synthesis of a novel tripeptide, methionyl-tyrosyl-lysine, from spinal cord and dorsal root ganglia of sheep, *Neuroscience* 5: 1437 (1980).

66. P. Panula, M. Hadjiconstantinou, H.-Y. T. Yang, and E. Costa, Immunohistochemical localization of bombesin/gastrin-releasing peptide and substance P in primary sensory neurons, *J Neurosci.* 3:2021 (1983).

67. K. Fuxe, L. F. Agnati, T. McDonald, V. Locatelli, T. Hökfelt, C.-J. Dalsgaard, N. Battistini, N. Yanaihara, V. Mutt, and A. C. Cuello, Immunohistochemical indications of gastrin releasing peptide - bombesin-like immunoreactivity in the nervous system of the rat. Codistribution with substance P-like immunoreactive nerve terminal systems and coexistence with substance P-like immunoreactivity in dorsal root ganglion cell bodies, *Neurosci. Lett.* 37:17 (1983).

68. J. D. Leah, A. A. Cameron, W. L. Kelly, and P. J. Snow, Coexistence of peptide immunoreactivity in sensory neurons of the cat, *Neuroscience* 16:683 (1985).

69. L. M. Kaplan, E. R. Spindel, K. J. Isselbacher, and W. W. Chin, Tissue-specific expression of the rat galanin gene, *Proc. Natl. Acad. Sci. USA* 85:1065 (1988).

70. A. Rökaeus and M. J. Brownstein, Construction of a porcine adrenal medullary cDNA library and nucleotide sequence analysis of two clones encoding a galanin precursor, *Proc. Natl. Acad. Sci. USA* 83:6287 (1986).

71. N. Itoh, K. Obata, N. Yanaihara, and H. Okamoto, Human preprovasoactive intestinal polypeptide contains a novel PHI-27-like peptide, PHM-27, *Nature (Lond.)* 304:547 (1983).

72. S. A. S. Shehab and M. E. Atkinson, Vasoactive intestinal polypeptide (VIP) increases in the spinal cord after peripheral axotomy of the sciatic nerve originate from primary afferent neurons, *Brain Res.* 372:37 (1986).

73. M. Vecsernyés, I. Jojart, J. Jojart, F. Laczi, and F. A. Laszlo, Presence of chromatographically identified oxytocin in hguman sensory ganglia, *Brain Res.* 414:153 (1987).

74. D. Nohr, H. J. Zentel, R. M. Arendt, and E. Weihe, Atrial natriuretic factor-like immunoreactivitiy in spinal cord and in primary sensory neurons of spinal and trigeminal ganglia: interrelation with tachykinin immunoreactivity, (submitted).

75. E. Weihe, R. M. Arendt, D. Nohr, B. C. Liebisch and A. Herz, Immunohistochemical and radioimmunological evidence for the presence of ANF-like immunoreactivity in sensory afferents of guinea- pig, *Neuroscience* 22, Suppl. S788 (1987).

76. R. Jozsa, H. W. Korf, I. Merchenthaler, Growth hormone-releasing factor (GRF)-like immunoreactivity in sensory ganglia of the rat, *Cell Tissue Res.* 247:441 (1987).

77. J. Szolcsnyi, Antidromic vasodilatation and neurogenic inflammation, *Agents and Actions* 23:4 (1988).

78. S. Saporta, Loss of spinothalamic tract neurons following neonatal treatment of rats with the neurotoxin capsaicin, *Somatosensory Res.* 4:153 (1986).

79. H. Kannan, H. Yamashita, K. Kouizumi, and C. McC. Brooks, Neuronal activity of the cat supraoptic nucleus is influenced by muscle small diameter afferent (groups III and IV) receptors, *Proc. Natl. Acad. Sci. USA* 85:5744 (1988).

80. A. Öhlén, L. Lindbom, W., Staines, T. Hökfelt, A. C. Cuello, J. A. Fischer, and P. Hedqvist, Substance P and calcitonin gene-related peptide: immunohistochemical localisation and microvascular effects in rabbit skeletal muscle, *Naun. Schmied. Arch. Pharmacol.* 336:87 (1987).

81. G. P. McGregor, S. J. Gibson, I. M. Sabate, M. A. Blank, N. D. Christofides, P. D. Wall, J. M. Polak, and S. R. Bloom, Effect of peripheral nerve section and nerve crush on spinal cord neuropeptides in the rat; increased VIP and PHI in the dorsal horn, *Neuroscience* 13: 207 (1984).

82. T. Hökfelt, Z. Wiesenfeld-Hallin, M. Villar, and T. Melander, Increase of galanin-like immunoreactivity in rat dorsal root ganglion cells after peripheral axotomy, *Neurosci. Lett.* 83:217 (1987).

83. K. Chung, W. T. Lee, and S. M. Carlton, The effects of dorsal rhizotomy and spinal cord isolation on calcitonin gene-related peptide-labeled terminals in the rat lumbar dorsal horn, *Neurosci. Lett.* 90:27 (1988).

84. T. Ogawa, I. Kanazawa, and S. Kimura, Regional distribution of substance P, neurokinin a and neurokinin β in rat spinal cord, nerve rootsand dorsal root ganglia, and the effects of dorsal root section or spinal transection, *Brain Res.* 359:152 (1985).

85. R. M. Lindsay and A. J. Harmar, Nerve growth factor regulates expression of neuropeptide genes in adult sensory neurons, *Nature* 337:362 (1989).

86. M. M. Oblinger and R. J. Lasek, Axotomy-induced alterations in the synthesis and transport of neurofilaments and microtubules in dorsal root ganglion cells, *J. Neurosci.* 8:1747 (1988).

87. D. Otto, K. Unsicker, and C. Grothe, Pharmacological effects of nerve growth factor and fibroblast growth factor applied to the transectioned sciatic nerve on neuron death in adult rat dorsal root ganglia, *Neurosci. Lett.* 83:156 (1987).

88. R. Heumann, D. Lindholm, C. Brandtlow, M. Meyer, M. J. Radeke, T. P. Misko, E. Shooter, and H. Thoenen, Differential regulation of mRNA encoding nerve growth factor and its receptor in rat sciatic nerve during development, degeneration and regeneration: role of macrophages, *Proc. Natl. Acad. Sci. USA* 84:8735 (1987).

89. D. B. MacLean, S. F. Lewis, and F. B. Wheeler, Substance P content in cultured neonatal rat vagal sensory neurons: the effect of nerve growth factor, *Brain Res.* 457:53 (1988).

90. Z. Wiesenfeld-Hallin, T. Hökfelt, J. M. Lundberg, W. G. Forssmann, M. Reinecke, F. A. Tschopp, and J. A. Fischer, Immunoreactive calcitonin gene-related peptide and substance P coexist in sensory neurons to the spinal cord and interact in spinal behavioral responses of the rat, *Neurosci. Lett.* 52:199 (1984).

91. R. A. Cridland and J. L. Henry, N- and C-terminal fragments of substance P: spinal effects in the rat tail flick test, *Brain Res. Bull.* 20:429 (1988).

92. A. Carlsson, Peptide neurotransmitters - redundant vestiges?, *Pharmacol. & Toxicol.* 62:241 (1988).
93. M. I. Sweeney and J. Sawynok, Evidence that substance P may be modulator rather than a transmitter of noxious mechanical stimulation, *Can. J. Physiol. Pharmacol.* 64:1324 (1986).
94. H. Frenk, D. Bossut, G. Urca, and D. J. Mayer, Is substance P a primary afferent neurotransmitter for nociceptive input? I. Analysis of pain-related behaviors resulting from intrathecal adminstration of substance P and 6 excitatory compounds, *Brain Res.* 455:223 (1988).
95. D. Bossut, H. Frenk, and D. J. Mayer, Is substance P a primary afferent neurotransmitter for nociceptive input? II. spinalization does not reduce and intrathecal morphine potentiates behavioral responses to substance P, *Brain Res.* 455:232 (1988).
96. H. Frenk, D. Bossut, and D. J. Mayer, Is substance P a primary afferent neurotransmitter for nociceptive input? III. Valproic acid and chlordiazepoxide decrease behaviors elicited by intrathecal injection of substance P and excitatory compounds, *Brain Res.* 455:240 (1988).
97. D. Bossut, H. Frenk, D. J . Mayer, Is substance P a primary afferent neurotransmitter for nociceptive input? IV. 2-Amino-5- phosphonovalerate (APV) and [D-Pro2, D-Trp7,9]-substance P exert different effects on behaviors induced by intrathecal substance P, strychnine and kaninic acid, *Brain Res.* 455:247 (1988).
98. G. Battaglia and A. Rustioni, Coexistence of glutamate and substance P in dorsal root ganglion neurons of the rat and monkey, *J. Comp. Neurol.* 277:302 (1988).
99. H. J. Cho and A. I. Basbaum, Increased staining of immunoreactive dynorphin cell bodies in the deafferented spinal cord of the rat, *Neurosci. Lett.* 84:125 (1988).
100. D. F. Cechetto and C. B. Saper, Neurochemical organization of the hypothalamic projection to the spinal cord in the rat, *J. Comp. Neurol.* 272:579 (1988).
101. D. L. McNeill, K. Chung, S. M. Carlton, and R. E. Coggeshall, Calcitonin gene-related peptide immunostained axons provide evidence for fine primary afferent fibers in the dorsal and dorsolateral funiculi of the rat spinal cord, *J. Comp. Neurol.* 272:303 (1988).
102 S. M. Carlton, D. L. McNeill, K. Chung, and R. E. Coggeshall, A light and electron microscopic level analysis of calcitonin gene-related peptide (CGRP) in the spinal cord of the primate: an immunohistochemical study, *Neurosci. Lett.* 82: 145 (1987).
103. O. Takahashi, R. J. Traub, and M. A. Ruda, Demonstration of calcitonin-gene-related peptide immunoreactive axons contacting dynorphin A (1- 8) immunoreactive spinal neurons in a rat model of peripheral inflammation and hyperalgesia, *Brain Res.* 475:168 (1988).
104. K. B. Thor and C. J. Helke, Serotonin- and substance P-containing projections to the nucleus tractus solitarii of the rat, *J. Comp. Neurol.* 265:275 (1987).
105. A. W. Duggan, C. R. Morton, Z. Q. Zhao, and I. A. Hendry, Noxious heating of the skin releases immunoreactive substance P in the substantia gelatinosa of the cat: a study with antibody microprobes, *Brain Res.* 403:345 (1987).
106. K. Noguchi, Y. Morita, H. Kiyama, K. Ono, and M. Tohyama, A noxious stimulus induces the preprotachykinin-A gene expression in the rat dorsal root ganglion: a quantitative study using in situ hybridization histochemistry, *Mol. Brain Res.* 4:31 (1988).
107. R. Oku, M. Sato, and H. Tagaki, Release of substance P from the spinal dorsal horn is enhanced in polyarthritic rats, *Neurosci. Lett.* 74:315 (1987).
108. C. Post, L. Alari, and T. Hökfelt, Intrathecal galanin increases the latency in the tail-flick and hot-plate tests in mouse, *Acta Physiol. Scand.* 132:583 (1988).
109. E. Weihe, D. Nohr, M. J. Millan, C. Stein, C. Gramsch, V. Höllt, A. Herz, Experimental mono- and polyarthritis differentially intensify immunostaining of multiple proenkephalin- and prodynorphin-opioid peptides in rat lumbosacral neurons. *Advances in the Biosciences,* Pergamon Press (in press) (1988).
110. E. Weihe, M. J. Millan, A. Leibold, D. Nohr, and A. Herz, Co-localization of proenkephalin- and prodynorphin-derived opioid peptides in laminae IV/V spinal neurons revealed in arthritic rats. *Neurosci. Lett.* 85:187 (1988).
111. M. J. Millan, M. H. Millan, A. Czlonkowski, V. Höllt, C. W. T. Pilcher, A. Herz, and F. C. Colpaert, A model of chronic pain in the rat: response of multiple opioid systems to adjuvant-induced arthritis, *J. Neurosci.* 6:899 (1986).
112. M. A. Ruda, M. J. Iadarola, L. V. Cohen, and W. S. Young III, In situ hybridization histochemistry and immunocytochemistry reveal an increase in spinal dynorphin biosynthesis in a rat model of peripheral inflammation and hyperalgesia, *Proc. Natl. Acad. Sci. USA* 85:622 (1988).
113. R. A. Cridland and J. L. Henry, Effects of intrathecal administration of neuropeptides on a spinal nociceptive reflex in the rat: VIP, galanin, CGRP, TRH, somatostatin and angiotensin II, *Neuropeptides* 11:23 (1988).

114. Z. Wiesenfeld-Hallin, Substance P and somatostatin modulate spinal excitability via physiologically different sensory pathways, *Brain Res.* 372:172 (1986).

115. Z. Wiesenfeld-Hallin, Intrathecal vasoactive intestinal polypeptide modulates spinal reflex excitability primarily to cutaneous thermal stimuli in rats, *Neurosci. Lett.* 80:293 (1987).

116. P. Mollenholt, C. Post, N. Rawal, J. Freedman, T. Hökfelt, and I. Paulsson, Antinociceptive and "neurotoxic" actions of somatostatin in rat spinal cord after intrathecal administration, *Pain* 32:95 (1988).

117. Y. Kuraishi, T Nanayama, H. Ohno, M. Minami, and M. Satoh, Antinociception induced in rats by intrathecal administration of antiserum against calcitonin gene-related peptide, *Neurosci. Lett.* 92:325 (1988).

118. I. L. Gibbins and J. L. Morris, Co-existence of neuropeptides in sympathetic, cranial autonomic and sensory neurons innervating the iris of the guinea-pig, *J. Autonom. Nerv. Syst.* 21:67 (1987).

119. E. Weihe, W. Hartschuh, Multiple peptides in cutaneous nerves: regulators under physiological conditions and a pathogenetic role in skin disease? *Seminars Dermatol.* 7: (in press).

120. H. C. Su, J. M. Wharton, J. M. Polak, P. K. Mulderry, M. A. Ghatei, S. J. Gibson, G. Terenghi, J. F. B. Morrison, J. Ballesta, and S. R. Bloom, Calcitonin gene-related peptide immunoreactivity in afferent neurons supplying the urinary tract: combined retrograde tracing and immunohistochemistry, *Neuroscience* 18:727 (1986).

121. D. Nohr, J. Krekel, and E. Weihe, Opioid peptides are present in nerve fibers of the respiratory tract, *Regulatory Peptides* 22:425 (1988).

122. S. Konrad, J. Zentel, and E. Weihe, Differential presence of opioid peptides in nerves supplying guinea-pig and rat urinary tract, *Regulatory Peptides* 22:414 (1988)

123. T. Fink, E. Weihe, Multiple neuropeptides in nerves supplying mammalian lymph nodes: messenger candidates for sensory and autonomic neuroimmunomodulation? *Neurosci. Lett.* 90:39 (1988).

124. E. Weihe, T. Fink, and S. Müller, Substance P in nerves suplying the immune system: A messenger role of tachykinins in sensory neuroimmunomodulation? *Regulatory Peptides* 22:186 (1988).

125. E. Weihe, S. Müller, T. Fink, H. J. Zentel, Peptides in nerves of the mammalian thymus: interactions with mast cells in autonomic and sensory neuroimmunomodulation? *Neurosci. Lett.* (in press).

126. M. R. Matthews, M. Connaughton, and A. C. Cuello, Ultrastructure and distribution of substance-P immunoreactive sensory collaterals in the guinea pig prevertebral sympathetic ganglia, *J. Comp. Neurol.* 258:28 (1987).

127. B. Lindh, T. Hökfelt, and L.-G. Elfvin, Distribution and origin of peptide-containing nerve fibers in the celiac superior mesenteric ganglion of the guinea-pig, *Neuroscience* 26:1037 (1988).

128. R. Amann, G. Skofitsch, and F. Lembeck, Species- related differences in the capsaicin-sensitive innervation of the rat and guinea-pig ureter, *Naun.-Schmied. Arch. Pharmacol.* 338:407 (1988).

129. D. Nohr and E. Weihe, Light microscopic immunohistochemistry reveals species-dependent presence of tachykinins in intrinsic neurons in the mammalian respiratory tract, *Regulatory Peptides* 22:425 (1988).

130. J. M. Lundberg, C.-R. Martling, and L. Lundblad, Cigarette smoke-induced irritation in the airways in relation to peptide-containing, capsaicin-sensitive sensory neurons, *Klin. Wochenschr.* 66 (Suppl. XI):151 (1988)

131. R. W. Fuller, J.-A. Karlsson, N. B. Choudry, and N. B. Pride, Effect of inhaled and sytemic opiates on responses to inhaled capsaicin in humans, *J. Appl. Physiol.* 65:1125 (1988).

132. A. I. Basbaum, D. Menetrey, R. Presley, and J. D. Levine, The contribution of the nervous system to experimental arthritis in the rat, in: "The arthritic rat as a model of clinical pain?," J. M. Besson and G. Guilbaud, eds., Excerpta Medica, Elsevier, Amsterdam, New York, Oxford (1988).

133. C. L. Kimberly and M. R. Byers, Inflammation of rat molar pulp and periodontium causes increased calcitonin gene-related and axonal sprouting, *Anat. Rec.* 222:289 (1988).

134. M. Büchler, E. Weihe, P. Malfertheimer, H. Friess, and H. G. Beger, Neurotransmitters in nerves in chronic pancreatitis, *Pancreas* (in press).

135. C. Stein, M. J. Millan, A, Yassouridis, and A. Herz, Antinociceptive effects of μ- and kappa-agonists in inflammation are enhanced by a peripheral opioid receptor-specific mechanism, *Eur. J. Pharm.* 155:255 (1988).

136. M. J. Millan, C. Stein, E. Weihe, D. Nohr, V. Höllt, A. Czlonkowski, and A. Herz, Dynorphin and kappa- receptors in the control of nociception: response to peripheral inflammation and the pharmacology of kappa-antinociception, in: "The arthritic rat as a model of clinical pain?", J. M. Besson and G. Guilbaud, eds., Excerpta Medica, Elsevier, Amsterdam, New York, Oxford (1988).

137. L. R. Steranka, D. C. Manning, C. J. De Haas, J. W. Ferkany, S. A. Borosky, J. R. Connor, R. J. Vavrek, J. M. Stewart, and S. H. Snyder, Bradykinin as a pain mediator: receptors are localized to sensory neurons and antagonists have analgesic actions, Proc. Natl. Acad. Sci. USA 85:3245 (1988).

138. D. M. White and M. Zimmermann, The bradykinin- induced release of subsatnce P from nerve fibre endings in the rat saphenous nerve neuroma is not related to electrophysiological excitation, Neurosci. Lett. 92:108 (1988).

139. N. E. S. Sibinga and A. Goldstein, Opioid peptides and opioid receptors in cells of the immune system, Ann. Rev. Immunol. 6:219 (1988).

140. H. M. Johnson and B. A. Torres, Immunoregulatory properties of neuroendocrine peptide hormones, in: "Progress in Allergy", P. Kallós, ed., S. Karger, Basel (1988).

141. J. M. Martin, M. B. Prystowski, and R. H. Angeletti, Preproenkephalin mRNA in T-cells, macrophages, and mast cells, J. Neurosci. Res. 18:82 (1987).

142. J. E. Taylor, J. P. Moreau, and F. V. DeFeudis, Small peptides and nerve growth: theraputic implications, Drug Development Res. 11:75 (1987)

144. J. P. Smith and T. E. Solomon, Effects of gastrin, and somatostatin on growth of human colon cancer, Gastroenterol. 95:1541 (1988).

145. T. C. Moore, C. H. Spruck and S. I. Said, In vivo depression of lymphocyte traffic in sheep by VIP and HIV (AIDS)-related peptides, Immunopharmacol. 16:181 (1988).

146. J. P. McGillis, M. L. Organist, and D. G. Payan, Substance P and immunoregulation, Fed. Proc. 46:196 (1987).

147. R. D. Helme, A. Eglezos, G. W. Dandie, P. V. Andrews, and R. L. Boyd, The effect of substance P on the regional lymph node antibody response to antigenic stimulation in capsaicin-pretreated rats, J. Immunol. 139:3470 (1987).

148. G. Nilsson and S. Ahlstedt, Altered lymphocyte proliferatiom of immunized rats after neurological manipulation with capsaicin Int. J. Immunopharmac. 10:747 (1988).

149. E. S. Kimball, F. J. Persico, and J. L. Vaught, Substance P, neurokinin A, and neurokinin B induce generation of IL-1 like activity in P388D1 cells: Possible relevance to arthritic disease, J. Immunol. 141:3564 (1988).

150. J. P. Robinson, G. B. Willars, D. R. Tomlinson, and P. Keen, Axonal transport and tissue contents of substance P in rats with long-term streptozotocin-diabetes. Effect of the aldolase reductase inhibitor 'Statil', Brain Res 426:339 (1987).

151. K. B. Walker, R. H. Serwonska, F. H. Valone, W. S. Harkonen, O.L. Frick, K. H. Scriven, W. D. Ratnoff, J. G. Browning, D. G. Payan, and E. J. Goetzl, J. Clin. Immunol. 8:108 (1988).

13

ON THE ORGANIZATION OF CUTANEOUS AND MUSCLE NERVE

PROJECTIONS IN THE RAT LUMBAR DORSAL HORN

C. Molander and G. Grant

Karolinska Institutet, Dept. of Anatomy

Stockholm, Sweden

INTRODUCTION

Central conduction of sensory information from the periphery includes two different functions: to specify the characteristics of the stimuli in terms of sensory modalities, and to inform about stimulus localization. Subpopulations of primary sensory neurons concerned with different types of stimuli and morphological prerequisites for stimulus localization created by the generally restricted peripheral innervations areas of these neurons are essential in fulfilling these tasks. Separation of different types of stimuli and stimulus localization is present also at higher levels: in the spinal cord, brainstem, thalamus and cerebral cortex. Thus, observations in physiological studies have indicated that afferent fibers for different sensory modalities are connected to separated groups of second order sensory neurons which differ with regard to their dorsoventral location within, for instance, the spinal gray matter (for reviews see Willis and Coggeshall, 1978; Brown, 1981). Furthermore, other physiological observations have suggested a somatotopic pattern in the localization of these second order neurons, where each point in the periphery is represented at a specific site along the mediolateral extent of the dorsal horn (cf., Wall, 1960; Bryan et al., 1973; Brown and Fuchs, 1975; Light and Durkovic, 1984; Wilson et al., 1986).

Until not long ago, the detailed anatomical knowledge of the central projections of primary sensory neurons was relatively limited, due to the lack of suitable neuroanatomical methods. Studies leading to such knowledge were much facilitated by the introduction of the method of transganglionic transport of horseradish peroxidase (DeGroat et al., 1978; Grant et al., 1979; Mesulam and Brushart, 1979). This method does not only allow investigation of the somatotopic organization of the projections of different peripheral sensory nerves to the dorsal horn (cf., Grant et al., 1979; Mesulam and Brushart, 1979; Koerber and Brown, 1980, 1982; Smith, 1983; Ygge and Grant, 1983; Nyberg and Blomqvist, 1985; Swett and Woolf, 1985) but also provides a method to study the laminar distribution of entire peripheral nerves (cf., Grant et al., 1979; Smith, 1983; Ygge and Grant, 1983). A related method is the technique of intracellular staining of different types of single, electrophysiologically characterized primary sensory neurons (Brown et al., 1977). This method provides information at the single unit level with regard to the detailed morphology of collateral branches and terminal arborizations in the dorsal horn (e.g., Brown, 1981; Sugiura et al., 1986; Woolf, 1987).

This chapter focuses on recent years' investigations in our laboratory of how central branches of primary sensory neurons originating in skin and muscle project to the dorsal horn at the lumbar level in the rat, using the method of transganglionic transport of horseradish peroxidase. When put into the context of the results of similar studies at other levels of the spinal cord and brainstem and in other animal species (Devor and Claman, 1980; Koerber and Brown, 1980, 1982; Arvidsson, 1982; Fitzgerald and Swett, 1983; Smith, 1983; Ygge and Grant, 1983; Nyberg and Blomqvist, 1985; Swett and Woolf, 1985; Woolf and Fitzgerald, 1986), it becomes clear that the findings (Molander and Grant, 1985, 1986, 1987) illustrate the general patterns on central projections of sensory nerves which innervate skin and muscle.

METHODS

Adult female rats were used. In those studies in which the projections from different nerves were investigated, plastic capsules filled with 50% horseradish peroxidase (HRP, Sigma type VI) were applied to the proximal end of the transected nerve. The following nerves were investigated: the tibial, medial plantar, lateral plantar, common peroneal, deep peroneal (a branch to the foot was excluded), saphenous, sural, lateral femoral cutaneous, obturator and the gastrocnemius nerves, and nerve branches to the quadriceps (from the femoral nerve) and hamstring muscles. Another set of animals was used to investigate the central distribution patterns, following application of 1% wheat germ agglutinin (WGA) -HRP (Sigma) or 50% HRP with 2% dimethylsulphoxide (DMSO), as described above, to the gastrocnemius nerves. The capsules were sealed off with an inert dental rubber impression material (President, Coltene Inc.), fixed to the surrounding tissue by Histoacryl (B. Brown, Melsungen A.G.) and left in the tissue during the postoperative survival period. In a third set of animals, various amounts of 1% WGA-HRP were injected intracutaneously with a Hamilton syringe in one or two circumscribed hindlimb foot areas. Following different survival times, ranging from 6 h to 10 days, the animals were perfused with an aldehyde mixture. Segments from the lower part of the spinal cord were removed and cut transversely in 80 μm thick sections. Every second or third section was incubated from the T12-L6 segments of animals used for somatotopic mapping and kept in serial order. Incubations were made with tetramethylbenzidine (TMB) according to Mesulam et al. (1980). The sections were examined in a light microscope. The distribution of labeling was related to the laminae, using the laminar scheme for the rat spinal cord of Molander et al. (1984). Special attention was paid to the distribution of labeling in lamina II (substantia gelatinosa), where the distance from the medial to the lateral border(s) of the labeled area(s) in each individual section was transferred to standard reconstruction schemes, so called dorsal view maps, where the curved

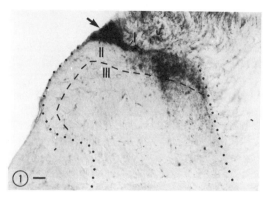

Figs. 1-4. Photomicrographs of parts of transverse sections of the lower spinal cord (1 and 2, L4; 3, T13). Labeling represents HRP reaction product. Bars: 50 μm.

Fig. 1. Labeling resulting from application of HRP to the tibial nerve. Medial right, dorsal above. Note the particularly dense labeling in Lissauers' tract (arrow), lamina II (II) and dorsal part of III (III), and the paler labeling more ventrally in the gray matter. Note also the more wide spread distribution of labeling in lamina I (I) compared to lamina II.

lamina was unrolled to be represented in one plane (see Devor and Claman, 1980; Koerber and Brown, 1980, 1982; Molander and Grant, 1986). The segmental boundaries used in these maps were normalized (see Molander and Grant, 1986).

RESULTS

A. LAMINAR DISTRIBUTION

1. Laminar distribution of labeling following application of free HRP to transected ends of mixed nerves.

The earliest labeling in the spinal cord following application of HRP to the tibial nerve was observed after 12 h, when it was found in the dorsal root entry zone, the tract of Lissauer and laminae I and IIo (outer part of lamina II). Longer survival times (24, 36, 48, and 72 h) resulted in labeling also in laminae IIi (inner part of lamina II) - VII and IX. The densest labeling was regularly found in lamina II and dorsal part of lamina III (Fig. 1). After survival times longer than 4.5 days, the density of labeling from the tibial nerve was lower.

A gradual appearance of labeling more ventrally in the dorsal horn after increasing survival times was found also by Grant et al. (1979), Ygge and Grant (1983) and Shigenaga et al. (1984), following application of HRP to peripheral sensory nerves, or injection of HRP in such nerves. Furthermore, also in accordance with previous findings (cf., Ygge and Grant, 1983; Shigenaga et al., 1984; Swett and Woolf, 1985) the results suggest that a postoperative survival time of 48-72 h should be optimal for demonstrating primary afferent projections to the dorsal horn by the method which was used.

Application of HRP to other mixed hindlimb nerves than the tibial nerve, such as the common peroneal or sural nerves, resulted in a laminar distribution of labeling which was similar to that of the tibial nerve described above.

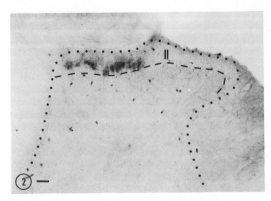

Fig. 2. Labeling primarily in lamina II (II) resulting from injection of WGA-HRP in digit I of the hindlimb. Note the restricted mediolateral distribution of the labeling.

2. Laminar distribution of labeling following application of free HRP to transected ends of cutaneous nerves or intracutaneous injections of WGA-HRP.

Application of HRP to cutaneous nerves resulted in labeling in laminae I-IV. The labeling was especially dense in lamina II and dorsal part of III, suggesting that these layers receive abundant cutaneous afferent projections. In contrast to the mixed nerves described above, labeling could not be confidently detected, however, in laminae V-VII and IX inclusive of Clarke's column, after application of HRP to these purely cutaneous nerves.

In agreement with this, Koerber and Brown (1980) and Swett and Woolf (1985) reported less intense labeling ventral to lamina III following application of HRP or WGA-HRP to hindlimb cutaneous nerves in the cat and rat, respectively. The lack of labeling in the more ventral laminae may not reflect a true lack of cutaneous afferent projections. It could be that the quantity of labeling in these laminae was below the limit of detection. Projection to lamina V of single physiologically identified coarse cutaneous afferents has been described in the cat (see Brown, 1981) and recently also in the rat (Woolf, 1987). Findings in other physiological studies in the cat (Pomeranz et al., 1968) and rat (Schouenborg, 1984) have given evidence that neurons in lamina V would receive unmyelinated and fine myelinated cutaneous afferents, as well. Another possibility would be that 2 or 3 days' survival time, which was used in our cases, was unsuitable for showing labeling ventral to lamina IV from the cutaneous nerves investigated. This is unlikely, however, since following application of HRP to the tibial nerve, which also contains cutaneous afferents, the density of labeling in these laminae is the same or weaker (see above) at survival times other than 2 or 3 days. Intracutaneous injections of 1% WGA-HRP in different part of the hindlimb foot resulted in transganglionic labeling primarily in laminae I, II and outer part of lamina III, with the densest labeling in lamina II (Fig. 2). This more restricted laminar distribution in comparison with that found after application of free HRP to transected cutaneous nerves may be due to limited uptake/transport by primary afferents which project to ventral lamina III and lamina IV (cf., Robertson and Arvidsson, 1985).

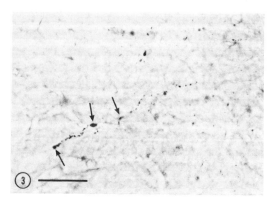

Fig. 3. Labeling in central lamina VII resulting from application of HRP to the gastrocnemius nerves. Note the axonal "swellings" (arrows).

3. Laminar distribution of labeling following application of HRP, WGA-HRP or HRP with DMSO to transected ends of muscle nerves

Application of free HRP to hindlimb muscle nerve branches resulted in transganglionic labeling primarily in laminae V-VII and IX. In some of the cases following application of free HRP to branches of the deep peroneal, femoral and hamstring muscle nerves, sparse labeling was found also in laminae I-IV. Although the nature of this labeling could not be established, its organization in rows of coarse labeling was suggestive of fibers passing from the dorsal root or dorsal column towards more ventral laminae and not of terminal ramifications. Similar findings and interpretations were made by Swett (1983) and Bakker (1984) in their studies of muscle afferent projections to the spinal cord in the cat. No labeling was observed in the dorsal laminae of the dorsal horn following application of free HRP to the gastrocnemius nerves. This was the case also after application of WGA-HRP or free HRP with DMSO to the gastrocnemius nerves, following 3 days' survival. The labeling in lamina V was dominated by fine granules, whereas that in laminae VI, VII and IX generally had a coarser appearance. The fine granular labeling, in contrast to the coarse labeling, largely disappeared after longer survival times. A similar decrease of labeling after longer survival times has been observed in lamina II following cutaneous injections of WGA-HRP and after application of HRP to nerves containing cutaneous afferents (Shigenaga et al., 1984; Swett and Woolf, 1985; Woolf and Fitzgerald, 1986). Since the primary afferent projection to lamina II seems to consist largely of fine calibered afferent fibers (see below), it is possible that the fine labeling in lamina V represents fine calibered afferents from muscle. This is supported by observations in physiological studies (Pomeranz et al., 1968), in which neurons were found in lamina V which could be excited by small caliber muscle afferents. The coarse labeling in laminae VI, VII and IX was sometimes arranged in rows with swelling suggestive of boutons "en passant" and "terminaux" (Fig. 3).

4. Methodological considerations

The validity of the description of primary afferent projections in the spinal cord in studies where transganglionic transport of HRP, or related substances has been employed, depends on the extent to what different populations of afferents are able to take up and transport the tracer. Thus, it is possible that the central projections visualized by peripheral application of free HRP, HRP with DMSO, or WGA-HRP reveals only certain populations of afferent neurons. No labeling in lamina II has been found following application of free HRP, HRP with DMSO or WGA-HRP to transected ends of muscle nerves (Ammann, et al., 1983; Jhaveri and Frank, 1983; Smith, 1983; Swett, 1983; Abrahams et al., 1984; Bakker et al., 1984; Nyberg and Blomqvist, 1984; Abrahams and Swett, 1986; Molander and Grant, 1987; see also Nishimori et al., 1986). Muscle injections, however (Brushart and Mesulam, 1980; Carson and Mesulam, 1982; Kalia, 1981), have been found to result in dense projections to lamina II. Provided that the results of these latter studies are not an effect of spread of injected tracer to the skin (see Craig and Mense, 1983), they would indicate the existence of lamina II projecting muscle afferents. In cases where WGA-HRP was used, the afferents would possibly be labeled more easily following uptake by adsorptive endocytosis (see Gonatas et al., 1979) directly from intact nerve terminals in muscle than after application of tracer to severed nerves. That there might be muscle afferents projecting to lamina II is supported by the findings in a physiological study that some slowly conducting muscle afferents project to the superficial part of the dorsal horn (McMahon and Wall, 1985). Furthermore, it is known that primary afferent neurons containing fluoride resistant acid phosphatase (FRAP) activity send their central branches primarily to lamina II (see Knyihár-Csillik and Csillik, 1981) and some FRAP containing dorsal root ganglion cells have been found to have their peripheral axons in muscle nerves (Molander et al., 1987). Failing uptake or transport of HRP, or WGA-HRP does not seem to have been reported, however, in any neuronal system so far. The only exception seems to have been a fiber population from tooth pulps in cats (see Johnson et al., 1983). In recent studies it has been shown, however, that also this fiber population is possible to label by HRP (Henry and Westrum, 1988).

The differential laminar distribution of labeling following HRP application to cutaneous versus muscle nerves is likely to reflect differences in fiber composition. The central distribution of single functionally identified myelinated afferent fibers has been studied with combined physiological and anatomical methods by Brown and coworkers (see Brown, 1981), Light and Perl (1979b), Ishizuka et al. (1979); Fyffe and Light (1984) and Woolf (1987). The central projections of unmyelinated afferent fibers are less known, although indirect evidence suggests that at least some project to lamina II (LaMotte, 1977; Light and Perl, 1977, 1979a; Réthelyi, 1977; Kumazawa and Perl, 1978; Schouenborg, 1984). Furthermore, a recent report by Sugiura et al. (1986) has demonstrated a C-fiber projection to laminae I-IV, using intracellular labeling of electrophysiologically identified skin afferents in the guinea pig.

The dorsal horn projection of a particular nerve may be found in different laminae at different rostrocaudal levels. This seems to be in principal agreement with observations made in the rat by Nagy and Hunt (1983), who studied the central projections of primary afferent neurons, following HRP injections in lumbar spinal ganglia and with those of Brown and collaborators in the cat (see Brown, 1981), who demonstrated that different functional types of afferent fibers differ both with regard to laminar distribution and rostrocaudal extent.

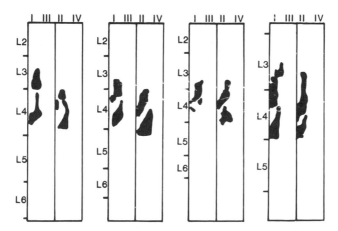

Fig. 4. Dorsal view maps of the lamina II projection resulting from cutaneous injections of WGA-HRP in digits I and III on one side and digits II and IV on the contralateral side. Each pair of columns represents the two sides from one case. Each single column is presented with medial to the left and rostral above. Note the different segmental lengths among different individual cases. The injections resulted in two rostrocaudally separated projections on each side, some of which were slightly interrupted. Note that the projections from digits II and IV are located more caudally than those of digits I and III. The more rostral of the two projections from digits II and IV had a location corresponding to the space in the contralateral dorsal horn, between the projections from digits I and III, although with some overlap. Reproduced from Molander and Grant, J. Comp. Neurol., 237:476-484, (1985).

These differential projections may reflect that second order neurons which are monosynaptically influenced by different functional types of afferents from a particular nerve are located at different rostrocaudal levels and in different laminae, and/or that they are contacted at different sites along their dendritic branches.

B.SOMATOTOPIC ORGANIZATION

Except for the gastrocnemius nerves, all nerves investigated, cutaneous as well as muscle nerves, gave rise to labeling in the medial 2/3 - 3/4 of *lamina I* in L1-6, although inconsistently for muscle nerves, i.e., the obturator and deep peroneal nerves, as well as for the nerve branches to the hamstring and quadriceps muscles. In most cases the labeling was suggestive of fibers of passage (see above). The projections in lamina I from the other nerves and from the skin areas investigated overlapped extensively. This is in accordance with the findings of Smith (1983), who investigated the spinal projections from rat thoracic spinal nerves and with those of Woolf and Fitzgerald (1986) at the lumbar level. Furthermore, the distribution of labeling in lamina I showed considerable inter-individual variations, both rostrocaudally and mediolaterally.

Discrete areas of labeling were seen in the medial 2/3 to 3/4 of *lamina II* in lumbar segments, from injected skin areas (Figs. 2, 4) as well as from the tibial, common peroneal, sural, saphenous and the lateral femoral cutaneous nerve, which all contain cutaneous afferents (see Fig. 1). The projections to this lamina from the different nerves and the skin areas investigated occupied compartments which could be compiled to general schematic somatotopic maps (Fig. 5). A comparison between the two maps in Fig. 5 shows that the lamina II projections from the lateral plantar nerve cover areas which are occupied by the plantar skin. The projection from the medial plantar nerve covers parts of the projection areas for the digital and plantar skin. The projection from the common peroneal nerve corresponds to part of the area occupied by the dorsal skin of the foot. In the case of the saphenous nerve, the projection covers the area occupied by the medial side of the foot as well as part of the plantar surface. The projection area from the sural nerve covers the area occupied by the lateral and the dorsal side of the foot. The maps do not show, however, possible overlap or inter-individual variations (see Fig. 4). In cases of intracutaneous injections of WGA-HRP, overlap and variations may in part have been due to diffusion at the injection sites, or to intraspinal transneuronal transfer of tracer. In cases where free HRP was applied to transected nerve ends, variations in uptake of tracer, in peripheral innervation patterns or presence of fibers of passage near the spinal termination areas may have been responsible. Finally, an even larger overlap than indicated (see Fig. 4) may be present, since some primary afferent fibers may have escaped labeling by the method used.

No *contralateral projections* were observed in our unilateral experiments. Such projections seem to be rare in the enlargements (Culberson et al., 1979). A small number of contralateral projecting fibers have been described in other studies, however, both in the rat and cat, at lumbar as well as at other levels (Light and Perl, 1977, 1979a, b; Réthelyi et al., 1979; Culberson et al., 1979; Smith, 1983).

In instances where labeling was found in lamina II following application of HRP to peripheral nerves, it was generally seen also in *laminae III and IV*. Its distribution, although less discernible medially in some instances, generally corresponded to the somatotopic organization in lamina II both rostrocaudally and mediolaterally. It did not always, however, extend as far laterally, but on the other hand it frequently extended more medially. This medially located labeling may in part have been due to fibers of passage, entering from the dorsal columns and taking a curving route dorsally to the area of dense labeling in lamina II. In the cat, a detailed somatotopic map of the projections to laminae III and IV from peripheral sensory nerves has been presented by Koerber and Brown (1980, 1982). It is possible that a somatotopic pattern in lamina III and IV is more readily distinguishable in the cat than in the rat. This could in part be due to differences in the cytoarchitectonic organization of the dorsal horn between the two species. Thus, whereas lamina IV in the cat has a transverse orientation, it has an oblique ventromedial orientation in its medial part in the rat. This might result in a more extensive passage of primary afferent fibers in medial lamina IV to more lateral parts of the dorsal horn in the rat than in the cat.

No clear somatotopic organization in the mediolateral direction similar to that found in laminae II-IV was found in laminae V-VII for the nerves investigated, but a somatotopic organization within these laminae can not be excluded. It may be that some of the fibers do project somatotopically to these layers but that the picture becomes distorted by fibers of passage. A tendency towards a somatotopic organization in the rostrocaudal direction seemed to be present,

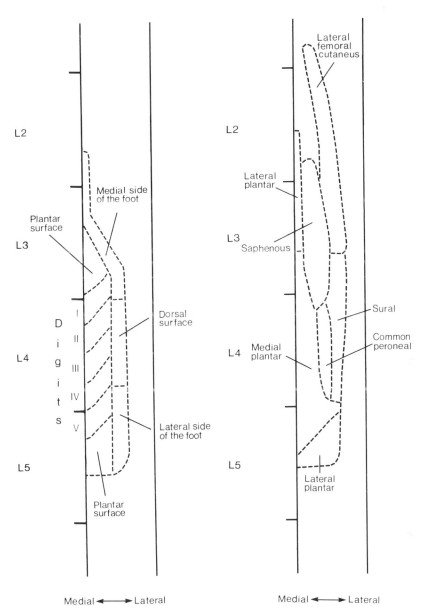

Fig. 5. Schematic dorsal view maps of lamina II depicting density centers of projections from various parts of foot skin (left column) and various hindlimb nerves (right column). The segmental lengths have been normalized to facilitate comparisons between the two columns.

although overlap was prominent. For instance, fine granular labeling in lamina V was common from the gastrocnemius nerves in L4-5. For the deep peroneal nerve it was found in L2-3, in addition.

The somatotopic organization found in the dorsal horn seems to be related to the position of the hindlimb during the embryonic period (Wall, 1960; Bryan et al., 1973; Brown and Fuchs, 1975). During this period the hindlimb is laterally rotated so that the medial side of the adult limb, including the medial side of the foot, is facing rostrally, and the lateral side caudally. The saphenous nerve, innervating the medial side of the adult leg and foot, including digit I projects more rostrally in the spinal cord than the sural nerve, which innervates the lateral side of the leg and foot. Furthermore, according to Bryan et al. (1973), the embryonically dorsal parts of the hindlimb, or extensor surface (i.e., anterior in the adult), are represented laterally and the embryonically ventral parts, or flexor surface, medially in the dorsal horn. The finding that the tibial nerve, innervating the plantar surface, projects medially in the dorsal horn, whereas the common peroneal nerve, which innervates the dorsal surface of the foot, projects laterally is in agreement with this.

GENERAL CONCLUSIONS

Sensory nerves which specifically innervate cutaneous or muscle tissue differ with regard to the laminar distribution of their central projections, as revealed by the method of transganglionic transport of horseradish peroxidase following application of tracer to transected hindlimb nerve ends. Thus, cutaneous afferents have been found to project primarily to laminae I-IV and muscle afferents primarily to laminae V-VII and IX.

The dorsal horn projections in laminae II-IV of hindlimb sensory nerves containing cutaneous afferents are organized in three dimensional compartments, which are somatotopically arranged. The somatotopic organization can be explained in the light of the embryonic development of the hindlimb. No clear somatotopic organization was found in laminae V-VII.

Acknowledgements: This study was supported by grants from the Swedish Medical Research Council, project no. 553, Karolinska Institutet and Svenska Sällskapet för Medicinsk Forskning.

REFERENCES

Abrahams, V. C., Richmond, F. J., and Keane, J., 1984, Projections from C2 and C3 nerves supplying muscles and skin of the cat neck: a study using transganglionic transport of horseradish peroxidase, *J. Comp. Neurol.*, 230:142-154

Abrahams, V. C., and Swett, J. E., 1986, The pattern of spinal and medullary projections from a cutaneous nerve and a muscle nerve of the forelimb in the cat: a study using the transganglionic transport of HRP, *J. Comp. Neurol.*, 246:70-84

Ammann, B., Gottschall, J., and Zenker, W., 1983, Afferent projections from the rat longus capitis muscle studied by transganglionic transport of HRP, *Anat. Embryol.*, 6:275-289

Arvidsson, J., 1982, Somatotopic organization of vibrissae afferents in the trigeminal sensory nuclei of the rat studied by transganglionic transport of HRP, *J. Comp. Neurol.*, 211:84-92

Bakker, D. A., Richmond, F. J. R., and Abrahams, V. C., 1984, Central projections from cat suboccipital muscles: a study using transganglionic transport of horseradish peroxidase, *J. Comp. Neurol.*, 228:409-421

Brown, A. G., 1981, "Organization in the spinal cord," Springer, Berlin

Brown, A. G., Rose, P. K., and Snow, J. P., 1977, The morphology of hair follicle afferent fibre collaterals in the spinal cord of the cat, *J. Physiol. (Lond.)*, 272:779-797

Brown, P. B., and Fuchs, J. L., 1975, Somatotopic representation of hindlimb skin in cat dorsal horn, *J. Physiol. (Lond.)*, 38:1-9

Brushart, T. M., and Mesulam, M.-M., 1980, Transganglionic demonstration of central sensory projections from skin and muscle with HRP-lectin conjugates, *Neurosci. Lett.*, 17:1-6

Bryan, R. N., Trevino, D. L., Coulter, J. D., and Willis, W. D., 1973, Location and somatotopic organization of the cells of origin of the spino-cervical tract, *Exp. Brain Res.*, 17:177-189

Carson, K. A., and Mesulam, M.-M., 1982, Ultrastructural evidence in mice that transganglionically transported horseradish peroxidase-wheat germ agglutinin conjugate reaches the intraspinal terminations of sensory neurons, *Neurosci. Lett.*, 29:201-206

Craig, A. D., and Mense, S., 1983, The distribution of afferent fibers from the gastrocnemius-soleus muscle in the dorsal horn of the cat, as revealed by the transport of horseradish peroxidase, *Neurosci. Lett.*, 41:233-238

Culberson, J. L., Haines,D. E., Kimmel, D. L., and Brown, P. B., 1979, Contralateral projection of primary afferent fibers to mammalian spinal cord, *Exp. Neurol.*, 64:83-97

DeGroat, W. C., Nadelhaft, I., Morgan, C., and Schauble, T., 1978, Horseradish peroxidase tracing of visceral efferent and primary afferent pathways in the cat's sacral spinal cord using benzidine processing, *Neurosci. Lett.*, 10:103-108

Devor, M., and Claman, D., 1980, Mapping and plasticity of acid phosphatase afferents in the rat dorsal horn, *Brain Res.*, 190:17-28

Fitzgerald, M., and Swett, J., 1983, The termination pattern of sciatic nerve afferents in the substantia gelatinosa of neonatal rats, *Neurosci. Lett.*, 43:149-154

Fyffe, R. E. W., and Light, A. R., 1984, The ultrastructure of group Ia afferent fiber synapses in the lumbosacral spinal cord of the cat, *Brain Res.*, 300:201-209

Gonatas, N. K., Harper, C., Mizutani, T.,and Gonatas, J. O., 1979, Superior sensitivity of conjugates of horseradish peroxidase with wheat germ agglutinin for studies of retrograde axonal transport, *J. Histochem. Cytochem.*, 37:728-734

Grant, G., Arvidsson, J., Robertson, B., and Ygge, J., 1979, Transganglionic transport of horseradish peroxidase in primary sensory neurons, *Neurosci. Lett.*, 12:23-28

Henry, M. A., and Westrum, L. E., 1988, Comparative central representation of maxillary and mandibular dental structures, *Soc. Neurosci. Abstr.*, 14:1164

Ishizuka, N., Mannen, H., Hongo, T., and Sasaki, S., 1979, Trajectory of group Ia afferent fibers stained with horseradish peroxidase in the lumbosacral spinal cord of the cat: three dimensional reconstructions from serial sections, *J. Comp. Neurol.*, 186:189-212

Jhaveri, S., and Frank, E., 1983, Central projections of the brachial nerve in bullfrogs: muscle and cutaneous afferents project to different regions of the spinal cord, *J. Comp. Neurol.*, 221:304-312

Johnson, L. R., Westrum, L. E., and Canfield, R. C., 1983, Ultrastructural study of transganglionic degeneration following dental lesions, *Exp. Brain Res.*, 52:226-234

Kalia, M., Mei, S. S., and Kao, F. F., 1981, Central projections from ergoreceptors (C-fibers) in muscle involved in cardiopulmonary responses to static exercise, *Circulat. Res.*, 48, Suppl. I, 48-62

Knyihár-Csillik, E., and Csillik, B., 1981, FRAP: Histochemistry of the primary nociceptive neuron, *Progress in Histochemistry and Cytochemistry*, 14,1

Koerber, R. H., and Brown, P. B., 1980, Projections of two hindlimb cutaneous nerves to cat dorsal horn, *J. Neurophysiol.*, 44:259-269

Koerber, R. H., and Brown, P. B., 1982, Somatotopic organization of hindlimb cutaneous nerve projections to cat dorsal horn, *J. Neurophysiol.*, 48:481-489

Kumazawa, T., and Perl, E. R., 1978, Excitation of marginal and substantia gelatinosa neurons in the primate spinal cord: Indications of their place in dorsal horn functional organization, *J. Comp. Neurol.*, 177:417-434

LaMotte, C., 1977, Distribution of the tract of Lissauer and the dorsal root fibers in the primate spinal cord, *J. Comp. Neurol.*, 172:529-562

Light, A. R., and Durkovic, R. G., 1984, Features of laminar and somatotopic organization of lumbar spinal cord units receiving cutaneous inputs from hindlimb receptive fields, *J. Neurophysiol.*, 52:449-458

Light, A. R., and Perl, E. R., 1977, Differential termination of large diameter and small diameter primary afferent fibers in the spinal dorsal gray matter as indicated by labeling with horseradish peroxidase, *Neurosci. Lett.*, 6:59-63

Light, A. R., and Perl, E. R., 1979a, Reexamination of the dorsal root projection to the spinal dorsal horn including observations on the differential termination of course and fine fibers, *J. Comp. Neurol.*, 186:117-132

Light, A. R., and Perl, E. R., 1979b, Spinal termination of functionally identified primary afferent neurons with slowly conducting myelinated fibers, *J. Comp. Neurol.*, 186:133-150

McMahon, S. B., and Wall, P. D., 1985, The distribution and central termination of single cutaneous and muscle unmyelinated fibers in the rat spinal cord, *Brain Res.*, 359:39-48

Mesulam, M.-M., and Brushart, T. M., 1979, Transganglionic and anterograde transport of horseradish peroxidase across dorsal root ganglia: a tetramethylbenzidine method for tracing central sensory connections of muscles and peripheral nerves, *Neurosci.*, 4:1107-1117

Mesulam, M.-M., Hegarty, E., Barbas, H., Carson, K. A., Gower, E. C., Knapp, A. G., Moss, M. B., and Mufson, E. J.,1980, Additional factors influencing sensitivity in the tetrametylbenzidine method for horseradish peroxidase neurohistochemistry, *J. Histochem. Cytochem.*, 28:1255-1259

Molander, C., Xu, Q., and Grant, G., 1984, The cytoarchitectonic organization of the spinal cord in the rat. I. The lower thoracic and lumbosacral cord, *J. Comp. Neurol.*, 230:133-141

Molander, C., and Grant, G., 1985, Cutaneous projections from the hindlimb foot to the substantia gelatinosa studied by transganglionic transport of WGA-HRP conjugate, *J. Comp. Neurol.*, 237:476-484

Molander, C., and Grant, G., 1986, Laminar distribution and somatotopic organization of primary afferent fibers from hindlimb nerves in the dorsal horn. A study by transganglionic transport of horseradish peroxidase in the rat, *Neurosci.*, 191:297-312

Molander, C., and Grant, G., 1987, Spinal cord projections from hindlimb muscle nerves in the rat studied by transganglionic transport using HRP, WGA-HRP, or HRP with DMSO, *J. Comp. Neurol.*, 260:246-255

Nagy, J. I., and Hunt, S. P., 1983, The termination of primary afferents within the rat dorsal horn: evidence for rearrangement following capsaicin treatment, *J. Comp. Neurol.*, 218:145-158

Nishimori, T., Sera, M., Suemune, S., Yoshida, A., Tsuru, K., Tsuiki, Y., Akisaka, T., Okamoto, T., Dateoka, Y., and Shigenaga, Y., 1986, The distribution of muscle primary afferents from the masseter nerve to the trigeminal sensory nuclei, *Brain Res.*, 372:375-381

Nyberg, G., and Blomqvist, A., 1984, The central projection of muscle afferent fibers to the lower medulla and upper spinal cord: An anatomical study in the cat with the transganglionic transport method, *J. Comp. Neurol.*, 230:99-109

Nyberg, G., and Blomqvist, A., 1985, The somatotopic organization of forelimb cutaneous nerves in the brachial dorsal horn: An anatomical study in the cat, *J. Comp. Neurol.*, 242:28-39

Pomeranz, B., Wall, P. D., and Weber, W. V., 1968, Cord cells responding to fine myelinated afferents from viscera, muscle and skin, *J. Physiol. (Lond.)*, 199:511-532

Réthelyi, M., 1977, Preterminal and terminal axon arborizations in the substantia gelatinosa of cat's spinal cord, *J. Comp. Neurol.*, 172:511-528

Réthelyi, M., Trevino, D. L., and Perl, E. R., 1979, Distribution of primary afferent fibers within the sacrococcygeal dorsal horn: An autoradiographic study, *J. Comp. Neurol.*, 185:603-622

Robertson, B., and Arvidsson, J., 1985, Transganglionic transport of wheat germ agglutinin-HRP and choleragenoid-HRP in rat trigeminal primary sensory neurons, *Brain Res.*, 348:44-51

Schouenborg, J., 1984, Functional and topographical properties of field potentials evoked in rat dorsal horn by cutaneous C-fiber stimulation, *J. Physiol. (Lond.)*, 356:169-192

Shigenaga, Y., Nishimori, T., Suemune, S., Chen, Y. C., Nasution, I. D., Sato, H., Okamoto, T., Sera, M., Tabuchi, K., Kagawa, K., and Hosoi, M., 1984, Laminar-related projection of primary trigeminal fibers in the caudal medulla demonstrated by transganglionic transport of horseradish peroxidase, *Brain Res.*, 309:341-345

Smith, C., 1983, The development and postnatal organization of primary afferent projections to the rat thoracic spinal cord, *J. Comp. Neurol.*, 220:29-43

Sugiura, Y., Lee, C. L., and Perl, E. R., 1986, Central projections of identified, unmyelinated (c) afferent fibers innervating mammalian skin, *Science*, 234:358-361

Swett, J. E., 1983, Few C-fibers from muscle terminate in lamina II of the spinal cord of the cat, *J. Physiol. (Lond.)*, 345:157P

Swett, J. E., and Woolf, C. J., 1985, The somatotopic organization of primary afferent terminals in the superficial dorsal horn of the rat spinal cord, *J. Comp. Neurol.*, 231:66-77

Wall, P. D., 1960, Cord cells responding to touch, damage and temperature of skin, *J. Neurophysiol.*, 23:197-210

Willis, W. D., and Coggeshall, R. E., 1978, "Sensory Mechanisms of the Spinal Cord," Wiley and Sons, Chichester

Wilson, P., Meyers, D. E. R., and Snow, P. J., 1986, The detailed somatotopic organization of the dorsal horn in the lumbosacral enlargement of the cat spinal cord, *J. Neurophysiol.*, 553:604-617

Woolf, C.-J., 1987, Central terminations of cutaneous mechanoreceptive afferents in the rat lumbar spinal cord, *J. Comp. Neurol.*, 261:105-119

Woolf, C.-J., and Fitzgerald, M., 1986, Somatotopic organization of cutaneous afferent terminals and dorsal horn neuronal receptive fields in the superficial and deep laminae of the rat lumbar spinal cord, *J. Comp. Neurol.*, 251:517-531

Ygge, J., and Grant, G., 1983, The organization of the thoracic spinal nerve projection in the rat dorsal horn demonstrated with transganglionic transport of horseradish peroxidase, *J. Comp. Neurol.*, 216:1-9

14

CENTRAL PROJECTIONS OF CERVICAL PRIMARY AFFERENTS
IN THE RAT. SOME GENERAL ANATOMICAL PRINCIPLES AND
THEIR FUNCTIONAL SIGNIFICANCE

Winfried L. Neuhuber, Wolfgang Zenker,
and Sergei Bankoul

Anatomisches Institut der Universität Zürich - Irchel
Zürich, Switzerland

INTRODUCTION

Since the classical studies of Longet (1845), Bárány (1906), de Kleijn (1921) and Magnus (1924) many scientific activities have been devoted to the exploration of the central mechanisms of body, head and eye movement coordination (for reviews see Biemond and deJong, 1969; Precht, 1974; Hülse, 1983; Mergner et al., 1983; Precht, 1983b). Proprioceptors in the skeleto-muscular system of the neck registrate position and movements of the head with respect to the trunk, and cooperate with the vestibular apparatus in posture control (Kornhuber, 1966; Guedry, 1974; Kasper and Thoden, 1981; Stampacchia et al., 1987). They have proved to be important for the tonic neck reflex apparatus (Magnus, 1924; Lindsay et al., 1976; Brink et al., 1985; for review see Wilson, 1988), for cervico-ocular reflexes (DeKleijn, 1921; Barmack et al., 1981), as well as for cervico-collic reflexes (Peterson et al., 1985; Chan et al., 1987). Furthermore, they are suspected to play a pivotal role in certain cervical syndromes (Decher, 1969; Hülse, 1983).

Information carried by both joint (McCouch et al., 1951; Hikosaka and Maeda, 1973; Wilson et al., 1976; Hirai et al., 1978; Brink et al., 1980) and muscle (Richmond and Abrahams, 1979; Brink et al., 1980; Abrahams, 1981; Mergner et al., 1982; Chan et al., 1987) afferents in rostral cervical spinal nerves is fed into a network consisting of the central cervical nucleus (CCN), spinal interneurons, the external cuneate (ECN) and cuneate (CU) nuclei, the vestibular nuclear complex including perihypoglossal nuclei (VNC), the medullary reticular formation, the superior colliculus, spinal motor and oculomotor nuclei, thalamus, cerebellum and cerebral cortex. A basis for somato-vestibulo-oculomotor cooperation may be seen in interactions of neck proprioceptive, vestibular, extraocular proprioceptive and visual impulses at various levels of this network. Specifically, these interactions have been shown to occur in the CCN (Hirai et al., 1978), on cervical interneurons (for review see Wilson, 1988), in the ECN (Jensen and Thompson, 1983), VNC (Fredrickson et al., 1965; Brink et al., 1980; Boyle and Pompeiano, 1981; Anastasopoulos and Mergner, 1982; Henn and Waespe, 1983; Ashton et al., 1988; Kasper et al., 1988a), reticular formation (Pompeiano et al., 1984), superior colliculus (Abrahams and Rose, 1975), on spinal motor neurons (Ezure et al., 1978; Peterson et al., 1985), abducens neurons (Hikosaka and Maeda, 1973; Thoden and Schmidt, 1979), in the thalamus (Deecke et al., 1977), as well as in cerebellar (Berthoz and Llinas, 1974; for review see Matsushita and Tanami, 1987) and cerebral (Becker et al., 1979; Barbas and Dubrovsky, 1981) cortices. Furthermore, tracing studies suggested convergence of neck and extraocular proprioceptive afferents in the ventrolateral CU and the pars interpolaris of the spinal trigeminal nucleus (Edney and Porter, 1986; Porter, 1986; Neuhuber and Zenker, 1989). This multisensory convergence is supposed to play a role also in recovery from labyrinthine lesions (Schaefer and Meyer, 1974; Precht, 1983a,b).

Of particular interest is a neck proprioceptive input to the VNC which has been shown electrophysiologically (Fredrickson et al., 1965; Wilson et al., 1976; Rubin et al., 1977; Boyle and Pompeiano, 1981; Anastasopoulos and Mergner, 1982; Mergner et al., 1982; Kasper et al., 1988a). Pathways for neck afferents to vestibular nuclei may be indirect via e.g. the CCN (Carleton and Carpenter, 1983; Mehler and Rubertone, 1985), or direct. Evidence for the latter possibility has been obtained by degeneration methods (Corbin and Hinsey, 1935; Escolar, 1948; Keller and Hand, 1970) and tracing techniques (Pfister and Zenker, 1984; Pfaller and Arvidsson, 1988; for review see Neuhuber and Zenker, 1989).

This chapter summarizes recent data from tracing experiments in neck primary afferents, and should provide a basis, enabling us to recognize

1. the general projection pattern of rostral cervical primary afferents as compared to afferents from other segments;

2. the contribution of neck proprioceptive afferents to the total projection from spinal ganglia C 2 - 4.

Moreover, principles governing the central course and collateralisation of *proprioceptive* neurons will be elaborated and compared with those discernible in neurons carrying *exteroceptive* informations. The question of direct proprioceptive input to various vestibular and related nuclei will receive particular consideration. Furthermore, some aspects of the connectivity of vestibular nuclei will be considered in the context of cervical primary afferent projections to the VNC.

The following descriptions are based on five types of experiments in rats. For technical details, the reader is referred to the original papers:

1. Injection of horseradish peroxidase (HRP) or wheat germ agglutinin-horseradish peroxidase conjugate (WGA-HRP) into *dorsal root ganglia C2, C3 and C4*, for comparison also into *ganglia C7, L5* (Neuhuber and Zenker, 1989) and *S1* (Neuhuber, unpublished);

2. Application of free HRP to the cut *greater occipital nerve*, as a representative for exteroceptive afferents (Scheurer et al., 1983; Neuhuber and Zenker, 1989);

3. Application of free HRP to the cut *sternomastoid nerve* (Mysicka and Zenker, 1981; Neuhuber and Zenker, 1989) and to the anastomosis between the dorsal rami C1 and C2 (*anastomosis C1/2*; Neuhuber and Zenker, 1989). These two nerves were regarded as prototypes for proprioceptive pathways. The anastomosis C1/2 in rats of the strain used in our studies was consistently present and, because of the absence of a spinal ganglion C1, had to carry proprioceptive fibers from all deep (suboccipital) neck muscles and from the cranial part of the splenius muscle. A branch from this anastomosis also to the atlanto-occipital joint could be detected by microdissection;

4. Iontophoretic WGA-HRP injections into *vestibular nuclei* to obtain supplementary evidence for a possible cervical primary afferent projection by retrograde labeling of spinal ganglion cells (Bankoul and Neuhuber, submitted);

5. Iontophoretic injections of WGA-HRP or *Phaseolus vulgaris* lectin (PHA-L, Gerfen and Sawchenko, 1984) into caudal and rostral divisions of the *medial vestibular nucleus* to study intrinsic, commissural, vestibulospinal and spinovestibular connections.

THE CENTRAL PROJECTIONS FROM THE ROSTRAL CERVICAL
SPINAL GANGLIA

After HRP or WGA-HRP-injection into spinal ganglia C2 - 4 (Neuhuber and Zenker, 1989), large amounts of reaction product were visible in the dorsal horn, in particular the substantia gelatinosa, in the segment of entry. Bundles of labeled collaterals arising from fibers in the cuneate fasciculus coursed ventromedially, bypassing the internal basilar nucleus (medial part of lamina IV; according to Torvik (1956) the caudal equivalent of the main cuneate nucleus), and arborized profusely in the central cervical nucleus (CCN; Cummings and Petras, 1977; Matsushita and Hosoya, 1979; Wiksten and Grant, 1983). From there, many axons continued to terminations in the ventral horn, in particular its medial part. Labeling in the superficial dorsal horn extended for about one segment rostrally and caudally to the segment of entry, whereas labeling of the CCN occurred along its whole rostro-caudal extent (C1 - C4). Labeled axons in dorsal columns, however, descended for 9 - 10 segments, and ascended within the cuneate fasciculus as far as up to pontine levels. Most of the ascending fibers terminated in the rostral part of the external cuneate nucleus (ECN); projections to the (main) cuneate nucleus (CU) were largely confined to its ventro-lateral portion.

From both descending and ascending axons in the dorsal funiculus, two sets of collaterals arose: medial ones and lateral ones. Both sets of collaterals distributed in a very characteristic way (Figs. 1 and 4):

Medial collaterals from *descending* fibers, like those in the segment of entry, bypassed the internal basilar nucleus and terminated lateral to the central canal in a region homologous to the CCN. Here they contacted cell groups consisting of medium sized and large neurons, which, between T1 and T5, most

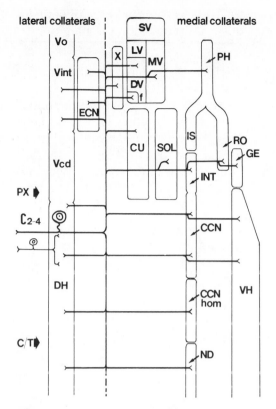

Fig. 1. Diagram summarizing the central projections of dorsal root ganglion neurons C2-4 as demonstrated by anterograde WGA-HRP tracing. Thick calibre afferents ascend to pontine, and descend to upper thoracic levels. Note the partition into a medial and lateral collateral system. Thin calibre afferents project to the superficial dorsal horn and hardly pass beyond the segment of entry and its vicinity. CCN-central cervical nucleus; CCNhom - area homologous to the CCN; C/T - border between cervical and thoracic spinal segments; CU - medial cuneate nucleus; DH - dorsal horn; DV - descending vestibular nucleus; ECN - external cuneate nucleus; f - cell group f; GE -geniohyoid nucleus; INT - intermediate nucleus; IS - intercalated nucleus of Staderini; LV - lateral vestibular nucleus; MV - medial vestibular nucleus; ND - Clarke's nucleus dorsalis; PH - nucleus prepositus hypoglossi; PX - pyramidal decussation; RO - Roller's nucleus; SOL - solitary nucleus; SV - superior vestibular nucleus; Vcd - spinal trigeminal nucleus, pars caudalis; Vint - sp. trig. nucl., pars interpolaris; Vo - sp. trig. nucl., pars oralis; VH - ventral horn; x - cell group x.

probably represented parts of Clarke's nucleus dorsalis. After C2/3 ganglion injections, the continuation of medial collaterals to the ventral horn could be followed down to segments C5/6, but almost not further caudally.

Medial collaterals from *ascending* fibers projected to rostral portions of the CCN and the ventral horn. At caudal brainstem levels, they bypassed the caudal CU to reach the intermediate nucleus which is considered the rostral continuation of the CCN (Torvik, 1956; Shriver et al., 1968; Ammann et al., 1983) and which merges rostrally with the intercalated nucleus of Staderini (Torvik, 1956). Labeled medial collaterals coursing in ventral direction beyond the intermediate nucleus, terminated in the supraspinal nucleus, in Roller's nucleus, and in the geniohyoid subnucleus (Krammer et al., 1979). Some of the medial collaterals at levels between the pyramidal decussation and the obex projected dorsome-

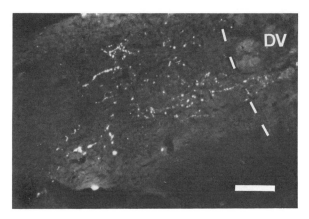

Fig. 2. Labeled fibers and terminals in the medial vestibular nucleus after WGA-HRP injection into ganglion C2. Dashed line indicates border between descending (DV) and medial vestibular nucleus. Bar: 100 µm.

dially to the solitary complex, mainly to the commissural nucleus.

The *most rostral medial collaterals* projected to vestibular and related nuclei (Fig. 1). They left the region of the rostral ECN, coursed medially and terminated in cell group x, the descending vestibular nucleus (DV), in particular cell group f (large neurons on its ventrolateral edge), the caudal half of the medial vestibular nucleus (MV, Fig. 2), and the nucleus praepositus hypoglossi. In the lateral and superior vestibular nuclei, labeled fibers were only occasionally seen. The caudal half of the MV appeared to be the *main termination* site. This was corroborated by results from experiments in which WGA-HRP was iontophoretically injected into this part of the MV (Bankoul and Neuhuber, submitted). Retrogradely labeled neurons were detected in spinal ganglia C2 and C3, but not at more caudal levels. The projection from ganglion C4, as seen with anterograde tracing, probably is not dense enough to result in retrograde labeling after a single small injection into the terminal field. The same may apply also to the projection from all three rostral cervical ganglia to the DV, since injections into that nucleus did not result in retrograde neuronal labeling in ganglia C2 - 4.

Lateral collaterals projected to lateral parts of the cervical and upper thoracic dorsal horns, mainly to laminae IV and V on one hand, and to trigeminal nuclei, particularly the pars interpolaris of the spinal trigeminal nucleus on the other hand. Some fibers were also found in the pars oralis and in the main sensory nucleus, as well as in the paratrigeminal nucleus.

In summary, dorsal column axons arising from neurons in rostral cervical spinal ganglia ascend up to pontine, and descend down to upper thoracic levels. Besides projections to the CU and ECN, collaterals are given off which display a characteristic geometry along the whole rostro-caudal extent: *Medial collaterals* pass to the CCN and its caudal and rostral homologues, as well as to vestibular and perihypoglossal nuclei. *Lateral collaterals* project to lateral dorsal horn laminae IV and V, and to the pars interpolaris of the spinal trigeminal nucleus.

PROPRIO- VERSUS EXTEROCEPTIVE PROJECTION ZONES

In the following section we shall try to analyze which of the labelings found after tracer injections into whole spinal ganglia can be related to projections of proprioceptive, and which to projections of exteroceptive primary afferents.

Exteroceptive projections

Transganglionic HRP tracing experiments on the greater occipital nerve (Scheurer et al., 1983; Neuhuber and Zenker, 1989) which was taken as a representative for exteroceptive modalities showed profuse projections of labeled fibers to the lateral third of dorsal horn laminae I-V, particularly to the substantia gelatinosa, whereas no reaction product could be found in the CCN and ventral horn. Labeled fibers descending in the dorsal columns as far as to T4, projected - via lateral collaterals only - to lateral parts of laminae IV and V. Fibers ascending in the cuneate fasciculus gave rise to terminals mainly in the ventrolateral CU. Again, only lateral collaterals projecting to the pars interpolaris, and sparsely to other trigeminal subnuclei were labeled (Fig. 3). This clearly shows that lateral collaterals represent an exteroceptive pathway. Only those of the medial collaterals were labeled which projected to the solitary nuclear complex. No exteroceptive projections were found in the ECN, and in vestibular and perihypoglossal nuclei.

Proprioceptive projections

Application of HRP to the sternomastoid nerve and to the anastomosis C1/2 resulted in labeling of proprioceptive, i.e., muscle and atlanto-occipital joint afferents (Mysicka and Zenker, 1981; Neuhuber and Zenker, 1989; for neck muscle afferents see also Ammann et al., 1983; Abrahams et al., 1984; Hirai et al., 1984; Pfister and Zenker, 1984; Bakker et al., 1985; Keirstead and Rose, 1988). In both cases transganglionic labeling occurred from pontine to upper thoracic spinal cord levels. Besides the prominent projections to the CCN and ventral horn, the ECN, and some terminals in the ventrolateral CU, terminations were found in all of those nuclei which - in experiments with tracer infiltration of whole spinal ganglia - were reached by labeled medial collaterals, except for the solitary complex. Within the brainstem, intermediate, supraspinal and Roller's nuclei, cell group x, ventrolateral parts of the DV (see also Pfister and Zenker, 1984) and the caudal part of the MV including the nucleus praepositus hypoglossi were found to contain labeled terminals (Fig. 3). More caudally, down to T4, labeled terminals in homologues of the CCN and in Clarke's nucleus dorsalis could be detected. These results show that medial collaterals represent a pathway for muscle afferents, although a contribution from joint afferents after labeling of the anastomosis C1/2 cannot be excluded. Common terminal areas for muscle and joint afferents have been recently described in the lumbar spinal cord (Craig et al., 1988).

In the dorsal horn, a projection of muscle afferents to laminae I, IV and V could be regularly observed. This labeling was limited to the segments of entry and their close vicinity and could be related to the central projections of fine muscle afferents (Mense, 1986). In this context, it is of interest to cast a glance upon the central projections of afferents from the hypoglossal nerve. After the

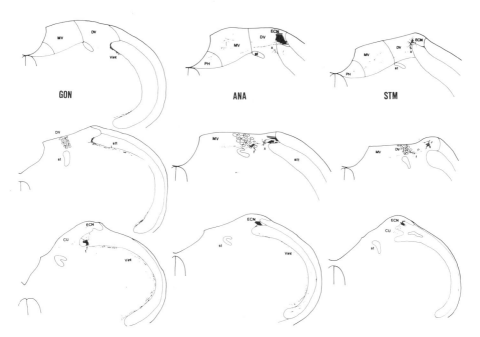

Fig. 3. Comparison of afferent labeling patterns after HRP application to the greater occipital nerve (GON), the anastomosis C1/2 (ANA) and the sternomastoid nerve (STM) on three selected levels from rostral (top) to caudal (bottom). Note medial collaterals to vestibular nuclei in ANA and STM, and their absence in GON. In contrast, lateral collaterals are labeled only in GON, and, less prominent, in ANA. st solitary tract; stt spinal trigeminal tract. Other abbreviations as in Fig.1.

branch to the geniohyoid muscle had been labeled, central projections followed a similar medial collateral pattern as was the case in sternomastoid or deep neck muscle afferents. The geniohyoid muscle contains muscle spindles (Maier, 1979) and therefore thick calibre (I and II) afferents. However, after tracer application to branches innervating the tongue musculature which does not contain muscle spindles in the rat (for references see Neuhuber and Mysicka, 1980), only fine afferents (III and IV) terminating in spinal laminae I, IV and V were labeled (Neuhuber and Fryscak-Benes, 1987).

Thus, hypoglossal afferents provide a good example for a *principle* which governs the central distribution of neck muscle afferents: fine afferents terminate in certain dorsal horn laminae, while large afferents project mainly to the CCN and the ventral horn in the spinal cord, as well as to the ECN, and to vestibular and perihypoglossal nuclei in the brainstem.

Attention should be paid to the consistent observation that several labeled lateral collaterals to the subnucleus interpolaris and to lateral spinal cord laminae IV and V were found after HRP application to the anastomosis C1/2, but not in sternomastoid nerve experiments (Fig. 3). No cutaneous branches from the anastomosis could be found by microdissection, and no labeling occured in spinal lamina II, indicating absence of skin afferents (cf. Abrahams et al., 1984). On the other hand, afferent fibers from the atlanto-occipital joint travel in the anastomosis, whereas no joint afferents are contained within the sternomastoid nerve. Thus, we presume that the labeling in these experiments was caused by proprioceptive fibers possibly from the atlanto-occipital joint.

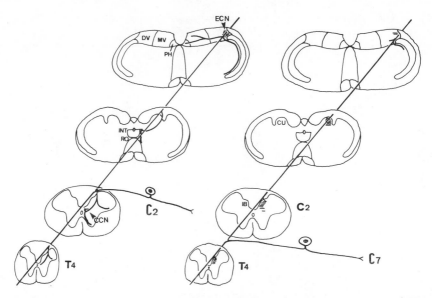

Fig. 4. Comparison of the trajectories of thick calibre afferents from spinal ganglia C2 and C7 at corresponding rostro-caudal levels. Note medial collaterals directed to the central canal and fourth ventricle in case of C2 afferents, whereas afferents from C7 are almost restricted to the internal basilar (IB) and cuneate nuclei. Note absence of a C7 projection to vestibular nuclei. Other abbreviations as in Fig. 1.

In summary, medial collaterals, except for those to the solitary complex, are proprioceptive, while lateral collaterals are exteroceptive, with a contribution from suboccipital proprioceptive fibers.

THE CENTRAL PROJECTIONS FROM GANGLIA C7, L5 AND S1

WGA-HRP injections into spinal ganglion C7 densely labeled the dorsal horn of the corresponding segment. Labeling in medial lamina IV corresponding to the internal basilar nucleus (Webster and Kemplay, 1987) extended rostrally up to C1, and caudally down to T8. From this region labeled axons coursed ventrolaterally to laminae VI, VII and to the ventral horn. Only a few fibers entered the CCN and homologous areas at more caudal levels. In the brainstem, dense labeling filled the lateral half of the CU, continuing the terminal field of the internal basilar nucleus in rostral direction. Terminals in the ECN were concentrated in more caudal parts of this nucleus than those from rostral cervical afferents. A few labeled axons reached the dorsal pars interpolaris of the spinal trigeminal nucleus. *No* projections to intermediate, vestibular and perihypoglossal nuclei could be detected, but for an occasional single fiber.

The most obvious *difference* between C7 and C2 - 4 afferent projections relates to the distribution pattern in the medial spinal grey and in homologous brainstem areas along the whole rostro-caudal extent (Fig. 4). Medial collaterals from C2 - 4 afferents bypass the internal basilar nucleus, and terminate profusely in the central cervical nucleus, whereas collaterals from C7 afferents terminate densely in the internal basilar nucleus, and project only sparsely to the central cervical nucleus. This difference in geometry is also evident in the medulla, where fibers from C2 - 4 ganglia bypass the cuneate nucleus, and project to intermediate, vestibular and perihypoglossal nuclei. Fibers from ganglion C7, on the other hand, are confined to the cuneate nucleus, and almost do not reach the more medially situated nuclei.

WGA-HRP injections into spinal ganglia L5 and S1 led to terminal labeling in dorsal and ventral horns in the segment of entry, in Clarke's nucleus dorsalis (L1 - T5) and in the gracile nucleus. No labeling was seen in CU, ECN, and in vestibular and perihypoglossal nuclei.

ANATOMICAL PECULARITIES OF NECK AFFERENTS AND THEIR FUNCTIONAL SIGNIFICANCE

A comparison of the labeling pattern after tracer injections into ganglia C2 - 4 with that of C7 and L5 afferents, and the differentiation into proprioceptive and exteroceptive neck afferents highlight the pecularities of the central distribution of rostral cervical primary afferents.

There is a profuse neck proprioceptive projection to the CCN and its caudal and rostral homologues. This *medial collateral geometry* is maintained, *mutatis mutandis,* in the brainstem, thus enabling neck proprioceptive afferents to contact vestibular and perihypoglossal nuclei. Exteroceptive fibers, as well as afferents from more caudal segments almost do not have direct access to these areas. Instead, they may reach them via secondary spinovestibular projections (Pompeiano and Brodal, 1957; Zemlan et al., 1978).

Afferent information from neck proprioceptors plays an important role in the control of posture and eye movements, as has been shown in numerous functional studies (Magnus, 1924; Lindsay et al., 1976; Thoden and Schmidt, 1979; Mergner et al., 1983; Precht, 1983b; Pettorossi and Petrosini, 1984; Dutia and Hunter, 1985; Stampacchia et al., 1987). The reason for this might be found in the central connections of these afferents, in particular to the CCN and the VNC.

The *CCN* not only relays neck afferent input to the cerebellum (Hirai et al., 1978, 1984), in particular the anterior vermis (Matsushita and Tanami, 1987; Wiksten, 1987), but also receives labyrinthine input via the vestibular nuclei (Hirai et al., 1978). This vestibular projection arises primarily from the MV (Carleton and Carpenter, 1983; Wiksten, 1987; Fig. 5a). Thus, the CCN appears to integrate neck proprioceptive and vestibular impulses and projects to the spinocerebellum, which independently receives information from the same sources via the ECN (Campbell et al., 1974; Jasmin and Courville, 1987) and the vestibular nuclei (for reviews see Gould, 1980; Matsushita and Tanami, 1987).

In tracing experiments, neck proprioceptive afferents had been shown to terminate in the caudal half of the *medial vestibular nucleus,* the *descending vestibular nucleus,* in particular its ventrolateral edge (cell group f), and in *perihypoglossal nuclei* (Neuhuber and Zenker, 1989). The concentration of terminals in the caudal MV could be confirmed by retrograde WGA-HRP tracing from this nucleus to rostral cervical spinal ganglia (Bankoul and Neuhuber, submitted). In these experiments, retrograde labeling was also detected in neurons of the contralateral CCN. The numbers of retrogradely labeled spinal

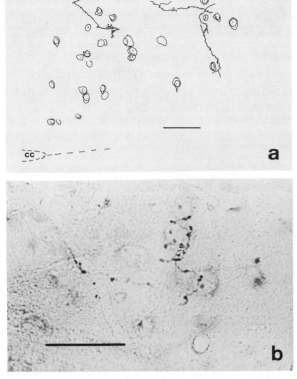

Fig. 5. Horizontal sections through segment C3 after PHA-L iontophoresis into the caudal medial vestibular nucleus. **A.** Camera lucida drawing of labeled fibers with terminal boutons in the central cervical nucleus. CC - central canal. **B.** Preterminal fiber with boutons in the ventral horn. Thionin counterstain. Bars: 100 μm.

ganglion and CCN neurons in the same segment were of the same order of magnitude, i.e., after a single microiontophoretic injection, about 30 cells were found in the spinal ganglion C2 and in the CCN at the same level, respectively. This underlines the significance of the neck primary afferent projection which is direct and ipsilateral, whereas the secondary afferent projection via the CCN is indirect and contralateral. Another difference between primary and secondary afferent projections concerns the distribution to the various parts of the VNC. Primary neck afferents are focused to the caudal MV and the ventrolateral DV, whereas secondary spinovestibular pathways from cervical and more caudal segments also project to parts of the VNC which receive only scattered or no primary afferent fibers, e.g., ventrolateral regions of the MV at levels of the facial genu, and the lateral and superior vestibular nuclei (for review see Mehler and Rubertone, 1985).

Neck primary afferent input *interacts* on vestibular neurons with afferents from other sources, in particular from the labyrinth (Fredrickson et al., 1965; Rubin et al., 1977; Anastasopoulos and Mergner, 1982; Kasper et al., 1988a). Many of those neurons have been identified as vestibulospinal (Brink et al., 1980; Boyle and Pompeiano, 1981; Kasper and Thoden, 1981). Thus, an integrated neck afferent-vestibular signal is transmitted to the spinal cord. In fact, the main termination area of neck afferents overlaps with the location of *vestibulospinal* neurons in the medial and descending vestibular nuclei from

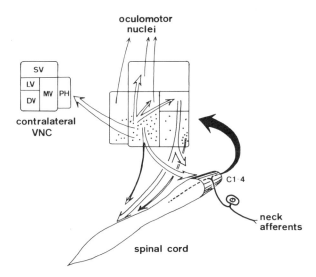

Fig. 6. Synopsis of neck proprioceptive terminal sites (dots) in the vestibular nuclear complex related to some efferent projections (open arrows) of vestibular neurons. Data for vestibulospinal neurons are taken from Brodal (1974), Peterson et al (1977), Leong et al (1984), and own results; for intrinsic and commissural neurons from Rubertone et al (1983), McCrea and Baker (1985), Epema et al (1988), and own results; for preoculomotor neurons from Brodal (1974), Lopez-Barneo et al (1981), McCrea and Baker (1985), and Mehler and Rubertone (1985). Abbreviations as in Fig.1.

where medial and caudal vestibulospinal tracts arise (Peterson et al., 1977; Leong et al., 1984; for review see Brodal, 1974; Fig. 6). Access of neck afferents to lateral vestibulospinal tract neurons (Brink et al., 1980; Boyle and Pompeiano, 1981; Kasper and Thoden, 1981; Kasper et al., 1988a) may be established via interneurons located in the caudal MV (Fig. 6), or via secondary projections from the CCN (Mehler and Rubertone, 1985). Results from anterograde tracing with WGA-HRP and PHA-L (Fig. 5b) in our laboratory indicate that many neurons in the caudal MV project primarily to rostral cervical segments, whereas neurons in the rostral MV which do not receive primary neck afferents project also farther caudally. Obviously, neck proprioceptive afferents are closely linked to vestibulospinal neurons projecting to corresponding levels of the spinal cord. These morphological findings support the concept of fast cervico-vestibulo-spinal loops controlling e.g. the fine adjustment of head position (Brink et al., 1981; Kasper and Thoden, 1981). These connections could contribute to the interaction of vestibulo-collic and cervico-collic reflexes (Dutia and Hunter, 1985; Peterson et al., 1985).

The relation of cervical primary afferents to *preoculomotor* neurons in the medial vestibular nucleus and the nucleus praepositus hypoglossi might be less direct since there is only marginal overlap between neck afferent terminal zones and the location of preoculomotor neurons (Mehler and Rubertone, 1985; Fig. 6). However, cervical afferents may contact intrinsic and commissural vestibular neurons which are abundant in the MV (Carleton and Carpenter, 1983; Rubertone et al., 1983; Epema et al., 1988). By this way, neck afferents could reach ipsi- or contralateral preoculomotor neurons in the medial or even superior vestibular nuclei, and in the nucleus praepositus hypoglossi (Fig. 6).

This pathway may supplement the secondary spinovestibular projection from the CCN (Mehler and Rubertone, 1985). Such indirect cervico-preoculomotor connections could be set into play after loss of labyrinthine function where the gain of the cervico-ocular reflex increases considerably (Dichgans et al., 1973; Bronstein and Hood, 1986). This reflex is weak or absent in normal subjects (Böhmer and Henn, 1983; Bronstein and Hood, 1986).

The extensive cranio-caudal spread of neck primary afferents must be seen in context with the close association of movements of the head, and of the shoulder and upper limb. Muscle spindle afferents from spinal ganglion C2 projecting to interneurons located in the medial ventral horn of midcervical segments play a crucial role in the tonic neck reflex on the forelimbs (Kasper et al., 1988b; Wilson, 1988). This concurs well with the anatomical demonstration of medial collaterals from C2/3 afferents to the ventral horn in segments C4 - 6. The functional significance of medial collateral projections from C2/3 ganglia to caudal homologues of the CCN and to Clarke's nucleus dorsalis is, however, unknown at present (for discussion see Pfaller and Arvidsson, 1988; Neuhuber and Zenker, 1989).

There is good reason to presume that the nervous apparatus which registrates and integrates head-on-trunk position and movements, as analyzed in anatomical and physiological animal studies, is similarly organized also in humans (Mergner et al., 1981; Hülse, 1983; Taylor and McCloskey, 1988).

In conclusion, the functional role played by neck afferents in the control of posture and eye movements emerges as a consequence from the anatomy of their central connections.

Acknowledgements: The authors are indebted to Mrs I. Drescher, R. Forster, C. Meyer and S. Richter for excellent technical assistance, and to Mrs H. Weber for expert photographical work. The studies in the author's laboratory as summarized in this article were supported by the EMDO - Stiftung, Zürich.

REFERENCES

Abrahams, V. C., 1981, Sensory and motor specialization in some muscles of the neck, *Trends Neurosci*, 4:24-27

Abrahams, V. C., Richmond, F. J., and Keane, J., 1984, Projections from C2 and C3 nerves supplying muscles and skin of the cat neck: A study using transganglionic transport of horse-radish peroxidase, *J. Comp. Neurol.*, 230:142-154

Abrahams, V. C., and Rose, P. K., 1975, Projections of extraocular, neck muscle and retinal afferents to the superior colliculus in the cat: Their connections to cells of origin of tectospinal tract, *J. Neurophysiol.*, 38:10-18

Ammann, B., Gottschall, J. and Zenker, W., 1983, Afferent projections from the rat longus capitis muscle studied by transganglionic transport of HRP, *Anat. Embryol.*, 166:275-289

Anastasopoulos, D., and Mergner, T., 1982, Canal-neck interaction in vestibular nuclear neurons of the cat, *Exp. Brain Res.*, 46:269-280

Ashton, J. A., Boddy, A., Dean, S. R., Milleret, C. and Donaldson, I. M. L., 1988, Afferent signals from cat extraocular muscles in the medial vestibular nucleus, the *nucleus praepositus hypoglossi* and adjacent brainstem structures, *Neuroscience*, 26:131-145

Bakker, D. A., Richmond, F. J. R., Abrahams, V. C., and Courville, J., 1985, Patterns of primary afferent termination in the external cuneate nucleus from cervical axial muscles in the cat, *J. Comp. Neurol.*, 241:467-479

Bankoul, S., and Neuhuber, W. L., Retrograde labeling of neurons in cervical spinal ganglia after wheat germ agglutinin-horseradish peroxidase conjugate (WGA-HRP) injection into the medial vestibular nucleus in the rat *(submitted)*

Bárány, R., 1906, Augenbewegungen durch Thoraxbewegungen ausgelöst, *Zbl. Physiol.*, 20:298-302

Barbas, H., and Dubrovsky, B., 1981, Excitatory and inhibitory interactions of extraocular and dorsal neck muscle afferents in the cat frontal cortex, *Exp. Neurology*, 74:51-66

Barmack, N. H., Nastos, M. A., and Pettorossi, V. E., 1981, The horizontal and vertical cervico-ocular reflexes of the rabbit, *Brain Res.*, 224:261-278

Becker, W., Deecke, L., and Mergner, T., 1979, Neuronal responses to natural vestibular and neck stimulation in the anterior suprasylvian gyrus of the cat, *Brain Res.*, 165:139-143

Biemond, A., and deJong, J. M. B. V., 1969, On cervical nystagmus and related disorders, *Brain*, 92:437-458

Boyle, R., and Pompeiano, O., 1981, Convergence and interaction of neck and macular vestibular inputs on vestibulospinal neurons, *J. Neurophysiol.*, 45:852-867

Böhmer, A., and Henn, V., 1983, Horizontal and vertical vestibulo-ocular and cervico-ocular reflexes in the monkey during high frequency rotation, *Brain Res.*, 277:241-248

Brink, E. E., Hirai, N., and Wilson, V. J., 1980, Influence of neck afferents on vestibulospinal neurons, *Exp. Brain Res.*, 38:285-285

Brink, E. E., Jinnai, K., Hirai, N., and Wilson, V. J., 1981, Cervical input to vestibulocollic neurons, *Brain Res.*, 217:13-21

Brink, E. E., Suzuki, I., Timerick, S. J. B., and Wilson, V. J., 1985, Tonic neck reflex of the decerebrate cat: A role for propriospinal neurons, *J. Neurophysiol.*, 54:978-987

Brodal, A., 1974, Anatomy of the vestibular nuclei and their connections, in: "Handbook of sensory physiology, Vol. VI/1, Vestibular system," H. H. Kornhuber, ed., Springer, Berlin:239-352

Bronstein, A. M., and Hood, J. D., 1986, The cervico-ocular reflex in normal subjects and patients with absent vestibular function, *Brain Res.*, 373:399-408

Campbell, S. K., Parker, T. D., and Welker, W., 1974, Somatotopic organization of the external cuneate nucleus in albino rats, *Brain Res.*, 77:1-23

Carleton, S. C., and Carpenter, M. B., 1983, Afferent and efferent connections of the medial, inferior and lateral vestibular nuclei in the cat and monkey, *Brain Res.*, 278:29-51

Chan, Y. S., Kasper, J., and Wilson, V. J., 1987, Dynamics and directional sensitivity of neck muscle spindle responses to head rotation, *J. Neurophysiol.*, 57:1716-1729

Corbin, K. B., and Hinsey, J. C., 1935, Intramedullary course of the dorsal root fibers of each of the first four cervical nerves, *J. Comp. Neurol.*, 63:119-126

Craig, A. D., Heppelmann, B., and Schaible, H.-G., 1988, The projections of the medial and posterior articular nerves of the cat's knee to the spinal cord, *J. Comp. Neurol.*, 276:279-288

Cummings, J. F., and Petras, J. M., 1977, The origin of spinocerebellar pathways. I. The nucleus cervicalis centralis of the cranial cervical spinal cord, *J. Comp. Neurol.*, 173:655-692

Decher, H., 1969, "Die zervikalen Syndrome in der Hals-Nasen-Ohren-Heilkunde," Thieme, Stuttgart.

Deecke, L., Schwarz, D. W. F., and Fredrickson, J. M., 1977, Vestibular responses in the rhesus monkey ventroposterior thalamus. II. Vestibulo-proprioceptive convergence at thalamic neurons, *Exp. Brain Res.*, 30:219-232

Dichgans, J., Bizzi, E., Morasso, P., and Tagliasco, V., 1973, Mechanisms underlying recovery of eye-head coordination following bilateral labyrinthectomy in monkeys, *Exp. Brain Res.*, 18:548-562

Dutia, M. B., and Hunter, M. J., 1985, The sagittal vestibulocollic reflex and its interaction with neck proprioceptive afferents in the decerebrate cat, *J. Physiol. (Lond.)*, 359:17-29

Edney, D. P., and Porter, J. D., 1986, Neck muscle afferent projections to the brainstem of the monkey: Implications for the neural control of gaze, *J. Comp. Neurol.*, 250:389-398

Epema, A. H., Gerrits, N. M., and Voogd, J., 1988, Commissural and intrinsic connections of the vestibular nuclei in the rabbit: a retrograde labeling study, *Exp. Brain Res.*, 71:129-146

Escolar, J., 1948, The afferent connections of the 1st, 2nd, and 3rd cervical nerves in the cat. An analysis by Marchi and Rasdolsky methods, *J. Comp. Neurol.*, 89:79-91

Ezure, K., Sasaki, S., Uchino, Y., and Wilson, V. J., 1978, Frequency-response analysis of vestibular-induced neck reflex in cat. II. Functional significance of cervical afferents and polysynaptic descending pathways, *J. Neurophysiol.*, 41:459-471

Fredrickson, J. M, Schwarz, D., and Kornhuber, H. H., 1965, Convergence and interaction of vestibular and deep somatic afferents upon neurons in the vestibular nuclei of the cat, *Acta Oto-Laryng. (Stockh.)*, 61:168-188

Gerfen, C. R., and Sawchenko, P. E., 1984, An anterograde neuroanatomical tracing method that shows the detailed morphology of neurons, their axons and terminals: Immunohistochemical localization of an axonally transported plant lectin, *Phaseolus vulgaris* leucoagglutinin (PHA-L), *Brain Res.*, 290:219-238

Gould, B. B., 1980, Organization of afferents from the brain stem nuclei to the cerebellar cortex in the cat, *Adv. Anat. Embryol. Cell Biol.*, 62:1-90

Guedry, F. E., Jr., 1974, Psychophysics of vestibular sensation, *in:* "Handbook of sensory physiology, Vol. VI/2, Vestibular system," H. H. Kornhuber, ed., Springer, Berlin:3-154

Henn, V., and Waespe, W., 1983, Visual-vestibular interactions: neurophysiological investigations in the monkey, *in:* "Multimodal convergences in sensory systems. Fortschritte der Zoologie, Bd. 28," E. Horn, ed., Fischer, Stuttgart: 261-273

Hikosaka, O., and Maeda, M., 1973, Cervical effects on abducens motoneurons and their interaction with vestibular-ocular reflex, *Exp. Brain Res.*, 18:512-512

Hirai, N., Hongo, T., and Sasaki, S., 1978, Cerebellar projection and input organizations of the spinocerebellar tract arising from the central cervical nucleus in the cat, *Brain Res.*, 341-345

Hirai, N., Hongo, T., Sasaki, S., Yamashita, M., and Yoshida, K., 1984, Neck muscle afferent input to spinocerebellar tract cells of the central cervical nucleus in the cat, *Exp. Brain Res.*, 55:286-300

Hülse, M., 1983, "Die zervikalen Gleichgewichtsstörungen," Springer, Berlin.

Jasmin, L., and Courville, J., 1987, Distribution of external cuneate nucleus afferents to the cerebellum: II. Topographical distribution and zonal pattern - an experimental study with radioactive tracers in the cat, *J. Comp. Neurol.*, 261:497-514

Jensen, D. W., and Thompson, G. C., 1983, Vestibular nerve input to neck and shoulder regions of lateral cuneate nucleus, *Brain Res.*, 280:335-338

Kasper, J., Schor, R. H., and Wilson, V. J., 1988a, Response of vestibular neurons to head rotations in vertical planes. II. Response to neck stimulation and vestibular-neck interaction, *J. Neurophysiol.*, 60:1765-1777

Kasper, J., Schor, R. H., Yates, B. J., and Wilson, V. J., 1988b, Three-dimensional sensitivity and caudal projection of neck spindle afferents, *J. Neurophysiol.*, 59:1497-1509

Kasper, J., and Thoden, U., 1981, Effects of natural neck afferent stimulation on vestibulospinal neurons in the decerebrate cat, *Exp. Brain Res.*, 44:401

Keirstead, S. A., and Rose, P. K., 1988, Structure of the intraspinal projections of single, identified muscle spindle afferents from neck muscles of the cat, *J. Neurosci.*, 8:3413-3426

Keller, J. H., and Hand, P. J., 1970, Dorsal root projections to nucleus cuneatus of the cat, *Brain Res.*, 20:1-17

Kleijn, A. de, 1921, Tonische Labyrinth- und Halsreflexe auf die Augen, *Pflügers Arch.*, 186:82-82

Kornhuber, H. H., 1966, Physiologie und Klinik des zentralvestibulären Systems, *in:* "Hals-Nasen-Ohrenheilkunde, III/3," J. Berendes, R. Link, and F. Zöllner, eds., Thieme, Stuttgart

Krammer, E. B., Rath, T., and Lischka, M., 1979, Somatotopic organization of the hypoglossal nucleus: a HRP study in the rat, *Brain Res.*, 170:533-537

Leong, S. K., Shieh, J. Y., and Wong, W. C., 1984, Localizing spinal-cord-projecting neurons in adult albino rats, *J.Comp. Neurol.*, 228:1-17

Lindsay, K. W., Roberts, T. D. M., and Rosenberg, J. R., 1976, Asymmetric tonic labyrinth reflexes and their interaction with neck reflexes in the decerebrate cat, *J. Physiol. (Lond.)*, 261:583-601

Longet, F. A., 1845, Sur les troubles qui surviennent dans l'équilibration, la station et la locomotion des animaux, apres la section des parties molles de la nuque, *Gaz. Med. Paris*, 13:565-567

Magnus, R., 1924, "Körperstellung," Springer, Berlin.

Maier, A., 1979, Occurrence and distribution of muscle spindles in masticatory and suprahyoid muscles of the rat, *Am. J. Anat.*, 155:483-506

Matsushita, M., and Hosoya, Y., 1979, Cells of origin of the spinocerebellar tract in the rat, studied with the method of retrograde transport of horseradish peroxidase, *Brain Res.*, 173:185-200

Matsushita, M., and Tanami, T., 1987, Spinocerebellar projections from the central cervical nucleus in the cat, as studied by anterograde transport of wheat germ agglutinin-horseradish peroxidase, *J. Comp. Neurol.*, 266:376-397

McCouch, G. P., Deering, I. D., and Ling, T. H., 1951, Location of receptors for tonic neck reflexes, *J. Neurophysiol.*, 14:191-195

Mehler, W. R., and Rubertone, J. A., 1985, Anatomy of the vestibular nucleus complex, *in:* "The rat nervous system. Vol. 2," G. Paxinos, ed., Academic Press, Sidney: 185-219

Mense, S., 1986, Slowly conducting afferent fibers from deep tissues: neurobiological properties and central nervous actions, *Progress in Sensory Physiology*, 6:139-219

Mergner, T., Deecke, L., and Becker, W., 1981, Pattern of vestibular and neck responses and their interaction: a comparison between cat cortical neurons and human psychophysics, *Ann. New York Acad. Sci.*, 374:361-372

Mergner, T., Anastasopoulos, D., and Becker, W., 1982, Neuronal responses to horizontal neck deflection in the group x region of the cat's medullary brainstem, *Exp. Brain Res.*, 45:196-206

Mergner, T., Deecke, L., Becker, W., and Kornhuber, H. H., 1983, Vestibular-proprioceptive interactions: neurophysiology and psychophysics, *in:* "Multimodal convergences in sensory systems. Fortschritte der Zoologie Bd. 28," E. Horn, ed., Fischer, Stuttgart: 241-252

Mysicka, A., and Zenker, W., 1981, Central projections of muscle afferents from the sternomastoid nerve in the rat, *Brain Res.*, 211:257-265

Neuhuber, W. L., and Fryscak-Benes, A., 1987, Die zentralen Projektionen afferenter Neurone des Nervus hypoglossus bei der Albinoratte, *Verh. Anat. Ges.*, 81:981-983

Neuhuber, W., and Mysicka, A., 1980, Afferent neurons of the hypoglossal nerve of the rat as demonstrated by horseradish peroxidase tracing, *Anat. Embryol.*, 158:349-360

Neuhuber, W. L., and Zenker, W., 1988, The central distribution of cervical primary afferents in the rat, with emphasis on proprioceptive projections to vestibular, perihypoglossal and upper thoracic spinal nuclei, *J. Comp. Neurol.*, 280:231-253

Peterson, B. W., and Coulter, J. D., 1977, A new long spinal projection from the vestibular nuclei in the cat, *Brain Res.*, 122:351-356

Peterson, B. W., Goldberg, J., Bilotto, G., and Fuller, J. H., 1985, Cervicocollic reflex: Its dynamic properties and interaction with vestibular reflexes, *J. Neurophysiol.*, 54:90-109

Pettorossi, V. E., and Petrosini, L., 1984, Tonic cervical influences on eye nystagmus following hemilabyrinthectomy: Immediate and plastic effects, *Brain Res.*, 324:11-19

Pfaller, K., and Arvidsson, J., 1988, Central distribution of trigeminal and upper cervical primary afferents in the rat studied by anterograde transport of horseradish peroxidase conjugated to wheat germ agglutinin, *J. Comp. Neurol.*, 268:91-108

Pfister, J., and Zenker, W., 1984, The splenius capitis muscle of the rat, architecture and histochemistry, afferent and efferent innervation as compared with that of the quadriceps muscle, *Anat. Embryol.*, 169:79-89

Pompeiano, O., and Brodal, A., 1957, Spino-vestibular fibers in the cat. An experimental study, *J. Comp. Neurol.*, 108:353-381

Pompeiano, O., Manzoni, D., Srivastava, U. C., and Stampacchia, G., 1984, Convergence and interaction of neck and macular vestibular inputs on reticulospinal neurons, *Neuroscience*, 12:111-128

Porter, J.D., 1986, Brainstem terminations of extraocular muscle primary afferent neurons in the monkey, *J. Comp. Neurol.*, 247:133-143

Precht, W., 1974, The physiology of the vestibular nuclei, *in:* "Handbook of sensory physiology, Vol. VI/1, Vestibular system," H. H. Kornhuber, ed., Springer, Berlin:353-416

Precht, W., 1983a, Panel discussion synthesis: Neurophysiological and diagnostic aspects of vestibular compensation, *Adv. Oto-Rhino-Laryng.*, 30:319-329

Precht, W., 1983b, The role of multisensory convergence in functional recovery after neural lesions, *in:* "Multimodal convergences in sensory systems. Fortschritte der Zoologie, Bd. 28," E. Horn, ed., Fischer, Stuttgart: 275-290

Richmond, F. J. R., and Abrahams, V. C., 1979, Physiological properties of muscle spindles in dorsal neck muscles of the cat, *J. Neurophysiol.*, 42:604-617

Rubertone, J. A., Mehler, W. R., and Cox, G. E., 1983, The intrinsic organization of the vestibular complex: evidence for internuclear connectivity, *Brain Res.*, 263:137-141

Rubin, A. M., Liedgren, S. R. C., Milne, A. C., Young, J. A., and Fredrickson, J. M., 1977, Vestibular and somatosensory interaction in the cat vestibular nuclei, *Pflügers Arch.*, 371:155-160

Schaefer, K. P., and Meyer, D. L., 1974, Compensation of vestibular lesions, *in:* "Handbook of sensory physiology, Vol. VI/2, Vestibular system," H. H. Kornhuber, ed., Springer, Berlin:463-490

Scheurer, S., Gottschall, J., and Groh, V., 1983, Afferent projections of the rat major occipital nerve studied by transganglionic transport of HRP, *Anat. Embryol.*, 167:425-438

Shriver, J. E., Stein, B. M., and Carpenter, M. B., 1968, Central projections of spinal dorsal roots in the monkey, *Am. J. Anat.*, 123:27-74

Stampacchia, G., Manzoni, D., Marchand, A. R., and Pompeiano, O., 1987, Convergence of neck and macular vestibular inputs on vestibulo-spinal neurons projceting to the lumbosacral segments of the spinal cord, *Arch. Ital. Biol.*, 125:201-224

Taylor, J. L., and McCloskey, D. I., 1988, Proprioception in the neck, *Exp. Brain Res.*, 70:351-360

Thoden, U., and Schmidt, P., 1979, Vestibular-neck interaction in abducens neurons, *in:* "Reflex control of posture and movement," R. Granit and O. Pompeiano, eds., Elsevier, Amsterdam: 561-566

Torvik, A., 1956, Afferent connections to the sensory trigeminal nuclei, the nucleus of the solitary tract and adjacent structures, *J. Comp. Neurol.*, 106:51-141

Webster, K. E., and Kemplay, S. K., 1987, Distribution of primary afferent fibres from the forelimb of the rat to the upper cervical spinal cord in relation to the location of spinothalamic neuron populations, *Neurosci. Lett.*, 76:18-24

Wiksten, B., 1987, Further studies on the fiber connections of the central cervical nucleus in the cat, *Exp. Brain Res.*, 67:284-290

Wiksten, B., and Grant, G., 1983, The central cervical nucleus in the cat. IV. Afferent fiber connections. An experimental anatomical study, *Exp. Brain Res.*, 51:405-412

Wilson, V. J., 1988, Convergence of neck and vestibular signals on spinal interneurons, *in:* "Progr. Brain Res., Vol. 76," O. Pompeiano and J. H. J. Allum, eds., Elsevier, Amsterdam:137-143

Wilson, V. J., Maeda, M., Franck, J. I., and Shimazu, H., 1976, Mossy fiber neck and second-order labyrinthine projections to cat flocculus, *J. Neurophysiol.*, 39:301-309

Zemlan, F. P., Leonard, C. M., Kow, L. M., and Pfaff, D. W., 1978, Ascending tracts of the lateral columns of the rat spinal cord: A study using the silver impregnation and horseradish peroxidase techniques, *Exp. Neurol.*, 62:298-334

15

VISCERAL AFFERENT PROJECTIONS TO THE THORACOLUMBAR SPINAL
CORD IN NORMAL AND CAPSAICIN-TREATED RATS

Winfried L. Neuhuber and Dominique Irminger

Anatomisches Institut der Universität Zürich-Irchel
Zürich, Switzerland

INTRODUCTION

Among the primary afferents projecting to the spinal cord, visceroafferent neurons represent a small but substantial population (Jänig and Morrison, 1986). Although their percentage, i.e., around 10% or even less of all afferent neurons in the main segments of entry, seems to be low as compared to somatic afferents (Cervero et al., 1984; Neuhuber et al., 1986), the effects elicited by visceroafferent stimulation are widespread and of well-known clinical significance (Jänig and Morrison, 1986; Procacci et al., 1986; Cervero, 1988; Ness and Gebhart, 1988).

Knowledge of visceroafferent termination sites in the spinal cord was limited until the advent of transganglionic tracing techniques (deGroat et al., 1978; Mesulam and Brushart, 1979), and was derived largely from electrophysiological data (Pomeranz et al., 1968; Gokin et al., 1977; Crousillat et al., 1979; Cervero, 1983). Transganglionic tracing studies with horseradish peroxidase (HRP) during the past few years have elucidated a typical visceroafferent projection pattern in the spinal cord: Afferents from thoracic, abdominal and pelvic organs distribute to laminae I, V, VII and X, sparing the remainder of the dorsal horn (for review see deGroat, 1986).

This chapter summarizes briefly the central distribution of visceral afferents in the thoracolumbar spinal cord of the rat. This "normal" distribution will be compared to that seen in rats treated neonatally with capsaicin. Since visceral afferents, besides small and a few large myelinated fibers, are composed largely of unmyelinated fibers (Mei, 1983) most of which are destroyed by capsaicin (for review see Jancsó et al., 1987), a significant reduction of visceroafferent terminals in the spinal cord was expected. However, it is unknown whether capsaicin-resistant and capsaicin-sensitive fibers of the visceroafferent pool terminate in the same or in different areas of the spinal cord. Since some visceral reflexes are abolished in capsaicin-treated rats (Cervero and McRitchie, 1982; Lembeck and Skofitsch, 1982), a distinction between termination sites of capsaicin-resistant and capsaicin-sensitive afferents might provide clues as to the location of second order neurons mediating these reflexes.

SPLANCHNIC VERSUS INTERCOSTAL AFFERENT PROJECTIONS IN NORMAL RATS

Afferent fibers travelling in the *greater splanchnic* and *intermesenteric nerves* enter the spinal cord at levels from T3 to L2(3) (Neuhuber, 1982; Neuhuber et al., 1986; Sepulveda and Nadelhaft, 1986; Nadelhaft and McKenna, 1987). The latter nerve represents the rat's equivalent to the abdominal aortic plexus and contains afferents from the hypogastric and lumbar colonic (inferior mesenteric plexus) nerves (for clarification of the anatomical situation and nomenclature see Baron et al., 1988). These afferents are conveyed further by the lumbar splanchnic nerves. Most of the fibers belonging to the A-delta and C classes are distributed by Lissauer's tract over 2-3 segments rostrocaudally. A few, presumably A-beta fibers enter the dorsal funiculus and reach the dorsal column nuclei. Collaterals of these fibers seem to enter the medial dorsal horn (Neuhuber et al., 1986).

Afferent fibers coursing in Lissauer's tract separate into two groups, a medial and a lateral collateral pathway (MCP and LCP) which are bundled at periodic rostro-caudal intervals (cf., Morgan et al., 1981; Neuhuber et al., 1986). Both groups of fibers embrace the dorsal horn in a forceps-like manner (Figs. 1b,3). Clusters of terminals are found in the most superficial layers of the dorsal horn equivalent to lamina I and possibly the outermost zone of lamina IIa (Neuhuber, 1982; Neuhuber et al., 1986), whereas the main part of lamina II, as well as laminae III and IV, are spared (Fig. 1b). Many fiber bundles of the LCP coursing through the dorsolateral funiculus give rise to terminals in the region of the nucleus of the dorsolateral funiculus (or lateral spinal nucleus, Fig. 3a).

The trajectories and terminal zones of the MCP and LCP correspond to the superficial band of fluoride-resistant acid phosphatase (FRAP) activity described by Csillik and Knyihar-Csillik (1986) as extensions of Lissauer's tract and "faisceau de la corne posterieure" of Cajal, respectively (see their figure 1.18b). In line with this, FRAP-positive spinal ganglion cells have been shown to constitute up to 50% of splanchnic afferents (Dalsgaard et al., 1984). In addition, thoracolumbar visceral afferents contain, as do afferents from other sources, substance P (SP) and calcitonin gene-related peptide (CGRP; Dalsgaard et al., 1982; Lindh et al., 1983; Green and Dockray, 1987; Molander et al., 1987; Su et al., 1987). These peptides are also abundant in visceroafferent terminal sites (Hökfelt et al., 1977; Rosenfeld et al., 1983). The main actions of these peptides might take place, however, in the periphery, since far greater amounts of them are transported from cell bodies to the peripheral branch of sensory neurons (for review see Dockray and Sharkey, 1986), and an influence of, e.g., SP on spinal second order neurons driven by splanchnic afferent stimulation has not been detected (Felpel and Huffman, 1986).

Both, the MCP and LCP, continue their pathways to the neck of the dorsal horn. Terminals in dorsal lamina V are primarily derived from fibers of the LCP,

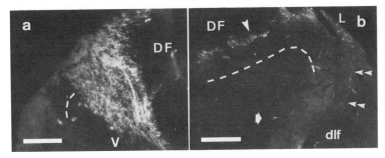

Fig. 1. Coronal sections through segment T10 after HRP application to the intercostal nerve (**A**), and the greater splanchnic nerve (**B**), showing the striking difference between somatic and visceral afferent projections to the dorsal horn. Intercostal afferents terminate densely in laminae I - dorsal V, whereas splanchnic afferent labeling is restricted to Lissauer's tract (L), lamina I (MCP, arrowhead), and some fibers (LCP, double arrowheads) in the dorsolateral funiculus (dlf), thus sparing laminae II-IV. Visceroafferent labeling in lamina V is almost absent in this section (arrow). Dashed lines indicate border between laminae II and III. DF - dorsal funiculus. Bars: 100 µm.

whereas the contribution of the MCP seems to be less pronounced (Fig. 3a). Again, many of these terminals contain SP (Sharkey et al., 1987) and presumably also CGRP (Rosenfeld et al., 1983; Carlton et al., 1988).

Some fibers of the MCP ultimately reach the ventral aspect of the central canal where they form a narrow rostro-caudally running bundle (Fig. 3a) which is part of the peptide-containing (Hökfelt et al., 1977; Jancsó et al., 1981) "fasciculus longitudinalis centroventralis" (Changgeng et al., 1983).

LCP fibers coursing through the dorsolateral funiculus may come in close contact to dendrites of sympathetic preganglionic neurons of the intermediolateral nucleus, thus suggesting monosynaptic connections of visceral primary afferent and visceral efferent neurons. However, the obvious association of the LCP with preganglionic neurons as seen in the lumbosacral cord of rats (Nadelhaft and Booth, 1984), and at all spinal levels in cats (for review see deGroat, 1986) cannot be demonstrated in the rat thoracolumbar spinal cord, possibly because of the scarcity of this projection, or because of insufficient HRP transport in very fine afferent fibers (for discussion see Neuhuber et al., 1986; Nadelhaft and McKenna, 1987).

Visceral afferent projections to the thoracolumbar spinal cord in rats as described above closely resemble those to the rat and monkey lumbosacral cord (Nadelhaft et al., 1983; Nadelhaft and Booth, 1984), and visceroafferent projections at all spinal levels in cats (Morgan et al., 1981; Kuo et al., 1983; Cervero and Connell, 1984a; Kuo et al., 1984; Kuo and deGroat, 1985; Morgan et al., 1986a,b). Thus, a typical visceroafferent projection pattern exists in mammals with a partition of the fibers into MCP and LCP, and terminal sites in laminae I, V, VII and X. Renal afferents in rats (in contrast to those in cats) seem to represent an exception since they follow the MCP only (Ciriello and Calaresu, 1983). In birds *(Gallus domesticus)*, pelvic visceral afferents form a LCP reaching the dorsal gray commissure (Ohmori et al., 1987). No MCP has been found in this species.

There is one obvious consequence of this distribution pattern: Visceral afferents do not project significantly to lamina II, the substantia gelatinosa proper (Cervero and Connell, 1984b), and to laminae III and IV. This sparing of lamina II is also evident in muscle afferents (Craig and Mense, 1983; Abrahams et al., 1984). Thus, lamina II appears to be concerned solely with processing of information from thin-calibre skin afferents.

The central distribution of *intercostal nerve* afferents and of afferents from *dorsal rami* which include a large number of fibers from the skin contrasts sharply with that of splanchnic afferents (Ygge and Grant, 1983; Neuhuber et al., 1986). They project densely to spinal laminae I-V, particularly to laminae II-IV (Fig. 1a). In addition, they show a clear somatotopic arrangement which is most obvious in lamina II, with ventral skin areas represented medially, and dorsal skin areas laterally in the dorsal horn (Ygge and Grant, 1983; Neuhuber et al., 1986). This pattern is also evident in the cat (Cervero and Connell, 1984a).

Overlap of visceral and skin afferents occurs in laminae I, V, and possibly VII and X. This concurs well with the reported location of convergent viscerosomatic spinal neurons in laminae I, V and VII, but almost any in II, III and IV (Pomeranz et al., 1968; Gokin et al., 1977; Cervero, 1983). Many of these neurons have ascending projections to supraspinal structures (Cervero, 1983). These anatomical and physiological findings support the convergence-projection theory for referred visceral pain (Ruch, 1947; Cervero, 1983).

The MCP/LCP duality of visceral afferents could have functional significance because MCP versus LCP fibers may differently contact different types of second order neurons in medial and lateral lamina I, in the lateral spinal nucleus, and in laminae V and X. Four classes of marginal neurons have been described which are distinguished not only on the basis of their perikaryal shape, dendritic tree and distribution along the mediolateral extent of lamina I, but also by their projections, peptide content and, presumably, also their functions (Lima and Coimbra 1983, 1985, 1986, 1988, 1989; see also Swett et al., 1985; Standaert et al., 1986; Leah et al., 1988; Miller and Seybold, 1989). Thus, spinobulboreticular cells are found in medial lamina I (Lima and Coimbra, 1985) and are likely to be contacted mainly by MCP fibers. Marginal pyramidal cells projecting to the periaquaeductal grey are distributed over the entire lamina I (Lima and Coimbra, 1989) and may relay information from both the MCP and LCP. On the other hand, fusiform cells in lateral lamina I projecting to the parabrachial area (Lima and Coimbra, 1989) may receive input from the LCP only. The parabrachial area is part of a network involved in the control of various autonomic, e.g., cardiovascular and respiratory functions (for reviews see Fulwiler and Saper, 1984; Cechetto et al., 1985) and may mediate systemic cardiovascular reflexes upon non-cardial visceroafferent stimulation (Lembeck and Skofitsch, 1982; Ness and Gebhart, 1988). Other brainstem areas participating in cardiovascular regulation are the ventrolateral medulla around the lateral reticular nucleus (Caverson et al., 1983), and the nucleus of the solitary tract (for review see Jordan and Spyer, 1986). Spinal neurons projecting to these nuclei are also found in positions where they can be contacted primarily by visceral afferents of the LCP (Menétrey et al., 1983; Menétrey and Basbaum, 1987). However, fibers of the MCP extending to lamina X may also have direct access to these

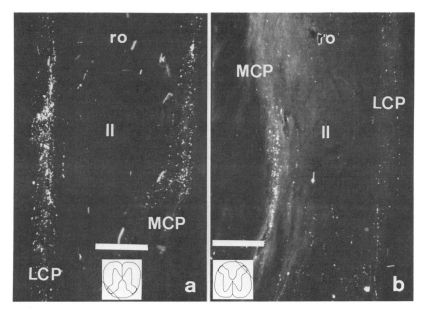

Fig. 2. Tangential sections through the superficial dorsal horn at segment T10 after HRP application to the greater splanchnic nerve. **A.** Note dense labeling of the LCP and MCP in the untreated animal. After neonatal capsaicin treatment (**B**), labeling in the LCP is reduced to an occasional axon, while there are relatively more HRP-positive structures in the MCP. II - lamina II. ro - rostral. Bars: 100 μm.

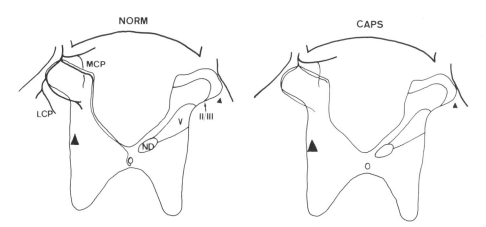

Fig. 3. Diagrams summarizing splanchnic afferent distribution in the thoracic spinal cord at levels T9-11 in normal (NORM) and neonatally capsaicin-treated (CAPS) rats. There is an overall reduction of fibers and terminals (as symbolized by thinner black lines) in neonatally capsaicin-treated rats which appears to be relatively more pronounced in the LCP. Note residual projection to lamina V, and the absence of an MCP projection to the ventral aspect of the central canal. Splanchnic afferent dorsal column fibers and their collaterals, as well as preganglionic sympathetic neurons (large triangles) are unaffected by capsaicin treatment. Small triangles represent lateral spinal nucleus. ND - Clarke's nucleus dorsalis.

neurons, particularly to those projecting to the nucleus of the solitary tract (Menétrey and Basbaum, 1987). Thus, although in earlier studies on spinoreticular (Menétrey et al., 1983) and spinomesencephalic (Menétrey et al., 1982;

Liu, 1983) neurons a tendency could be recognized for these cells to be clustered laterally in the lower thoracic spinal cord, coincident with trajectories of the LCP, the suggestion of the LCP being the link for supraspinal, and the MCP the link for propriospinal connections of splanchnic afferents (Neuhuber et al., 1986) appears as an naive oversimplification. Nevertheless, LCP and MCP fibers may still specifically contact different projection neurons and interneurons at least in lateral and medial lamina I since most marginal neurons show a rather limited mediolateral spread of their dendrites (Lima and Coimbra, 1986, 1988, 1989).

As to direct connections of splanchnic afferents with spinothalamic neurons, the scarcity of such cells reported for the thoracic spinal cord (Giesler et al., 1979) led us to suspect that visceroafferent signals in rats might be transmitted in first line to the lower brainstem (Neuhuber et al., 1986), and perhaps from there to the thalamus (cf., Granum, 1986). However, more recent studies using sensitive tracers (Granum, 1986; Lima and Coimbra, 1988) have demonstrated substantial numbers of spinothalamic neurons also in the lower thoracic cord, in particular in lamina I and the lateral spinal nucleus. Spinothalamic cells in lamina I are located in its intermediate third (Lima and Coimbra, 1988) and may receive input from both the LCP and MCP. Thus, splanchnic afferent information may be relayed by spinothalamic neurons to comparable extents in rats, cats and primates (for references see Willis, 1986).

SPLANCHNIC AFFERENT PROJECTIONS IN CAPSAICIN-TREATED RATS

In adult rats which had been treated with 50 mg/kg capsaicin on their second postnatal day, HRP was applied to the greater splanchnic nerve in three animals, and to the intermesenteric nerve in another rat (Irminger, 1989). There was a drastic reduction of labeled cell bodies in spinal ganglia (see Table 1), and of labeled afferent fibers and terminals in the spinal cord. Although the MCP/LCP pattern was maintained, the reduction of labeling was most pronounced in the LCP (Figs. 2,3). Only single fibers could be traced along the lateral border of the dorsal horn to dorsal lamina V, and no fibers and terminals were seen in in the area of the lateral spinal nucleus. In the MCP, however, there was a denser residual labeling in lamina I occurring at about the same rostro-caudal periodicity of the "labeling clusters" seen in normal rats. Instead of the mixed fine and coarse granular labeling present normally, only scattered coarse labeled structures were seen. Fibers in the dorsal columns and their collaterals were still present. There was no residual labeling ventrally to the central canal. Labeling of preganglionic sympathetic neurons was unimpaired.

These results are in agreement with the well-known toxic action of capsaicin on C and possibly A-delta afferent neurons (for reviews see Jancsó et al., 1987; Lawson, 1987). They are also comparable to the drastically reduced labeling of gastropancreatic (Sharkey et al., 1984) and urinary bladder afferents (Jancsó and Maggi, 1987) after neonatal capsaicin treatment, and concur very well with the abolition of certain visceral reflexes in capsaicin-treated rats (Cervero and McRitchie, 1982; Lembeck and Skofitsch, 1982). However, the non-uniform reduction of visceroafferent terminal labeling, i.e., more residual labeling in the MCP and lamina V in contrast to complete loss of labeling in the area of the

TABLE 1: Numbers of labeled spinal ganglion cells after HRP application to the greater splanchnic nerve in normal (NORM) and neonatally capsaicin-treated (CAPS) rats

	T 9	T 10	T 11
NORM (n=3)	214±33	307±87	222±71
CAPS (n=3)	38±30	83±48	85±55

(mean ± standard deviation)

lateral spinal nucleus and ventrally to the central canal, deserves attention. The disappearance of visceroafferent labeling ventrally to the central canal agrees with the capsaicin-sensitive content of SP and other peptides within these fibers (Jancsó et al., 1981). The almost complete loss of visceroafferent labeling along the LCP, particularly in the region of the lateral spinal nucleus, and the complete loss in lamina X might indicate that second-order neurons located there could mediate visceral reflexes known to be lost in capsaicin-treated animals, e.g., the depressor reflex upon intestinal distension (Lembeck and Skofitsch, 1982). The notion that this reflex depends on supraspinal pathways may be seen in context with the location of neurons within lateral lamina I, the lateral spinal nucleus and lamina X which project to the ventrolateral medulla (Menétrey et al., 1983), the nucleus of the solitary tract (Menétrey and Basbaum, 1987), and the parabrachial area (Lima and Coimbra, 1989; see also preceding section).

The present results and those from other studies (e.g., Sharkey et al., 1984) indicate that visceral afferents include capsaicin-sensitive as well as capsaicin-resistant neurons. The terminal fields of these two populations do not completely overlap since significant residual labeling was only found in medial lamina I and lamina V in our experiments. Besides some capsaicin-resistant small dark neurons emitting C or A-delta fibers (for review see Lawson, 1987), large light spinal ganglion cells connected to A-beta axons appear as a probable source for some of the "surviving" terminals, particularly in medial lamina I. However, because of the scarcity of these neurons and their collaterals to the dorsal horn (Neuhuber et al., 1986), they hardly can account for all the residual terminal labeling in capsaicin-treated animals, although the possibility of a compensatory sprouting of surviving fibers to capsaicin-denervated terminal areas (Nagy and Hunt, 1983; Rethelyi et al., 1986) has to be considered.

CONCLUSION

Transganglionic HRP tracing has revealed a typical visceroafferent projection pattern in the mammalian spinal cord. The bulk of visceroafferent fibers divides at the apex of the dorsal horn into a medially and laterally directed set of fiber bundles, the medial and lateral collateral pathways. They embrace the dorsal horn in a forceps-like manner, giving rise to terminals in lamina I, and at the neck and base of the dorsal horn, while sparing laminae II-IV. Overlap of these visceroafferent projections with somatoafferent terminal fields in laminae I and V provide an anatomical basis for viscerosomatic convergence on second-order neurons. The partition of visceroafferent fibers into a MCP and LCP may result in the formation of contacts with a variety of neurons in medial versus lateral lamina I and the lateral spinal nucleus whose projections and hence functional significance differ. Tracing of visceroafferent projections in capsaicin-treated rats can provide information about terminal sites of capsaicin-sensitive versus capsaicin-resistant fibers and might be useful in further elucidating the relationship between anatomy and physiology of visceroafferent neurons and their central connections.

Acknowledgements: The authors are indebted to Dr. M. Müntener for providing capsaicin-treated rats, to Mrs I. Drescher, C. Meyer and R. Forster for excellent technical assistance, and to Mrs H. Weber for expert photographical work. The studies in the authors' laboratory as summarized here were supported by the EMDO-Stiftung Zürich.

REFERENCES

Abrahams, V.C., Richmond, F.J., and Keane, J., 1984, Projections from C2 and C3 nerves supplying muscles and skin of the cat neck: A study using transganglionic transport of horseradish peroxidase, *J. Comp Neurol.*, 230:142-154

Baron, R., Jänig, W., and Kollmann, W., 1988, Sympathetic and afferent somata projecting in hindlimb nerves and the anatomical organization of the lumbar sympathetic nervous system of the rat, *J. Comp. Neurol.*, 275:460-468

Carlton, S.M., McNeill, D.L., Chung, K., and Coggeshall, R.E., 1988, Organization of calcitonin gene-related peptide-immunoreactive terminals in the primate dorsal horn, *J. Comp. Neurol.*, 276:527-536

Caverson, M.M., Ciriello, J., and Calaresu, F.R., 1983, Direct pathway from cardiovascular neurons in the ventrolateral medulla to the region of the intermediolateral nucleus of the upper thoracic cord: an anatomical and electrophysiological investigation in the cat, J. Autonom. Nerv. Syst., 9:451-475

Cechetto, D.F., Standaert, D.G., and Saper, C.B., 1985, Spinal and trigeminal dorsal horn projections to the parabrachial nucleus in the rat, *J. Comp. Neurol.*, 240:153-160

Cervero, F., 1983, Somatic and visceral inputs to the thoracic spinal cord of the cat: Effects of noxious stimulation of the biliary system, *J. Physiol. (Lond.)*, 337:51-67

Cervero, F., 1988, Neurophysiology of gastrointestinal pain, *Bailliere's Clin. Gastroenterol.*, 2:183-199

Cervero, F., and Connell, L.A., 1984a, Distribution of somatic and visceral primary afferent fibers within the thoracic spinal cord of the cat, *J. Comp. Neurol.*, 230:88-98

Cervero, F., and Connell, L.A., 1984b, Fine afferent fibers from viscera do not terminate in the substantia gelatinosa of the thoracic spinal cord, *Brain Res.*, 294:370-374

Cervero, F., Connell, L.A., and Lawson, S.N., 1984, Somatic and visceral primary afferents in the lower thoracic dorsal root ganglia of the cat, *J. Comp. Neurol.*, 228:422-431

Cervero, F., and McRitchie, H.A., 1982, Neonatal capsaicin does not affect unmyelinated efferent fibers of the autonomic nervous system: Functional evidence, *Brain Res.*, 239:283-288

Changgeng, Z., Zenker, W., and Celio, M.R., 1983, Substance P-positive structures in rat spinal cord. A longitudinal bundle ventral to the central canal, *Anat. Anz.*, 154:193-203

Ciriello, J., and Calaresu, F.R., 1983, Central projections of afferent renal fibers in the rat: an anterograde transport study of horseradish peroxidase, *J. Autonom. Nerv. Syst.*, 8:273-285

Craig, A.D., and Mense, S., 1983, The distribution of afferent fibers from the gastrocnemius-soleus muscle in the dorsal horn of the cat, as revealed by the transport of horseradish peroxidase, *Neurosci. Lett.*, 41:233-238

Crousillat, J., Ranieri, F., and Perrin, J., 1979, Interactions entre afférences somatiques et viscérales le long de la voie sensitive splanchnicque, *Arch. Ital. Biol.*, 117:186-188

Csillik, B., and Knyihar-Csillik, E., 1986, "The protean gate", Akademiai Kiadó, Budapest

Dalsgaard, C.-J.,Hökfelt, T., Elfvin, L.-G., Skirboll, L., and Emson, P., 1982, Substance P-containing primary sensory neurons projecting to the inferior mesenteric ganglion: evidence from combined retrograde tracing and immunohistochemistry, *Neuroscience*, 7:647-654

Dalsgaard, C.-J., Ygge, J., Vincent., S.R., Ohrling, M., Dockray, G.J., and Elde, R., 1984, Peripheral projections and neuropeptide coexistence in a subpopulation of fluoride-resistant acid phosphatase reactive spinal primary sensorx neurons, *Neurosci. Lett.*, 51:139-144

DeGroat, W.C., 1986, Spinal cord projections and neuropeptides in visceral afferent neurons, in: "Visceral sensation: Progress in Brain Research, Vol. 67," F. Cervero and J.F.B. Morrison, eds., Elsevier, Amsterdam:165-187

DeGroat, W.C., Nadelhaft, I., Morgan, C., and Schauble, T., 1978, Horseradish peroxidase tracing of visceral efferent and primary afferent pathways in the cat's sacral spinal cord using benzidine processing, *Neurosci. Lett.*, 10:103-108

Dockray, G.J., and Sharkey, K.A., 1986, Neurochemistry of visceral afferent neurons, *in:* "Visceral sensation; Progress in Brain Research, Vol. 67," F. Cervero and J.F.B. Morrison, eds., Elsevier, Amsterdam:133-148

Felpel, L.P., and Huffman, R.D., 1986, Response of splanchnic-driven neurons to substance P and eledoisin-related peptide, *J. Autonom. Nerv. Syst.*, 15:269-274

Fulwiler, C.F., and Saper, C.B., 1984, Subnuclear organization of the efferent connections of the parabrachial nucleus in the rat, *Brain Res. Rev.*, 7:229-259

Giesler, Jr., G.J., Menetrey, D., and Basbaum, A.I., 1979, Differential origins of spinothalamic tract projections to medial and lateral thalamus in the rat, *J. Comp. Neurol.*, 184:107-126

Gokin, P., Kostyuk, P.-G., and Preobrazhensky, N.-N., 1977, Neuronal mechanisms of interactions of high-threshold visceral and somatic afferent influences in spinal cord and medulla, *J. Physiol. (Paris)*, 73:319-333

Granum, S.L., 1986, The spinothalamic system of the rat. I. Locations of cells of origin, *J. Comp. Neurol.*, 247:159-180

Green, T., and Dockray, G.J., 1987, Calcitonin gene-related peptide and substance P in afferents to the upper gastrointestinal tract in the rat, *Neurosci. Lett.*, 76:151-156

Hökfelt, T., Johansson, O., Kellerth, J.-O., Ljungdahl, A., Nilsson, G., Nygards, A., and Pernow, B., 1977, Immunohistochemical distribution of substance P, *in:* "Substance P," U.S. von Euler and B. Pernow, eds., Raven Press, New York:117-145

Irminger, D.C., 1989, Zentrale Verteilung viszeraler Primärafferenzen in neonatal capsaicin-behandelten Ratten, Inaugural-Dissertation, Universität Zürich

Jancsó, G., Hökfelt, T., Lundberg, J.M., Kiraly, E., Halasz, N., Nilsson, G., Terenius, L., Rehfeld, J., Steinbusch, H., Verhofstad, A., Elde, R., Said, S., and Brown, M., 1981, Immunohisto-chemical studies on the effect of capsaicin on spinal and medullary peptide and monoamine neurons using antisera to substance P, gastrin/CCK, somatostatin, VIP, enkephalin, neurotensin and 5-hydroxytryptamine, *J. Neurocytol.*, 10:963-980

Jancsó, G., Kiraly, E., Such, G., Joo, F., and Nagy, A., 1987, Neurotoxic effect of capsaicin in mammals, *Acta Physiol. Hung.*, 69:295-313

Jancsó, G., and Maggi, C.A., 1987, Distribution of capsaicin-sensitive urinary bladder afferents in the rat spinal cord, *Brain Res.*, 418:371-376

Jänig, W., and Morrison, J.F.B., 1986, Functional properties of spinal visceral afferents supplying abdominal and pelvic organs, with special emphasis on visceral nociception, *in:* "Visceral sensation; Progress in Brain Research, Vol. 67," F. Cervero and J.F.B. Morrison, eds., Elsevier, Amsterdam:87-114

Jordan, D., and Spyer, K.M., 1986, Brainstem integration of cardiovascular and pulmonary afferent activity, *in:* "Visceral sensation; Progress in Brain Research, Vol. 67," F. Cervero and J.F.B. Morrison, eds., Elsevier, Amsterdam:295-314

Kuo, D.C., and DeGroat, W.C., 1985, Primary afferent projections of the major splanchnic nerve to the spinal cord and gracile nucleus of the cat, *J. Comp. Neurol.*, 231:421-434

Kuo, D.C., Nadelhaft, I., Hisamitsu, T., and DeGroat, W.C., 1983, Segmental distribution and central projections of renal afferent fibers in the cat studied by transganglionic transport of horseradish peroxidase, *J. Comp. Neurol.*, 216:162-174

Kuo, D.C., Oravitz, J.J., and DeGroat, W.C., 1984, Tracing of afferent and efferent pathways in the left inferior cardiac nerve of the cat using retrograde and transganglionic transport of horseradish peroxidase, *Brain Res.*, 321:111-118

Lawson, S.N., 1987, The morphological consequences of neonatal treatment with capsaicin on primary afferent neurones in adult rats, *Acta Physiol. Hung.*, 69:315-321

Leah, J., Menetrey, D., and dePommery, J., 1988, Neuropeptides in long ascending spinal tract cells in the rat: evidence for parallel processing of ascending information, *Neuroscience*, 24:195-207

Lembeck, F., and Skofitsch, G., 1982, Visceral pain reflex after pretreatment with capsaicin and morphine, *Naunyn Schmiedeberg's Arch. Pharmacol.*, 321:116-122

Lima, D., and Coimbra, A., 1983, The neuronal population of the marginal zone (lamina I) of the rat spinal cord. A study based on reconstructions of serially sectioned cells, *Anat. Embryol.*, 167:273-288

Lima and Coimbra, A., 1985, Marginal neurons of the rat spinal cord at the origin of a spinobulboreticular projection, *Neurosci. Lett. Suppl.* 22:S9

Lima, D., and Coimbra, A., 1986, A golgi study of the neuronal population of the marginal zone (lamina I) of the rat spinal cord, *J. Comp. Neurol.*, 244:53-71

Lima, D., and Coimbra, A., 1988, The spinothalamic system of the rat: Structural types of retrogradely labelled neurons in the marginal zone (lamina I), *Neuroscience*, 27:215-230

Lima, D., and Coimbra, A., 1989, Morphological types of spinomesencephalic neurons in the marginal zone (lamina I) of the rat spinal cord, as shown after retrograde labelling with cholera toxin subunit B, *J. Comp. Neurol.*, 279:327-339

Lindh, B., Dalsgaard, C.-J., Elfvin, L.-G., Hökfelt, T., and Cuello, A.C., 1983, Evidence of substance P immunoreactive neurons in dorsal root ganglia and vagal ganglia projecting to the guinea-pig pylorus, *Brain Res.*, 269:365-369

Liu, R.P.C., 1983, Laminar origins of spinal projection neurons to the periaqueductal gray of the rat, *Brain Res.*, 264:118-122

Mei, N., 1983, Sensory structures in the viscera, in: "Progress in Sensory Physiology, Vol. 4," D. Ottoson, ed., Springer, Berlin:1-42

Menétrey, D., and Basbaum, A.I., 1987, Spinal and trigeminal projections to the nucleus of the solitary tract: A possible substrate for somatovisceral and viscerovisceral reflex activation, *J. Comp. Neurol.*, 255:439-450

Menétrey, D., Chaouch, A., Binder, D., and Besson, J.M., 1982, The origin of the spinomesencephalic tract in the rat: An anatomical study using the retrograde transport of horseradish peroxidase, *J. Comp. Neurol.*, 206:193-207

Menétrey, D., Roudier, F., and Besson, J.M., 1983, Spinal neurons reaching the lateral reticular nucleus as studied in the rat by retrograde transport of horseradish peroxidase, *J. Comp. Neurol.*, 220:439-452

Mesulam, M.-M., and Brushart, T.M., 1979, Transganglionic and anterograde transport of horseradish peroxidase across dorsal root ganglia: A tetramethylbenzidine method for tracing central sensory connections of muscles and peripheral nerves, *Neuroscience*, 4:1107-1117

Miller, K.E., and Seybold, V.S., 1989, Comparison of met-enkephalin, dynorphin A, and neurotensin immunoreactive neurons in the cat and rat spinal cords: II. Segmental differences in the marginal zone, *J. Comp. Neurol.*, 279:619-628

Molander, C., Ygge, J., and Dalsgaard, C.-J., 1987, Substance P-, somatostatin-, and calcitonin gene-related peptide-like immunoreactivity and fluoride resistant acid phosphatase-activity in relation to retrogradely labeled cutaneous, muscular and visceral primary sensory neurons in the rat, *Neurosci. Lett.*, 74:37-42

Morgan, C., DeGroat, W.C., and Nadelhaft, I., 1986a, The spinal distribution of sympathetic preganglionic and visceral primary afferent neurons that send axons into the hypogastric nerves of the cat, *J. Comp. Neurol.*, 243:23-40

Morgan, C., Nadelhaft, I., and DeGroat, W.C., 1981, The distribution of visceral primary afferents from the pelvic nerve to Lissauer's tract and the spinal gray matter and its relationship to the sacral parasympathetic nucleus, *J. Comp. Neurol.*, 201:415-440

Morgan, C., Nadelhaft, I., and DeGroat, W.C., 1986b, The distribution within the spinal cord of visceral primary afferent axons carried by the lumbar colonic nerve of the cat, *Brain Res.*, 398:11-17

Nadelhaft, I., and Booth, A.M., 1984, The location and morphology of preganglionic neurons and the distribution of visceral afferents from the rat pelvic nerve: A horseradish peroxidase study, *J. Comp. Neurol.*, 226:238-245

Nadelhaft, I., and McKenna, K.E., 1987, Sexual dimorphism in sympathetic preganglionic neurons of the rat hypogastric nerve, *J. Comp. Neurol.*, 256:308-315

Nadelhaft, I., Roppolo, J., Morgan, C., and DeGroat, W.C., 1983, Parasympathetic preganglionic neurons and visceral primary afferents in monkey sacral spinal cord revealed following application of horseradish peroxidase to pelvic nerve, *J. Comp. Neurol.*, 216:36-52

Nagy, J.I., and Hunt, S.P., 1983, The termination of primary afferents within the rat dorsal horn: Evidence for rearrangement following capsaicin treatment, *J. Comp. Neurol.*, 218:145-158

Ness, T.J., and Gebhart, G.F., 1988, Colorectal distension as a noxious visceral stimulus: physiologic and pharmacologic characterization of pseudoaffective reflexes in the rat, *Brain Res.*, 450:153-169

Neuhuber, W., 1982, The central projections of visceral primary afferent neurons of the inferior mesenteric plexus and hypogastric nerve and the location of the related sensory and preganglionic sympathetic cell bodies in the rat, *Anat. Embryol.*, 164:413-425

Neuhuber, W.L., Sandoz, P.A., and Fryscak, T., 1986, The central projections of primary afferent neurons of greater splanchnic and intercostal nerves in the rat. A horseradish peroxidase study, *Anat. Embryol.*, 174:123-144

Ohmori, Y., Watanabe, T., and Fujioka, T., 1987, Projections of visceral and somatic primary afferents to the sacral spinal cord of the domestic fowl revealed by transganglionic transport of horseradish peroxidase, *Neurosci. Lett.*, 74:175-179

Pomeranz, B., Wall, P.D., and Weber, W.V., 1968, Cord cells responding to fine myelinated afferents from viscera, muscle and skin, *J. Physiol. (Lond.)*, 199:511-532

Procacci, P., Zoppi, M., and Maresca, M., 1986, Clinical approach to visceral sensation, *in:* "Visceral sensation; Progress in Brain Research, Vol. 67," F. Cervero and J.F.B. Morrison, eds., Elsevier, Amsterdam:21-28

Rethelyi, M., Salim, M.Z., and Jancsó, G., 1986, Altered distribution of dorsal root fibers in the rat following neonatal capsaicin treatment, *Neuroscience,* 18:749-761

Rosenfeld, M.G., Mermod, J.J., Amara, S.G., Swanson, L.W., Sawchenko, P.E., Rivier, J., Vale, W.W., and Evans, R.M., 1983, Production of a novel neuropeptide encoded by the calcitonin gene via tissue-specific RNA processing, *Nature,* 304:129-135

Ruch, T.C., 1947, Visceral sensation and referred pain, *in:* "Howell's Textbook of Physiology, 15th ed.," J.F. Fulton, ed., Saunders, Philadelphia:385-401

Sepulveda, L., and Nadelhaft, I., 1986, Preganglionic and sensory neurons of the rat greater splanchnic nerve: distribution and morphology, *Neurosci. Abstr.,* 12:1174

Sharkey, K.A., Sobrino, J.A., and Cervero, F., 1987, Evidence for a visceral afferent origin of substance P-like immunoreactivity in lamina V of the rat thoracic spinal cord, *Neuroscience,* 22:1077-1083

Sharkey, K.A., Williams, R.G., and Dockray, G.J., 1984, Sensory substance P innervation of the stomach and pancreas. Demonstration of capsaicin-sensitive sensory neurons in the rat by combined immunohistochemistry and retrograde tracing, *Gastroenterology,* 87:914-921

Standaert, D.G., Watson, S.J., Houghten, R.A., and Saper, C.B., 1986, Opioid peptide immunoreactivity in spinal and trigeminal dorsal horn neurons projecting to the parabrachial nucleus in the rat, *J. Neurosci.,* 6:1220-1226

Su, H.C., Bishop, A.E., Power, R.F., Hamada, Y., and Polak, J.M., 1987, Dual intrinsic and extrinsic origins of CGRP- and NPY-immunoreactive nerves of rat gut and pancreas, *J. Neurosci.,* 7:2674-2687

Swett, J.E., McMahon, S.B., and Wall, P.D., 1985, Long ascending projections to the midbrain from cells of lamina I and nucleus of the dorsolateral funiculus of the rat spinal cord, *J. Comp. Neurol.,* 238:401-416

Willis, Jr., W.D., 1986, Visceral inputs to sensory pathways in the spinal cord, *in:* "Visceral sensation; Progress in Brain Research, Vol. 67", F. Cervero and J.F.B. Morrison, eds., Elsevier, Amsterdam:207-225

Ygge, J., and Grant, G., 1983, The organization of the thoracic spinal nerve projection in the rat dorsal horn demonstrated with transganglionic transport of horseradish peroxidase, *J. Comp. Neurol.,* 216:1-9

16

RELATIONSHIP BETWEEN FUNCTIONAL AND MORPHOLOGICAL

PROPERTIES IN SINGLE PRIMARY AFFERENT NEURONS

S.Mense

Institut für Anatomie und Zellbiologie
Universität Heidelberg, Heidelberg, F.R.G.

INTRODUCTION

This report presents data obtained from single primary afferent neurons at two different levels, namely in the dorsal root ganglion (DRG) and in the spinal dorsal horn where the terminal arborizations are formed. The morphology of DRG cells has been the subject of anatomical studies for almost a century (for a review, see Lieberman, 1976). However, many questions remained unsolved, e.g., those concerning the quantitative correlation between soma dimension and axonal conduction velocity, or the presence of axonal branching in or in the vicinity of the ganglion. Since the advent of the technique of intracellular recording and staining of individual cells, these questions have gained new interest and several investigations dealing with these problems have appeared recently.

The same technique can be applied for identifying and visualizing single axons that enter the dorsal funiculus and form synaptic connections via segmental collaterals. It is thus possible to study the relationship between the physiological properties of a primary afferent fibre and the pattern of its terminations in the dorsal horn.

The present study aimed at the following questions:1) DRG level: Is the correlation between soma dimension in the DRG and axonal conduction velocity so tight, that the fibre calibre can be inferred from the soma size? Are the postulated axonal branching points (in addition to the bifurcation) close to the ganglion numerous enough that they can be considered to form the neuroanatomical basis for a common clinical type of pain, namely referred pain? 2) Spinal cord level: Where in the dorsal horn do slowly conducting afferent fibres from deep tissues (muscle, tendon, joint, bone) terminate? Is there a relation between the functional properties of an afferent fibre and the shape and location of its terminal arborizations in the dorsal horn?

Since a considerable amount of data has been accumulated concerning fast conducting afferent units (cf., Brown, 1981) the present investigation concentrates on neurons with slowly conducting (non-myelinated and thin myelinated) afferent fibres.

METHODS

The experimental data were obtained from cats anaesthetized with chloralose. Access to the lumbar spinal cord or cervical DRGs was gained by laminectomy, and the gastrocnemius-soleus (GS) muscle together with its nervous and vascular supplies was surgically exposed. A stimulating electrode was placed around the sciatic or forelimb nerves for the application of electrical searching stimuli. Glass micropipettes filled with a solution of 5-10% horseradish peroxidase (HRP, Boehringer) in 0.05 M Tris buffer and 0.5 M KCl (pH 7.3) were lowered into the spinal cord or DRG until unitary electrical activity was encountered. The tip of the electrode was assumed to have an intracellular position if a negative resting potential was present and the action potentials were positive-going with a negative afterpotential. From the latency of the action potential following electrical stimulation the conduction velocity of the afferent unit was calculated; its functional properties were determined with the use of a variety of non-noxious and noxious mechanical stimuli (Fig. 1).

The primary afferent neurons were neurophysiologically classified by the conduction velocity of the peripheral axon of DRG cells or - in the case of recordings from intraspinal branches - of the entire afferent unit. The limits were set as follows: Group IV units, <2.5 m/s; Group III units, 2.5-30 m/s; Group II units, 30-72 m/s; Group I units, >72 m/s (modified after Lloyd, 1943 and Paintal, 1967). After identification of a cell HRP was iontophoretically injected

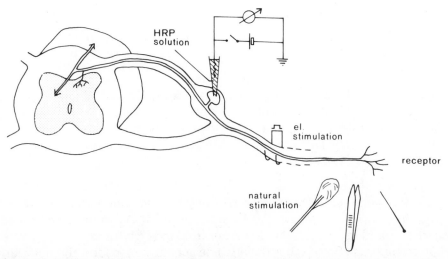

Fig. 1. Experimental set-up for identification and intracellular staining of single neurones. The electrical activity of the cells [in this case dorsal root ganglion (DRG) cells] is recorded with glass micropipettes filled with a horseradish peroxidase (HRP) solution. Electrical stimulation of a peripheral nerve is used as a searching stimulus. The neurones encountered are functionally identified by natural stimulation of their receptive field (in this case touch, pressure, pinch, prick).

using positive current pulses of 5-20 nA amplitude for 3-15 min. After a survival time of 2-30 h the animals were perfused through the carotid artery with saline at 37 °C followed by fixative containing 0.5% paraformaldehyde and 2.5% glutaraldehyde at room temperature, and finally 10% phosphate-buffered sucrose solution (pH 7.4, 4°C). The DRGs and the spinal cord were removed and stored in 30% phosphate-buffered sucrose for 3-10 h at 4°C. Serial frozen sections were cut at 60 μm and processed with diaminobenzidine after a modified protocol by Graham and Karnovsky (1966).

The shape and course of the stained units were reconstructed from serial sections in camera lucida drawings at a final magnification of 1060x. The size of DRG somata was measured as cross-sectional area with an electronic planimeter. Since the nucleus or nucleolus was not visible in heavily stained somata, the cells were focused in that plane where they had the largest cross-sectional area. For describing the projections of single afferent fibers in the dorsal horn, the lamination scheme of Rexed (1952) was used. Usually the tissue was not counterstained, therefore the laminae were determined in darkfield illumination, where laminae I,II and V can be easily recognized. The other layers had to be defined by interpolation. Control experiments in which the tissue was counterstained with thionin showed that this method was sufficiently correct to allow statements to be made on the main projection areas of a primary afferent unit.

RESULTS

DRG cells

This section concentrates on neurons having slowly conducting afferent fibres (Groups III and IV) since the discrepancies in the literature are greatest for these units and the branching of axons has been reported to occur predominantly in fibres of small diameter (Chung and Coggeshall, 1984). Some faster conducting units were also studied for reasons of comparison.

Thirty-five out of 36 stained neurones had the typical appearance of pseudo-unipolar cells with the stem axon bifurcating in a T-shaped fashion in a central and a peripheral axon. In all Group IV units there was a conspicuous asymmetry in the diameters of both branches, the central axon having the appearance of an extremely thin collateral of the stem axon (Fig. 2). Within Group III units, the thickness of the axons and the degree of myelination varied with the conduction velocity of the peripheral axon. Slowly conducting Group III units possessed peripheral and central axons without visible nodes of Ranvier, i.e., both branches were non-myelinated. The appearance of these axons was similar to that of the Group IV units except that the central axon was thicker and thus the difference in diameter between central and peripheral branch smaller. Group III units with an intermediate conduction velocity had nodes of Ranvier in the peripheral branch only, and fast conducting Group III cells showed clear nodes on both sides of the bifurcation. In this respect, the latter ones resembled Group II and Group I units. It thus appears that with increasing conduction velocity fibres within the Group III population show the following morphological featu-res: increase in diameter of the non-myelinated central axon; myelin sheath

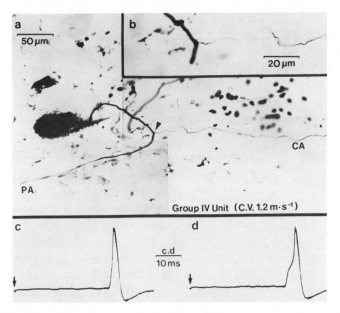

Fig. 2. Morphology and electrical activity of a cat dorsal root ganglion cell with non-myelinated axons (Group IV unit). Conduction velocity (C.V.) of the peripheral axon 1.2 m/ s. **A.** Composite light micrograph. The central axon (CA) originates as an extremely thin collateral from the stem axon (arrowhead). PA, peripheral axon. **B.** Enlarged micrograph of the bifurcation (see arrowhead in a). **C.** Action potential of the cell shown in A and B. **D.** Action potential recorded directly after the end of the iontophoresis. The spike is delayed because of slower invasion of the soma. In C and D the arrow marks the time of stimulation of the peripheral nerve (from Hoheisel and Mense, 1987).

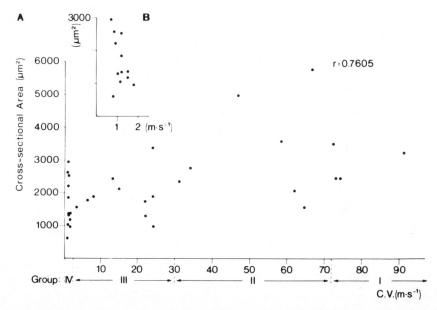

Fig. 3. Relationship between the cross-sectional area of cat DRG cell somata and the conduction velocity of the peripheral axon. **A.** diagram showing the data of 12 Group IV, 10 Group III, 7 Group II, and 4 Group I units. r, correlation coefficient for all units. **B.** diagram showing the data of the 12 Group IV units at an expanded scale of the abscissa. Note the great scatter of the soma sizes in each cell group (from Hoheisel and Mense, 1987).

around the peripheral branch; myelinated axons on both sides of the bifurcation.

As to the correlation between soma size and conduction velocity of the peripheral axon (Fig. 3), the results were heterogeneous. Although the correlation coefficient was significant if calculated for all the stained units the scatter of the data points was considerable. Further, there was no recognizable correlation within the Group III units, and within Group IV units the correlation was negative (see inset B in Fig. 3). For these reasons it is impossible to infer with reasonable probability the diameter of the peripheral axon from the soma size of an individual cell.

The only cell that did not follow the rule that DRG neurones possess a single peripheral and a single central branch was a Group III unit which had one central and two peripheral axons (Fig. 4). In this unit, the stem axon issued - in addition to the usual bifurcation - a second peripheral axon, which ran parallel to the first one and left the ganglion at its distal pole. Since only one out of 36 cells showed such an additional branching the overall number of branched fibres close to the ganglion is probably small. This is particularly true for axons in the dorsal root which could be followed until they entered the spinal cord. None of these fibres branched.

Termination pattern in the spinal cord

In this series of experiments the central branches of the DRG cells were impaled with micropipettes in the dorsal root entry zone and HRP was injected intra-axonally after physiological identification of a given unit. A survival time of several hours was found to be sufficient for the enzyme to move into the terminal arborizations and thus to visualize the spinal projection areas of a fibre. The investigation was confined to slowly conducting myelinated afferent units (Group II and III) from the deep tissues of the cat hindlimb since no information

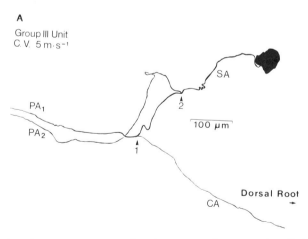

Fig. 4. Camera lucida reconstruction of a Group III afferent unit with two peripheral axons, peripheral conduction velocity 5 m/s (only one action potential could be elicited by electrical stimulation of the peripheral nerve). Note that in addition to what appears to be the original bifurcation (arrowhead 1) an additional bifurcation (arrowhead 2) exists from which a second peripheral axon (PA 2) originates. PA 1, original peripheral axon, CA, central axon; SA, stem axon (from Hoheisel and Mense, 1987).

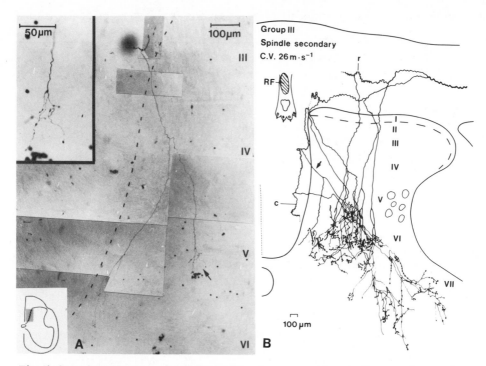

Fig. 5. Spinal termination of an afferent fibre from a muscle spindle secondary ending, C.V. 26 m/s. **A.** Composite micrograph of a collateral terminating with many presynaptic boutons in ventral lamina V. The dashed line indicates the border between the medial dorsal horn and the dorsal columns; the lower insert shows the area of the spinal cord (hatched) which contained the collateral. An enlargement of the terminal arborization (arrow) can be seen in the upper insert. **B.** Same unit as in A. Camera lucida reconstruction of all stained collaterals. In the upper right hand corner of the figure, the afferent fibre can be seen entering the spinal cord. It then divides into a rostral (r) and caudal (c) branch both of which issue collaterals. The receptive ending was located in a muscle underneath the sole of the hindpaw, it could be activated by stretching the toes and by slightly pressing upon a restricted area of the sole (RF).

on this fibre system is available in the literature. Group IV units could not be studied because the technique of intra-axonal ionophoresis is not (yet) applicable to non-myelinated fibres.

Neurophysiological data was obtained from 84 fibres with conduction velocities ranging from 18-40 m/s. Out of these, 32 units were recovered from the histological material. With the use of graded mechanical stimuli each unit could be sorted into one of the following types: 1) Afferents from muscle spindle secondary endings. These units responded to muscle stretch with a non-adapting discharge and low dynamic sensitivity. Their receptive fields to local pressure were situated in skeletal muscle outside joints. 2) Low-threshold mechanosensitive (LTM) units. They responded strongly to weak local pressure stimulation and - if located in joint capsules - also to moving joints. They could be distinguished from musle spindles by their adapting discharge and the irregular intervals between action potentials. 3) High-threshold mechanosensitive (HTM) units. These neurones did not respond to weak mechanical stimulation but required noxious local pressure or movements of joints in the noxious range to be activated.

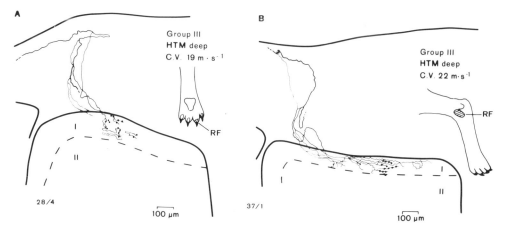

Fig. 6. Termination patterns of two Group III afferent fibres from high-threshold mechanosensitive (HTM) receptors in deep tissues of the cat hindlimb. Both units required noxious pinching of the receptive field (RF) to be activated. Note the exclusive termination with many boutons in lamina I. The approximate border between lamina I and II is marked by the dashed line.

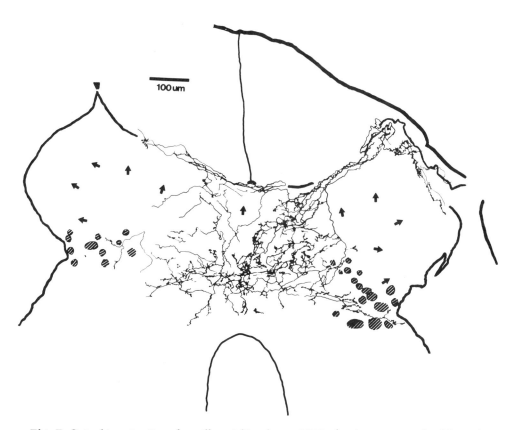

Fig. 7. Spinal termination of an afferent fibre from a HTM receptor in a muscle of the cat's tail, C.V. 46 m/s. Camera lucida reconstruction of all stained collaterals rostral to its point of entry into the spinal cord. Arrows indicate approximate border between lamina II and III. Note the numerous contralateral branches and the strong projection to lamina X (from Mense et al., 1981).

Within this population of group II and III afferent units there were clear differences in the spinal termination pattern. Afferent fibres from muscle spindle secondary endings had no terminal arborizations dorsal to lamina IV in the dorsal horn but projected heavily with many boutons to laminae VI/VII and into the ventral horn (Fig. 5). The other extreme form of termination pattern was that of some HTM units which terminated exclusively in lamina I (Fig. 6). Both types had such a typical appearance that the function of the neurone could be inferred from the termination pattern. Other HTM units had terminal arborizations in both lamina I and IV/V. The LTM units differed from the HTM units in that they did not project to lamina I, and from muscle spindle secondaries in that they often had terminals in lamina II and lacked the many boutons in laminae VI/VII.

The comparison of these data with those of a previous investigation on primary afferent neurons from the deep tissues of the cat's tail yielded several interesting differences: The HTM units from the tail had numerous contralateral branches which were not encountered in afferent units from the hindlimb, and they projected strongly to lamina X (Fig. 7). Again, HTM units terminated predominantly in lamina I and IV/V, while LTM units possessed no projections to lamina I. With one exception none of the stained afferent fibres from deep tissues had boutons in lamina III. Neurones in this layer appear to receive no primary afferent input from deep structures.

DISCUSSION

The results obtained from DRG cells show that it is an oversimp-lification to state that small cells belong to non-myelinated axons in a peripheral nerve (Dogiel, 1897; Cajal, 1906; Ha, 1970; for reviews, see Lieberman, 1976; Mei, 1983). Peripheral Group IV fibres can be connected to quite large DRG neurones and, conversely, somata having non-myelinated peripheral branches are not restricted to the lower extreme of the size spectrum of DRG cells. Although the correlation between soma size and peripheral conduction velocity was found to be significant if calculated for all cells, the scatter of the data is so great that it is not possible to infer the axonal conduction velocity (and hence fibre diameter) from the size of the soma in histological sections. The great scatter in the soma size-conduction velocity correlation diagram might indicate that factors other than the diameter of the peripheral axon determine the dimension of the soma. It is conceivable that the overall length of the afferent unit (cf., Cameron et al., 1984) or the mean frequency of discharge (Lee et al., 1986) are more important factors. This does not necessarily mean that electrical activity and soma size are positively correlated. Since small cells have a high surface-to-volume ratio and consequently a faster exchange of oxygen and metabolites they could theoretically be better suited to meet the requirements of high electrical activity.

In the literature, there is some discrepancy concerning the relation between soma size and axon calibre in Group IV units. Our results agree best with those of Lee et al. (1986) who likewise found a negative correlation between soma size and axonal conduction velocity in an in vitro preparation of feline DRG cells. Cameron and coworkers (1986), however, reported a positive correlation between these parameters in feline Group IV DRG cells. Besides the fact that these authors used a different marker (Lucifer Yellow) no explanation can be given for these conflicting results. Harper and Lawson (1985) likewise found a positive corrrelation between soma size and conduction velocity within Group IV DRG cells. Since their data were obtained from immature rats a comparison between their results and ours is difficult to make.

As to the question of dichotomizing axons near the ganglion, the cell shown in Fig. 4 appears to be the first one for which the existence of a third axon has been demonstrated by directly visualizing the branching point in an individually stained neurone. Up to now, the available evidence in favour of axonal branching was based on more indirect data from electrophysiological and anatomical experiments using the collision technique (Bahr et al., 1981; Pierau et al, 1982), double labelling methods (Taylor and Pierau, 1982) or comparisons between the numbers of DRG somata and dorsal root fibres (Langford and Coggeshall, 1981; Chung and Coggeshall, 1984). In general, these studies yielded higher proportions of branched axons than those found in the present investigation. Although additional branching points in distal portions of the peripheral nerve cannot be excluded, the overall figure of neurones with dichotomizing axons is probably small (Devor et al., 1984). This raises the question as to whether the existence of such a small number of additional branches can explain the rather common phenomenon of referred pain.

The data concerning the spinal terminations of primary afferent fibres demonstrate that a close relationship exists between the function of an afferent unit and the location of its terminals in the spinal gray matter. The HTM - presumably nociceptive - units projected predominantly to lamina I and in addition to laminae IV/V. It is well known that in these layers cells of origin of ascending tracts are situated which are thought to be involved in the mediation of pain, e.g., the spinothalamic (Carstens and Trevino, 1978; Craig and Kniffki, 1985) and the spino-cervical tract (Brown and Franz, 1969; Bryan et al., 1973; Craig, 1976). This arrangement suggests that a monosynaptic pathway for deep pain may exist at the spinal level.

From the data obtained from muscle spindle secondary endings it appears that the slowly conducting units of the present study exhibit the same pattern of spinal termination as the ones having a faster conduction velocity (cf., Fyffe, 1979). There are also many parallels to the morphology of Ia fibres, e.g., the course of the collaterals through the middle and medial dorsal horn and the projection into the ventral horn (cf., Brown, 1981).

The different patterns of spinal termination demonstrate that the projection areas of afferent units from the deep tissues of the hindlimb are determined by their functional properties. Since all the different functional types studied had similar conduction velocities in the Group II /III range, the conduction velocity and hence fibre diameter cannot be considered to be a decisive factor for the spinal projections of the units. Conversely, if a given type of receptor has a broad spectrum of afferent fibres - as is the case in muscle spindle secondary endings the afferent fibres of which have conduction velocities ranging from about 20 m/s to 72 m/s - the pattern of spinal termination will be the same and characteristic for the receptor type.

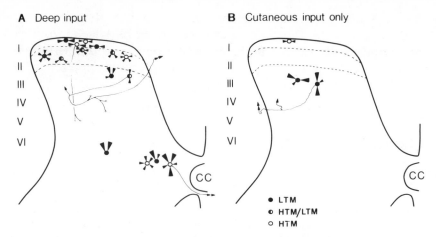

Fig. 8. Location of intracellularly stained dorsal horn neurones processing input from deep tissues (**A**) and the skin (**B**) of the cat's hindlimb. The soma is indicated by a filled circle if the neurone had a low mechanical threshold (LTM) and by an open circle if the neurone had HTM properties. The arrowheads mark the predominant orientation of the dendrites. Stained axons are indicated by thin lines; a single arrowhead marks a descending, a double arrowhead an ascending branch. The small open circles attached to the thin lines represent segmental collaterals with boutons. The horizontal dashed lines indicate the approximate borders of lamina II. Note that the majority of the neurones with deep input (**A**) are located in the superficial and deep dorsal horn while the intermediate layers (particularly laminae IV/V) are free from such cells. CC, central canal.

Topical relationship between primary afferent fibres and dorsal horn cells

The conspicuous lack of terminations of deep primary afferent fibres in lamina III raises the question as to whether this layer is excluded from the processing of information from deep tissues. In order to answer this question, dorsal horn cells with input from deep tissues were stained intracellularly using the same method as described above. When the locations of all the stained cells are plotted on a standardized cross section of the dorsal horn (Fig. 8) it is obvious that there is no gap in lamina III. Since there was no projection from deep afferents to this layer, lamina III cells cannot receive their deep input monosynaptically. (Theoretically, the deep input could reach the cells via monosynaptic connections with non-myelinated fibres, but there is no evidence indicating that such afferents from deep tissues terminate in lamina III, cf., Mense and Craig, 1988). On the other hand, the cells in lamina I with deep HTM input could well have monosynaptic connections with the HTM afferent fibres shown in Fig. 6.

An interesting finding was that laminae IV and V had the least density of somata with deep input in spite of the fact that both deep HTM and deep LTM fibres had massive terminations in these layers. Apparently, the terminations are not effective in eliciting action potentials in cells the somata of which are located in laminae IV/V. The function of these terminals is unknown; possible explanations are that the fibres form axo-axonal synapses, are inhibitory in nature, or elicit subthreshold membrane potential changes in second order cells. The last hypothesis is of particular interest, since it constitutes a possible mechanism for changes in the excitability of dorsal horn neurons. Such changes have been observed in cells with multiple receptive fields following noxious stimulation of skeletal muscle (Hoheisel and Mense, 1989).

REFERENCES

Bahr, R., Blumberg, H., and Jänig, W., 1981, Do dichotomizing afferent fibers exist which supply visceral organs as well as somatic structures? A contribution to the problem of referred pain, *Neurosci. Lett.*, 24:25

Brown, A. G., 1981, "Organization in the spinal cord. The anatomy and physiology of identified neurones," Springer, Berlin, Heidelberg, New York.

Brown, A. G., and Franz, D. N., 1969, Responses of spinocervical tract neurones to natural stimulation of identified cutaneous receptors, *Exp. Brain Res.*, 7:231

Bryan, R. N., Trevino, D. L., Coulter, J. D., and Willis, W. D., 1973, Location and somatotopic organization of the cells of origin of the spino-cervical tract, *Exp. Brain Res.*, 17:177

Cajal, S. R. y, 1906, Die Struktur der sensiblen Ganglien des Menschen und der Tiere, *Anat. Hefte, 2. Abtlg.*, 16: 177

Cameron, A. A., Leah, J. D., and Snow, P. J., 1986, The electrophysiological and morphological characteristics of feline dorsal root ganglion cells, *Brain Res.*, 362:1

Carstens, E., and Trevino, D. L., 1978, Anatomical and physiological properties of ipsilaterally projecting spinothalamic neurons in the second cervical segment of the cat's spinal cord, *J. Comp. Neurol.*, 182:167

Chung, K., and Coggeshall, R. E., 1984, The ratio of dorsal root ganglion cells to dorsal root axons in sacral segments of the cat, *J. Comp. Neurol.*, 225:24

Craig, A. D., 1976, Spinocervical tract cells in the cat and dog, labeled by the retrograde transport of horseradish peroxidase, *Neurosci. Lett.*, 3:173

Craig, A. D., and Kniffki, K.-D., 1985, Spinothalamic lumbosacral lamina I cells responsive to skin and muscle stimulation in the cat, *J. Physiol. (Lond.)*, 365:197

Devor, M., Wall, P. D., and McMahon, S. B., 1984, Dichotomizing somatic nerve fibers exist in rats but they are rare, *Neurosci. Lett.*, 49:187

Dogiel, A. S., 1897, Zur Frage über den feineren Bau der Spinalganglien und deren Zellen bei Säugetieren, *Int. Monatsschrift für Anat. u. Physiologie*, 14:73

Fyffe, R. E. W., 1979, The morphology of group II muscle afferent fibre collaterals, *J. Physiol. (Lond.)*, 296:39P

Graham, R. C., and Karnovsky, M. J., 1966, The early stages of absorption of injected horseradish peroxidase in the proximal tubules of mouse kidney: ultrastructural cytochemistry by a new technique, *J. Histochem. Cytochem.*, 14:291

Ha, H., 1970, Axonal bifurcation in the dorsal root ganglion of the cat: a light and electron microscopic study, *J. Comp. Neurol.*, 140:227

Harper, A. A., and Lawson, S. N., 1985, Conduction velocity is related to morphological cell type in rat dorsal root ganglion neurones, *J. Physiol. (Lond.)*, 359:31

Hoheisel, U., and Mense, S., 1987, Observations on the morphology of axons and somata of slowly conducting dorsal root ganglion cells in the cat, *Brain Res.*, 423:269

Hoheisel, U., and Mense, S., 1989, Long-term changes in discharge behaviour of cat dorsal horn neurones following noxious stimulation of deep tissues, *Pain* (in press)

Langford, L. A., and Coggesshall, R. E., 1981, Branching of sensory axons in the peripheral nerve of the rat, *J. Comp. Neurol.*, 203:745

Lee, K. H., Chung, K., Chung, J. M., and Coggeshall, R. E., 1986, Correlation of cell body size, axon size, and signal conduction velocity for individually labelled dorsal root ganglion cells in the cat, *J. Comp. Neurol.*, 243:335

Lieberman, A. R., 1976, Sensory ganglia, in: "The Peripheral Nerve," D. N. Landon (ed.), Wiley, New York

Lloyd, D. P. C., 1943, Neuron patterns controlling transmission of ipsilateral hind limb reflexes in cat, *J. Neurophysiol.*, 6:293

Mei, N., 1983, Sensory structures in the viscera, *Progr. Sensory Physiol.*, 4:1

Mense, S., and Craig, A. D., 1988, Spinal and supraspinal terminations of primary afferent fibers from the gastrocnemius-soleus muscle in the cat, *Neuroscience*, 26:1023

Mense, S., Light, A. R., and Perl, E. R., 1981, Spinal terminations of subcutaneous high-threshold mechanoreceptors, in: "Spinal cord sensation," A. G. Brown and M. Réthelyi (eds.), Scottish Academic Press, Edinburgh

Paintal, A. S., 1967, A comparison of the nerve impulses of mammalian non-medullated nerve fibres with those of the smallest diameter medullated fibres, *J. Physiol. (Lond.)*, 193:523

Pierau, F.-K., Taylor, D. C. M., Abel, W., and Friedrich, B., 1982, Dichotomizing peripheral fibres revealed by intracellular recording from rat sensory neurones, *Neurosci. Lett.*, 31:123

Rexed, B., 1952, The cytoarchitectonic organization of the spinal cord in the cat, *J. Comp. Neurol.*, 96:415

Taylor, D. C. M., and Pierau, F.-K., 1982, Double fluorescence labelling supports electrophysiological evidence for dichotomizing peripheral sensory nerve fibres in rats, *Neurosci. Lett.*, 33:1

17

ULTRASTRUCTURE OF PRIMARY AFFERENT TERMINALS

IN THE SPINAL CORD

M. Réthelyi

Second Department of Anatomy, Semmelweis
University Medical School, Budapest, Hungary

INTRODUCTION

Primary afferent fibers (PAF) arborize and terminate in the spinal gray matter with characteristic arborizations and terminal structures. The arborizations blend into the neuropil, i.e., they are tangential in lamina I, sagittally oriented sheets in laminae II and III, discs oriented in the transverse plane in the intermediate zone as well as in the ventral horn, and narrow rostro-caudally oriented bundles in Clarke's column.

Primary sensory neurons establish synaptic contacts with the spinal cord neurons through *en passant* and terminal enlargements of the terminal axon branches. The presynaptic enlargements contact mainly dendrites and less frequently perikarya.

Historical overview

Synapses established by PAFs were studied on electron micrographs of intact spinal cord and/or obtained from the spinal cord after various experimental interferences.

Terminals of PAFs could be identified with a high degree of confidence only in certain specific territories of the grey matter, where they could be identified on the basis of one or several characteristic properties. For example, large caliber PAFs were found to terminate in Clarke's column as multiple, giant *en passant* synapses (Szentágothai and Albert, 1955) running along the major longitudinally oriented dendrites of the Clarke neurons. Their large size boutons could be identified as of primary afferent origin in subsequent ultrastructural studies (Réthelyi, 1970; Saito, 1974).

Terminals of PAFs could be "labeled" in various ways. For a long period, dorsal rhizotomy and the ensuing anterograde axonal degeneration was the only way to select PAFs from the thousands of fibers of other origins. Degeneration studies were instrumental in defining the termination patterns of PAFs in the spinal gray matter. Terminal degeneration at the ultrastructural level often yielded equivocal results because of various factors. First, different types of PAFs degenerated at different rates. At the moment of sacrifice of the animal, some terminals might still resist degeneration and therefore might look normal, others more sensitive to being disconnected from the parent cell body were already engulfed and digested by glial terminals. Theoretically and practically, it was only a fraction of terminals that showed the signs of degeneration while still being in contact with the postsynaptic component of the connection. It is by now of historical interest to mention the discrepancies encountered in the terminations of PAFs in the substantia gelatinosa. It was only the drastic reduction of survival period to 24-48 hours (Heimer and Wall, 1968; Réthelyi and Szentágothai, 1969) that made it possible to show clearly the degenerating terminals in that enigmatic portion of the spinal gray matter. Another source of difficulties in the interpretation laid in the various cytoplasmic alterations in the axon terminals that follow the disconnection from the perikaryon. Some types of terminals became dark (dark-type degeneration), while others lost their cytoplasmic components (electron lucent-type swelling). In a third group degenerating terminals became filled with tangles or whorls of neurofilaments (neurofilamentous-type of degeneration). Furthermore, degenerating axon terminals usually shrunk and were surrounded by glial processes and ultimately they were removed from the postsynaptic component. All these alterations caused serious difficulties in the clear observation of synaptic structures labeled by the process of anterograde (Wallerian) degeneration. In spite of these difficulties, the degeneration technique was a powerful tool in studying the connections between various groups of neurons.

Labeling axon terminals with the axoplasmic transport of radioactive proteins (Lasek et al.,1968) resulted in intact ultrastructure of the terminal arborization. Although the technique was often used to study the distribution of PAFs in the spinal gray matter (LaMotte, 1977; Réthelyi et al., 1979), the long exposure time and the interpretational problems due to background effects did not make it a popular technique for labeling terminal boutons of PAFs in the spinal cord (but see Ralston and Ralston, 1979a and b).

With the introduction of immunohistochemistry with electron dense precipitate (PAP technique, Sternberger, 1986), the terminal synaptic arrangement of peptide containing PAFs could be studied with the electron microscope.

The real breakthrough in studying the synaptic structure of PAFs was accomplished by the iontophoretic injection of tracer substances like horseradish peroxidase (HRP) and more recently *Phaseolus vulgaris* leukoagglutinin (PHAL). Bundles of PAFs within a dorsal root filament were stained with HRP, or single PAFs were impaled by a tracer-filled microelectrode. After having identified the physiological properties of the fiber, the tracer was injected and transported into the terminal arborization. HRP was widely used for studying the central arborization of primary sensory neurons with myelinated axons, whereas PHAL gave excellent results in tracing the central arborizations of neurons with unmyelinated axons. With both tracer substances the labeling of the terminals was unequivocal, the histochemically and/or immuno-histochemically detected tracer molecules did not obscure the cytoplasmic structure of the terminals and the structure of the neuropil, neither were the connections between pre- and postsynaptic components significantly altered.

A combination of the HRP and the degeneration techniques could be used to label neurons and fiber tracts and to study the synaptic connections between labeled structures (Maxwell et al., 1984).

DISTRIBUTION OF PRIMARY SENSORY FIBERS IN THE SPINAL GRAY MATTER

Degeneration and tracer studies showed that the bulk of PAFs terminate in the upper four laminae of Rexed (1952). This area which could be separated from the rest of the gray matter upon architectural principles (Scheibel and Scheibel, 1968) was defined as the dorsal horn in stricter sense (Réthelyi, 1976). PAFs could also be traced to the intermediate region of the gray matter (central core of Réthelyi, 1976), to the ventral horn and to Clarke's column. Long ascending branches of PAFs terminate also in the dorsal column nuclei.

1. Terminal structures in the dorsal horn (Laminae I-IV)

PAFs terminate in all four laminae of the dorsal horn forming predominantly longitudinally oriented preterminal arborizations (Schimert [Szentágothai], 1939; Sprague and Ha, 1964; Sterling and Kuypers, 1967; Imai and Kusama, 1969; LaMotte, 1977; Réthelyi et al., 1979). Recent studies revealed that individual laminae or parts thereof receive PAFs of defined physiological characteristics (Light and Perl, 1979; Brown, 1981).

Synaptic enlargements of PAFs were identified as boutons with spherical synaptic vesicles (R-type), boutons with large densecore vesicles (LGV-type, Fig. 1a) and centrally located boutons synapsing with several postsynaptic and presynaptic components in the form of synaptic glomeruli (C-type, Figs 1b and 1c) in both cat (Ralston, 1968; Beattie et al., 1978) and monkey (Ralston and Ralston, 1979a; Knyihár et al., 1982).

1. 1 Lamina I. Fine myelinated and unmyelinated PAFs terminate in this narrow superficial layer of the gray matter.

Fine myelinated PAFs with high threshold mechanoreceptors (Burgess and Perl, 1967) in the cat possessed 2-3 μm large *en passant* and terminal enlargements that synapsed with several smaller profiles (Réthelyi et al., 1982). The axon enlargements contained spherical synaptic vesicles. Dense core vesicles rarely occurred in these boutons. The smaller profiles (dendritic protrusions, dendritic shafts more rarely axons) were often partially compressed into the centrally located axon terminal thus forming a synaptic glomerulus (C-type boutons). Few dendritic protrusions contained synaptic vesicle-like structures and seemed to be presynaptic to the large bouton (Fig. 1d). Also the small size axon profiles were in presynaptic location with respect to the large axon. Terminals of high threshold mechanoreceptor fibers of the C-type boutons were found also in the monkey dorsal horn (Réthelyi et al., 1982). They synapsed with numerous dendrites and were contacted by small size axon terminals (Fig. 1e). Vesicle containing dendrites were never observed, though, in these complex synapses of the monkey.

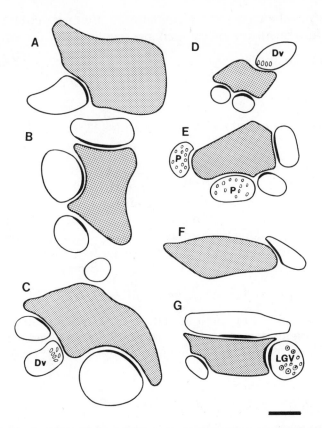

Fig. 1. Schematic drawings of synaptic arrangements in the superficial dorsal horn. The drawings are the copies of electron microscopic pictures taken from published works. The primary afferent boutons are stippled, dendrites in postsynaptic position have an empty contour, vesicle containing profiles represent either axons presynaptic to the primary afferent bouton (P), or vesicle containing dendrites (Dv). For easy comparison all drawings were made on scale. **A.** LGV-type terminal in lamina I; **B.** C-type terminal in lamina I; **C.** C-type terminal in lamina II; **D.** Glomerulus around the bouton of a fiber with high threshold mechanoreceptor in the cat; **E.** Glomerulus around the bouton of a fiber with high threshold mechanoreceptor in the monkey; **F.** VIP positive axon enlargement; **G.** Glomerulus around the bouton of an unmyelinated fiber, a desmosome-like specialization connects the primary afferent bouton to a LGV-type terminal. Bar: 1 μm.

Unmyelinated (C) PAFs of various receptor properties were found to terminate in lamina I in the guinea pig (Sugiura et al., 1986). The terminal arborization of these fibers showed numerous *en passant* and terminal arborizations of 1 to 7 μm in diameter. The ultrastructural analysis of these terminals will give the first direct evidence for the synaptic connections made by C-type PAFs.

An indirect approach in the study the synaptology of C-fibers is to search for immunohistochemically detectable peptides in axon enlargements. Vasoactive intestinal peptide (VIP) immunostaining in the cat spinal cord is the best in fulfilling the criteria for labeling axon enlargements that belong exclusively to peripheral and unmyelinated fibers. VIP positive axon enlargements were densely filled with round synaptic and dense core vesicles 90-120 nm in diameter (LGV-type boutons). They formed synaptic connections only infrequently, in which they contacted dendritic shafts (Fig. 1f). No glomerulus formation was seen, and no axo-axonic synapses were found in connection with the VIP positive enlargements (Honda et al., 1982). Practically all other known peptides which occur in fibers in the superficial dorsal horn are distributed either also in descending fibers and/or in local interneurons (substance-P, cholecystokinin, somatostatin) or also in primary sensory neurons with myelinated axons (calcitonin gene-related peptide, CGRP).

In a fortunate experiment a physiologically unidentified C-fiber was labeled iontophoretically with HRP in a monkey spinal cord (Réthelyi and Light, unpublished). The fiber ended with a rich arborization and numerous large size boutons in laminae I and in the subjacent part of lamina II. The boutons contained round synaptic vesicles and numerous dense-core vesicles (LGV-type boutons) and synapsed with several dendritic shafts and protrusions. Large size unlabeled boutons filled similarly with round synaptic and dens-core vesicles were seen to be in direct apposition to the labeled terminal (Fig. 1g). Symmetrical, desmosome-like contacts could be detected between the two profiles. In a recent study Carlton et al. (1988) detected CGRP positive boutons in the monkey spinal cord. Their immunoreactive profiles shared all ultrastructural features with the HRP-stained terminal structures (size of the boutons, combination of round and dense-core vesicles, desmosome-like connections between CGRP positive terminals). Thus it appears very probable that the unidentified C-fiber corresponded to a CGRP-containing fiber which arborized richly and terminated with multiple boutons in the superficial dorsal horn.

1.2 Lamina II. This lamina could be subdivided into an outer (IIo) and inner (IIi) sublaminae. IIo resembles lamina I with respect to termination of PAFs, whereas IIi seems to be the exclusive termination site of C-fibers (Réthelyi, 1977; Jancsó and Kiraly, 1980). Laminae I and II together form the superficial dorsal horn.

Large scalloped axon terminals (Gobel, 1974) in the center of synaptic glomeruli in lamina IIi in various species proved to be of primary sensory origin (Ralston, 1968, 1979; Réthelyi and Szentágothai, 1969; Knyihár et al., 1974, Coimbra et al., 1974). They were filled with regular round synaptic vesicles, dense-core vesicles could be seen only occasionally (C-type bouton). In these complex synapses the *en passant* and terminal enlargements of C-fibers synapsed mainly with dendritic protrusions (occasionally with dendritic shafts), while they received synaptic connections from vesicle containing dendrites and small axon terminals. Synapses might occur between the dendritic protrusions and small axon terminals resulting in triadic synaptic complexes. Large scalloped axon terminals in the rodent's lamina II showed fluoride resistent acid phosphatase (FRAP) activity (Coimbra et al., 1974; Knyihár et al., 1974).

1.3 Laminae III. and IV. Coarse and fine myelinated PAFs of cutaneous origin arborize and terminate in these two laminae. Each physiological type of these fibers were labeled with HRP injection, therefore both the distribution (see Brown, 1981) and synaptic connections of individual fibers are now well known. Coarse myelinated fibers terminate in general with R-type boutons.

Hair follicle afferent fibers. The size of the *en passant* and terminal boutons in the cat varied between 1 and 4 μm, they synapses mainly with dendritic shafts

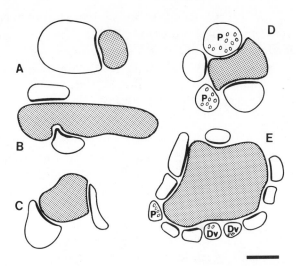

Fig. 2. Schematic drawings of synaptic arrangements in laminae III and IV. The drawings are the copies of electron microscopic pictures taken from published works. The primary afferent boutons are stippled, dendrites in postsynaptic position have an empty contour, vesicle containing profiles represent either axons presynaptic to the primary afferent bouton (P), or vesicle containing dendrites (Dv). For easy comparison all drawings were made on scale. **A.** A simple synapse made by a coarse hair follicle afferent fiber in the cat; **B.** Synapses made by a coarse hair follicle afferent fiber in the monkey; **C.** Synapse made by a Type I slowly adapting fiber; **D.** Glomerular synapse made by a rapidly adapting fiber; **E.** Large size synaptic glomerulus around the bouton of a fine caliber D-hair afferent fiber. Bar: 1 μm.

with asymmetrical synapses (Fig. 2a). Synaptic boutons and also the unmyelinated preterminal axon segments received axo-axonic connections (Maxwell et al., 1982). In another study (Ralston et al., 1984) boutons of G2-type hair follicle afferents in cat and monkey were seen to synapse with both dendritic spines and shafts (Fig. 2b). They were postsynaptic to axon terminals in both species. In addition, boutons in the cat participated in dendro-axonic synaptic connections.

Slowly adapting Type I afferents. The 1-2 μm large boutons form simple axo-dendritic synapses in the cat; glomerular synapses were only occasionally found (Fig. 2c). Vesicle containing structures were seen to contact the primary afferent boutons (Semba et al., 1983). - The same type of axon enlargements were described as contacting 1 to 5 dendritic profiles and frequently receiving synapses from vesicle containing profiles (Bannatyne et al., 1984). In both cat and monkey, slowly adapting Type I afferents terminated in relatively simple axo-dendritic and axon-spine synapses with additional presynaptic components (Ralston et al., 1984).

Slowly adapting Type II afferents. Both the size and synaptic articulation of the boutons were similar to those seen at the Type I fibers; 40% of the terminals received presynaptic boutons. Triadic arrangements were only occasionally seen (Bannatyne et al., 1984).

Rapidly adapting afferents. Descriptions of the ultrastructure of central arborization of PAFs connected with Pacinian corpuscles agreed in three independent studies (Maxwell et al., 1984; Ralston et al., 1984; Semba et al., 1984) in that the 1-2 μm large boutons synapsed with a small number of dendrites and were contacted by vesicle containing profiles (axons and vesicle containing dendrites). Rapidly adapting afferents with receptors other than Pacinian corpuscles (Fig. 2d) ended with similar synaptic structures (Maxwell et al., 1984; Semba et al., 1985).

D-hair follicle afferents. The dense terminal arborization of this type of fine myelinated PAFs arborized in lamina III with boutons ranging from 1.5 to 6 μm (Réthelyi et al., 1982). Characteristically, the labeled boutons were entirely surrounded by dendritic spines and small size axon terminals forming thus typical synaptic glomeruli (Fig. 2e). Among the dendritic spines some contained pleomorphic synaptic vesicles and synapsed with adjacent dendrites. The labeled bouton was often postsynaptic to small size axon terminals containing flattened synaptic vesicles.

2. Terminal structures in the central core

The central core, a cylindrical structure encompassing the intermediate zone of the spinal gray matter (laminae V to VII; see Réthelyi, 1976), is the termination area of three kinds of PAFs. Some fibers in addition to arborizing extensively in the dorsal horn, send collateral branches into the dorsal region of the central core (e.g., rapidly adapting fibers with Pacinian corpuscles and fine myelinated fibers with high threshold mechanoreceptors). Some other fibers, like a large portion of PAFs from muscles arborize and terminate in the central core (see

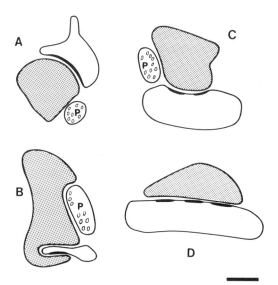

Fig. 3. Schematic drawings of synaptic arrangements in the central core (**A** and **B**) and in the ventral horn (**C** and **D**). The drawings are the copies of electron microscopic pictures taken from published works. The primary afferent boutons are stippled, dendrites in postsynaptic position have an empty contour, vesicle containing profiles represent either axons presynaptic to the primary afferent bouton (P), or vesicle containing dendrites (Dv). For easy comparison all drawings were made on scale. **A.** Synapse made by the deep collateral of a fiber with high threshold mechanoreceptor; **B.** Synapse made by a collateral of a Group Ia muscle fiber; **C.** and **D.** Synapses of Group Ia muscle fibers with motoneuron dendrites. Bar: 1 μm.

Brown, 1981). A third group of PAFs, the Group Ia muscle afferents traverse the central core issuing collateral branches.

Synaptic structure of PAFs with Pacinian corpuscles did not show laminar differences, thus boutons in laminae V and VI have the same connections as those located in the more dorsal laminae (Semba et al., 1984).

In contrast, collateral branches of the fine myelinated fibers with high threshold mechanoreceptors arborizing in lamina V did not terminate in glomeruli, as the parent fibers did in lamina I. Instead, the boutons synapsed mainly with dendritic shafts, or occasionally with irregular dendritic appendages (Fig. 3a). The labeled boutons were often contacted with small size axon terminals with flattened synaptic vesicles (Réthelyi et al., 1982).

Collateral branches of Group Ia muscle afferents terminated in the central core with relatively large size boutons (average diameter is 2.1 µm). They contacted mostly perikarya or dendrites (Fig. 3b), and their majority (up to 92%) received a small size axon terminals which were presynaptic to the labeled boutons (Maxwell and Bannatyne, 1983; Fyffe and Light, 1984).

3. Terminal structures in the ventral horn

Neurons in the ventral horn of the cat's spinal cord receive synapses from different types of axon terminals (Bodian, 1966). In a systematic study in the cat Conradi (1969) identified two types of boutons of presumed primary afferent origin: boutons containing spherical synaptic vesicles (S-type boutons) and large boutons with multiple release sites (M-type boutons). Both the large S-type and all M-type boutons were of large size (1.5 x 3.5-5 µm) and synapsed on the proximal part of the motoneuron dendrites. Conradi's results were substantiated by McLaughlin (1972), while Ralston and Ralston (1979b) have found in an electron microscopic autoradiographic study labeled S-type boutons which synapsed with small diameter dendrites in the monkey ventral horn.

Single HRP-labeled Group Ia muscle afferents were found to terminate in the ventral horn with medium to large size terminals establishing multiple (2-5) synaptic sites with large size dendrites (Figs 3c and d). The boutons contained round synaptic vesicles (S-type boutons) and received synaptic contacts from small size, so called P-terminals (Conradi et al., 1983; Fyffe and Light, 1984).

4. Terminal structures in the CLARKE's column

PAFs terminate in Clarke's column in giant synaptic bulbs which were regularly discernible on silver impregnated light microscopical preparations (Szentágothai and Albert, 1955). Huge, generally elongated boutons up to 4-5 to 8-10 µm in length were found in Clarke's column with later electron microscopical analyses (Réthelyi, 1970; Saito, 1974). They contained round synaptic vesicles and established multiple synaptic contacts mostly with large dendritic shafts (Fig. 4a) and occasionally with perikarya. The giant boutons were often contacted by small size profiles containing pleomorphic synaptic vesicles.

These early observations were confirmed by Walmsley et al. (1985, 1987) who stained Group Ia and Ib muscle sensory fibers individually with HRP and traced

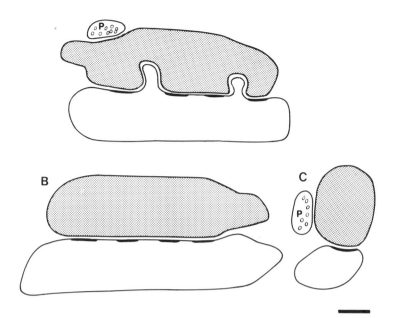

Fig. 4. Schematic drawings of synaptic arrangements in Clarke's column. The drawings are the copies of electron microscopic pictures taken from published works. The primary afferent boutons are stippled, dendrites in postsynaptic position have an empty contour, vesicle containing profiles represent either axons presynaptic to the primary afferent bouton (P), or vesicle containing dendrites (Dv). For easy comparison all drawings were made on scale. **A.** Giant axon terminal synapsing with a dendrite; **B.** Giant axon terminal of a Group Ia muscle afferent; **C.** Cross section of a giant terminal of a Group Ib muscle afferent. Bar: 1 μm.

the collateral branches into Clarke's column. Both types of fibers possessed several large size *en passant* boutons (up to 11.5 μm in length) which synapsed with the dendrites (Fig. 4b). Each bouton formed multiple synaptic sites with the postsynaptic components, and regularly received an axo-axonic synapse (Fig. 4c).

TERMINAL STRUCTURES IN THE DORSAL COLUMN NUCLEI

Terminals of PAFs were described as large boutons synapsing with several dendrites and receiving synaptic connections from vesicle containing profiles. These latter were assumed to be partly axon terminals with flattened synaptic vesicles, partly dendritic profiles with pleomorphic synaptic vesicles (Walberg, 1966; Ellis and Rustioni, 1981). More recently, HRP labeled hair follicle afferent fibers were seen to terminate in the cuneate nucleus with simple axo-dendritic synapses (Fyffe et al., 1985).

CONCLUSIONS

Classical and recent ultrastructural studies revealed that the bulbous enlargements along the branches of PAFs are filled with round synaptic vesicles and at these sites the fibers contacted one or more postsynaptic structures. These latter were usually dendrites or dendritic spines and occasionally perikarya. The numerical ratio between pre- and postsynaptic components at a given synapse showed territorial differences . It seems to depend also on the diameter and peripheral distribution of the parent fiber.

In the central core, ventral horn and in Clarke' column one bouton synapses with one postsynaptic component. In the ventral horn and in the Clarke's column there are several (repeated) synaptic sites between the pre- and postsynaptic components. This latter arrangement can be interpreted as multiple release sites. It is matter of speculation whether or not the probability of transmitter release is identical or different at the release sites within one bouton (see Walsmley et al., 1985). Large diameter PAFs of muscle origin terminate predominantly in the above mentioned three regions of the grey matter.

In the dorsal horn and in the dorsal column nuclei the complex (glomerular) synapses occur intermingled with simple synapses. Large diameter cutaneous PAFs have a tendency to terminate with simple synapses, whereas the terminals of many fine myelinated and unmyelinated fibers form synaptic glomeruli. Altogether it seems reasonable to assume that complex synaptic arrangements are more common in connection with slowly conducting sensory fibers, whereas fast fibers terminate with simple synapses. The significance of this correlation is not clear.

Many primary afferent boutons receive synapses from small size axon terminals and also from vesicle containing dendrites. Small size axon terminals contact the terminals of PAFs in both simple and glomerular synapses, while vesicle containing dendrites were described in the glomerular synapses only. In all probability the small axon terminals belong to neurons with locally arborizing axon trees. Indeed, small size GABA-ergic axon terminals were seen in the dorsal horn terminating on axon terminals (Barber et al., 1978; Maxwell and Noble, 1987). Unfortunately, neither of the single neuron studies furnished any evidence about the cellular origin of the small axon terminals that contact the boutons of PAFs. Neurons with vesicle containing dendrites were located in the superficial dorsal horn (Todd, 1988), but their dendritic processes did not participate in synaptic glomeruli.

After the initial ultrastructural characterization of primary afferent boutons, the more recent studies, using various tracer substances, furnished a wealth of valuable new information. It can be foreseen that by increasing the number of samples and by the application of computer reconstruction of serial sections in the future we may better understand the significance of the structure in both the impulse transmission and presynaptic control at the "first synapse".

REFERENCES

Bannatyne, B. A., Maxwell, D. J., Fyffe, R. E., and Brown, A. G., 1984, Fine structure of primary afferent axon terminals of slowly adapting cutaneous receptors in the cat, *Quart. J. Exp. Physiol.*, 69:547-557

Barber, R. P., Vaughn, J. E., Saito, K., McLaughlin, B. J., and Roberts, E., 1978, GABAergic terminals are presynaptic to primary afferent terminals in the substantia gelatinosa of the rat spinal cord, *Brain Res.*, 141:35-55

Beattie, M. S., Bresnahan, J. C., and King, J. S., 1978, Ultrastructural identification of dorsal root primary afferent terminals after anterograde filling with horseradish peroxidase, *Brain Res.*, 153:127-134

Bodian, D, 1966, Synaptic types on spinal motoneurons: an electron microscopic study, *Bull. Johns Hopkins Hosp.*, 119:16-45

Brown, A. G., 1981, "Organization in the spinal cord," Springer Verlag, Berlin-Heidelberg New York.

Burgess, P. R., and Perl, E. R., 1967, Myelinated afferent fibres responding specifically to noxious stimulation of the skin, *J. Physiol. (Lond.)*, 190:541-562

Carlton, S. M., McNeill, D. L., Chung, K., and Coggeshall, R.E., 1988, Organization of calcitonin gene-related peptide-immunoreactive terminals in the primate dorsal horn, *J. Comp. Neurol.*, 276:527-536

Coimbra, A., Sodre-Borges, B. P., Magalhaes M. M., 1974, The substantia gelatinosa olandi of the rat. Fine structure, cytochemistry (acid phosphatase) and changes after dorsal root section , *J. Neurocytol.*, 3:199-217

Conradi, S., 1969, Ultrastructure and distribution of neuronal and glial elements on the motoneuron surface in the lumbosacral spinal cord of the adult cat, *Acta Physiol. Scand.*, Suppl. 332:5-48

Conradi, S., Culheim, S., Gollvik D., and Kellerth, J. O., 1983, Electron microscopic observations on the synaptic contacts of group Ia muscle spindel afferents in the cat lumbosacral spinal cord, *Brain Res.*, 265:31-39

Gobel, S., 1974, Synaptic organization of the substantia gelatinosa glomeruli in the spinal trigeminal nucleus of the adult cat, *J. Neurocytol.*, 3:219-243

Ellis, L. C., and Rustioni, A., 1981, A correlative HRP, Golgi and EM study of the intrinsic organization of the feline dorsal column nuclei, *J. Comp. Neurol.*, 197:341-367

Fyffe, R. E. W., Cheema, S. S., Light, A. R., and Rustioni, A., 1985, The organization of neurons and afferent fibers in the cat cuneate nucleus, *in:* Development, Organization and Processing in Somatosensory Pathways, M. Rowe and W. D. Willis, Jr., ed., Alan R. Liss, New York, pp 163-173

Fyffe, R. E. W., and Light, A. R., 1984, The ultrasructure of group Ia afferent fiber synapses in the lumbosacral spinal cord of the cat, *Brain Res.*, 300:201-209

Heimer, L., and Wall, P. D., 1968, The dorsal root distribution to the substantia gelatinosa of the rat with a note on the distribution in the cat, *Exp. Brain Res.*, 6:89-99

Honda, C. N., Réthelyi, M., and Petrusz, P., 1982, Preferential immunohistochemical localization of vasoactive intestinal polypeptide (VIP) in the sacral spinal cord of the cat: light and electron microscopic observations, *J. Neuroscience*, 3:2183-2196

Imai, Y., and Kusama, T., 1969, Distribution of the dorsal root fibers in the cat. An experimental study with the Nauta method, *Brain Res.*, 13:338-359

Jancsó, G., and Kiraly, E., 1980, Distribution of chemosensitive primary sensory afferents in the central nervous system of the rat, *J. Comp. Neurol.*, 190:781-792

Knyihár, E., Laszlo I., and Tornyos S., 1974, Fine structure and fluoride resistant acid phosphatase activity of electron dense sinusoid terminals in the substantia gelatinosa Rolandi of the rat after dorsal root transection, *Exp. Brain Res.*, 19:529-544

Knyihár-Csillik, E., Csillik, B., and Rakic P., 1982, Periterminal synaptology of primary afferents in the primate substantia gelatinosa, *J. Comp. Neurol.*, 210:376-399

LaMotte, C., 1977, Distribution of the tract of Lissauer and the dorsal root fibers in the primate spinal cord, *J. Comp. Neurol.*, 172:529-561

Lasek, R., Joseph B. S., and Whitlock D. G., 1968, Evaluation of a radioautographic neuroanatomical tracing method, *Brain Res.*, 8:319-336

Light, A. R., and Perl, E. R., 1979, Spinal termination of functionally identified primary afferent neurons with slowly conducting myelinated fibers, *J. Comp. Neurol.*, 186:133-150

Maxwell, D. J., and Bannatyne, B. A., 1983, Ultrastructure of muscle spindle afferent terminations in lamina VI of the cat spinal cord, *Brain Res.*, 288:297-301

Maxwell, D. J., Bannatyne, B. A., Fyffe R. E. W., and Brown, A. G., 1982, Ulrastructure of hair follicle afferent fiber terminations in the spinal cord of the cat, *J. Neurocytol.*, 11:571-582

Maxwell, D. J., Bannatyne, B. A., Fyffe R. E., and Brown, A. G., 1984a, Fine structure of primary afferent axon terminals projecting from rapidly adapting mechanoreceptors of the toe and foot pads of the cat, *Quart. J. Exp. Physiol.*, 69:381-392

Maxwell, D. J., Fyffe, ER. E. W., and Brown, A. G., 1984b, Fine structure of normal and degenerating primary afferent boutons associated with characterized spinocervical tract neurons in the cat, *Neuroscience*, 12:151-163

Maxwell, D. J., and Noble, R., 1987, Relatiosnships between hairfollicle afferent terminations and glutamic acid decarboxyalse -containing boutons in the cat's spinal cord, *Brain Res.*, 408:308-312

McLaughlin, B. J., 1972, Dorsal root projections to the motor nuclei in the cat spinal cord, *J. Comp. Neurol.*, 144:461-474

Ralston, H. J. III., 1968, The fine structure of neurons in the dorsal horn of the cat spinal cord, *J. Comp. Neurol.*, 132:275-301

Ralston, H. J. III., 1979, The fine structure of laminae I, II and III of the Macaque spinal cord, *J. Comp. Neurol.*, 184:619-641

Ralston, H. J. III., Ralston, D. D., 1979a, The distribution of dorsal root axons in laminae I, II and III of the Macaque spinal cord: a quntitqtive electron microscopic study, *J. Comp. Neurol.*, 184:643-684

Ralston, H. J., and Ralston, D. D., 1979b, Identification of dorsal root synaptic terminals on monkey ventral horn cells by electron microscopic autoradiography, *J. Neurocytol.*, 8:151-166

Ralston H. J. III., Light, A. R., Ralston D. D., and Perl, E. R., 1984, Morphology and synaptic relationships of physiologically identified low-threshold dorsal root axons stained with intraaxonal horseradish peroxidase in the cat and monkey, *J. Neurophysiol.*, 51:777-792

Réthelyi, M., 1970, Ultrastructural synaptology of Clarke's column, *Exp. Brain Res.*, 11:159-174

Réthelyi, M., 1976, Central core in the spinal grey matter, *Acta morph. Acad. Sci. hung.*, 24:63-70

Réthelyi, M., 1977, Preterminal and terminal axon arborizarions in the substantia gelatinosa of cat's spinal cord, *J. Comp. Neurol.*, 172:511-528

Réthelyi, M., Light A. R., and Perl, E.R., 1982, Synaptic complexes formed by functionally defined primary afferent units with fine myelinated fibers, *J. Comp. Neurol.*, 207:381-393

Réthelyi, M., and Szentágothai, J., 1969, The large synaptic complexes of the substantia gelatinosa, *Exp. Brain Res.*, 7:258-274

Réthelyi, M., Trevino, D. L., and Perl, E. R., 1979, Distribution of primary afferent fibers within the sacrococcygeal dorsal horn: An autoradiographic study, *J. Comp. Neurol.*, 185:603-622

Rexed, B., 1952, The cytoarchitectonic organization of the spinal cord in the cat, *J. Comp. Neurol.*, 96:415-466

Saito, K., 1974, The synaptology and cytology of the Clarke cell in nucleus dorsalis of the cat: An electron microscopic study, *J. Neurocytol.*, 3:179-197

Scheibel, M. E., Scheibel, A. B., 1968, Terminal axonal patterns in cat spinal cord. II. The dorsal horn, *Brain Res.*, 9:32-58

Schimert, J., 1939, Das verhalten des Hinterwurzelkollateralen im Ruckenmark, *Z. Anat. Entwickl.*, 109:665-687

Semba, K., Masarachia, P., Malamed, S., Jacquin, M., Harris, S., Yang G., and Egger, M.D., 1983, An electron microscopic study of primary afferent terminals from slowly adapting type I receptors in the cat, *J. Ccom. Neurol.*, 221:466-481

Semba, K,. Masarachia, P., Malamed, S., Jacquin, M., Harris, S., Yang G., and Egger, M. D., 1984, Ultrastructure of pacinian corpuscle primary afferent terminals in the cat spinal cord, *Brain Res.*, 302:135-150

Semba, K., Masarachia, P., Malamed, S., Jacquin, M., Harris, S., Yang G., and Egger, M. D., 1985, An electron microscopic study of terminals of rapidly adapting mechanoreceptive afferent fibers in the cat spinal cord, *J. Comp. Neurol.*, 232:229-240

Sprague, J. M., Ha, H., 1964, The terminal fields of dorsal root fibers in the lumbosacral spinal cord of the cat and the dendritic organization of the motor nuclei *in*: Organization of the spinal Cord, J. C. Eccles, J. P. Schade, ed., Progr. Brain Res., 11:120-152

Sterling, R., Kuypers, H. G. J. M., 1967, Anatomical organization of the brachial spinal cord of the cat. I. The distribution of dorsal root fibers, *Brain Res.*, 4:1-15

Sternberger, L. A., 1986, "Immunocytochemistry," Churchill Livingstone, New York

Sugiura, Y., Lee, C. L., and Perl, E. R., 1986, Central projections of identified, unmyelinated (C) afferent fibers innervating mammalian skin, *Science*, 234:358-361

Szentágothai, J., and Albert, A., 1955, The synaptology of Clarke's column. *Acta Morph. Acad. Sci. Hung.*, 5:43-51

Todd, A. J., 1988, Electron microscope study of Golgi-stained cells in lamina II of the rat spinal dorsal horn, *J. Comp. Neurol.*, 275:145-157

Walberg, F., 1966, The fine structure of the cuneate nucleu in normal cats and following interruption of afferent fibers. An electron microscopical study with particular refernce to findings made in Glees and Nauta sections, *Exp. Brain Res.*, 2:107-128

Walmsley, B., Wieniawa-Narkiewicz, E., and Nicol, M. J., 1985, The ultrastructural basis for synaptic transmission between primary muscle afferents and neurons in Clarke's column of the cat, *J. Neurosci.*, 5:2095-2106

Walmsley, B., Wieniawa-Narkiewitz, E., and Nicol, M. J., 1987, Ultrasructural evidence related to presynaptic inhibition of primary muscle afferents in Clarke's column of the cat, *J. Neurosci.*, 7:236-243

18

STRUCTURAL, FUNCTIONAL AND CYTOCHEMICAL PLASTICITY OF

PRIMARY AFFERENT TERMINALS IN THE UPPER DORSAL HORN

Elizabeth Knyihár-Csillik and Bert Csillik

Department of Anatomy, Albert Szent-Györgyi
Medical University, Szeged, Hungary

ABSTRACT

Structural and functional plasticity of central terminals of primary sensory neurons is regulated by nerve growth factor supplied by retrograde axoplasmic transport to dorsal root ganglion cells. Two important aspects of this regulatory system: transganglionic degenerative atrophy and regenerative synaptoneogenesis are reviewed in view of electron microscopic histochemical and immunocytochemical studies.

INTRODUCTION

Structural and functional plasticity is one of the genuine properties of the neuron; however, the extent and the organizational level of this plasticity varies considerably amongst different kinds of nerve cells. In this respect, pseudounipolar primary sensory nerve cells in dorsal root ganglia (and in the analogous sensory ganglia of cranial nerves) are exceptionally characteristic since blockade of the retrograde axoplasmic transport in their peripheral processes induces spectacular biodynamic alterations of their central terminals. Such degenerative alterations, most conspicuous in the substantia gelatinosa Rolandi (and in homologous structures of the brain stem) after transection of a peripheral sensory nerve, were reported as early as more than one hundred years ago (see historical data in Csillik and Knyihár, 1978a; Csillik and Knyihár-Csillik, 1986; Knyihár-Csillik and Csillik, 1981); but the plasticity of the upper dorsal horn was revealed in full extent only by means of cytochemical, electron microscopical and electrophysiological techniques performed in the course of the last two decades. In this paper, details, causes and neurobiological consequences of "transganglionic degenerative atrophy" (TDA) and "regenerative proliferation" (RP) will be summarized. The second part of this review will be devoted to the enzyme histochemical and immunocytochemical processes that result in the re-establishment of synaptic connectivity lost in the course of TDA. It will be shown that this process involves a straightforward synapto-neogenesis which offers new insights into the mechanism of the still elusive problem of central nervous regeneration.

Premises

According to the law of neuronal trophism, as outlined at the turn of the century, axons and axon terminals are dependent on the trophic substance(s) manufactured by the cell bodies or perikarya (Cajal, 1918). Thus, whenever an axon is transected, the distal stump undergoes a series of destructive cytological and electrophysiological changes called Wallerian degeneration. According to this concept, often referred to as the trophic entity of the neuron, neuroproteins synthesized in the perikaryon are carried distally by axoplasmic flow, so as to replace proteins degraded in the course of the cell's metabolic activity. Therefore, the axon cannot survive long when separated from the perikaryon. At the same time, it has been implicitly assumed that, if the connection of the terminal with the perikaryon was intact, it had no reason to undergo degeneration.

On the basis of light- and electron cytochemical investigations started in our laboratory in the early '70-es (Csillik and Knyihár, 1977b; Knyihár and Csillik, 1976c; Knyihár and Csillik, 1976b; Knyihár and Csillik, 1976a; Knyihár and Csillik, 1977) and confirmed by studies in numerous other laboratories, however, it became obvious that primary sensory neurons represent a remarkable exception from this general rule. It turned out that whenever the anatomical continuity of the peripheral axon of a dorsal root ganglion cell is interrupted (by a mechanical injury, like transection, crush or ligature), histochemical reactions characterizing axon terminals of primary sensory neurons in the upper dorsal horn, like fluoride resistant acid phosphatase (FRAP), thiamine monophosphatase (TMPase; Csillik et al., 1986; Knyihár-Csillik et al., 1986), substance P (SP; Léránth et al., 1984), and lectin-binding carbohydrate epitopes (Fischer and Csillik, 1985; Fischer et al., 1985; Tajti et al., 1988) are depleted in the course of days, weeks or months; depletion is accompanied by fine structural alterations of central primary sensory terminals that bear close resemblance to a slowly proceeding Wallerian degeneration. Therefore, the term transganglionic degenerative atrophy (TDA) was coined to this process, in order to emphasize (i) its transganglionic character and (ii) its similarity to, but distinct difference from, Wallerian degeneration. Also it became obvious that TDA is a transient effect; as soon as the anatomical continuity of the peripheral axon is re-established by successful regeneration, histochemical marker substances are slowly replenished in the Rolando substance, accompanied by a *lege artis* restitution of the original synaptic circuitry, by means of an axonal regeneration and synapto-neogenesis which is unique in the mammalian central nervous system.

MATERIALS AND METHODS

Experiments were performed on CFY (R-Amsterdam) albino rats of both sexes; body weight 200-250 g, and on *Macacus rhesus* monkeys (both sexes; body weight 8-12 kg). A buffered glutaraldehyde-paraformaldehyde solution was used for transcardial fixation for enzyme histochemical studies; Zamboni's picric acid-formalin for immunohistochemical studies; neutral 4% formalin for lectin histochemical investigations and Rakic's aldehyde fixative for electron microscopic investigations. FRAP was demonstrated in serial frozen sections by a slightly modified Gömöri technique (Knyihár-Csillik and Csillik, 1981).

For light microscopic demonstration of TMPase, 40 micron thick serial frozen sections, obtained on a Reichert freezing microtome, were incubated at 37°C for 1 hr in serological agglutination plates, using the formula of Knyihár-Csillik et al. (Knyihár-Csillik et al., 1986), i.e.,: Thiamine monophosphate chloride (Sigma), 1.25% 10 ml; Tris-maleate buffer, pH 5.0 10 ml; $Pb(NO_3)_2$, 0.2% 20 ml; Distilled water 10 ml. Sections were developed in 1% Na_2S to obtain PbS as a visible end-product of the reaction. Evaluation of the extent of the TMPase reaction in substantia gelatinosa Rolandi was performed by ocular micrometry of the rectilinear projections of the reactive areas; data were consequently fed into a Commodore Type 64 personal computer (Bezzegh et al., 1986). Tricolored maps generated this way were photographed from the computer's monitor.

For electron microscopic cytochemical purposes, 35 micron thick Vibratome slices were incubated in a TMPase medium as above; however, instead of development in Na_2S, they were transferred directly into 1% s-collidine-buffered osmic acid where they were postfixed for 1 hr at 4°C. Samples dehydrated and processed according to the usual electron microscopic technique, were embedded in Durcupan ACM and sectioned on a Reichert Ultrotome, using glass or diamond knives. Silver interference color sections were collected on Formvar-coated 100-mesh copper grids, stained with lead citrate and studied on Tesla 100 B or Opton (Zeiss) electron microscopes at 60 kV or 80 kV accelerating voltage.

For the immunocytochemical visualization of SP and vasoactive intestinal polypeptide (VIP), the standard technique developed in this laboratory, based on earlier studies, consisted of the following steps:

(1) transcardial flush with isotonic saline, in order to remove blood from the vessels;

(2) fixation by transcardial perfusion of the fixative solution: Zamboni's fixative (for SP) or 4% paraformaldehyde (for VIP), followed by dissection of the lumbosacral spinal cord;

(3) post-fixation for 1-3 hrs in Bouin's fixative;

(4) rinsing in PBS for 4 hrs;

(5) serial sectioning on a Reichert freezing microtome.

Sections were processed according to the "sessile-drop" system (Nadelhaft, 1984), using Parafilm-coated slides for support. As a rule, 8 sections were processed on a single slide, amounting to 96 serial sections on 12 slides.

Application of primary and secondary sera and development of the reaction with H_2O_2-DAB was performed with Finpipe automatic pipettes. Antisera for SP and VIP were obtained from Amersham Inc., England; normal sera were purchased from HUMAN, Hungary.

Before the evaluation of the immunocytochemical reaction, sections were serially photographed at low power. The intensity of the immunocytochemical reaction was assessed by the 10-40 Electronic Image Analyser, using the Shade Comparison System. Lengths of the rectilinear projections of normally reacting areas (80-100%), those of decreased reaction (20-80%) and those of depletion (0-20%) were measured either by ocular micrometry or by using the same Image Analyser System. The values were fed into a Commodore Type 64 personal computer, equipped with a MUDH program to build up three-coloured maps of the lumbar dorsal horn.

Experimental surgery was performed under Nembutal and/or ether anesthesia in rats, and in Halothan anesthesia in monkeys.

Unilateral transection of the lumbar dorsal roots was performed intrathecally, after laminectomy. The sciatic nerve was transected at midthigh level, at about 6 mm below the sciatic notch. Crushing of the sciatic nerve (axocompression) was done with watchmaker's tweezers, applying 3 powerful compressions 3 times in succession, which results in a flattened and translucent portion of the nerve, about 3 mm long. Utmost care was taken not to disrupt the anatomical continuity of the nerve; if this inadvertently occurred, the animal was discarded. After surgery, the skin was sutured and the hind legs were painted daily by 1% chinine chloride solution in order to prevent self-mutilation.

RESULTS

1. Transganglionic Degenerative Atrophy

Transganglionic degenerative atrophy (TDA) is a recently (re-) discovered trophic phenomen peculiar to primary sensory neurons that apparently contradicts the maxims of the neuron theory (Csillik, 1980). When a peripheral sensory axon is damaged (transected, crushed or ligated), dramatic alterations occur not only in the distal stump resulting in Wallerian degeneration, but also in the central terminals of the affected primary sensory neuron that undergo structural, cytochemical and functional impairment (Fig. 1). Obviously related to the metabolic alterations within the perikaryon (manifested in the "axon reaction" or chromatolysis of the Nissl substance), TDA, similar in many respects to a slowly proceeding Wallerian degeneration (Arvidsson, 1979; Arvidsson and Grant, 1979; Gobel, 1984; Gobel and Binck, 1977; Grant and Ygge, 1981; Johnson and Westrum, 1980; Johnson et al., 1983; Risling et al., 1983; Westrum and Canfield, 1977; Westrum et al., 1984), is characterized by the following signs:

(1) Vesiculolysis (Fig. 2) which results in an electron lucent terminal axoplasm (light type of TDA), starts on the 4th postoperative day (Knyihár-Csillik and Csillik, 1981). The swollen electron lucent terminals form, in later stages, intricate axonal labyrinths that consist of spirally wound, onion-sheet-like, flattened axoplasmic profiles (Fig. 3) sometimes alternating with flattened

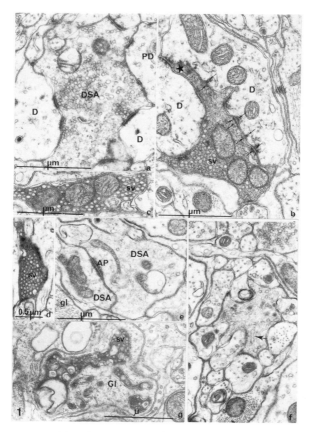

Fig. 1. A. Dense sinusoid (DSA) axon terminal in the substantia gelatinosa at the L6 level of the spinal cord in a normal adult rhesus monkey. Note numerous larger-than-usual synaptic vesicles in this type of dorsal root ending. D: dendrite; PD: presynaptic dendrite. **B.** Early stage of the electron-dense type of TDA. Note increased density of the terminal. Synaptic vesicles (sv) remain preserved only in the lefthand portion of the terminal. The rest of the terminal is empty although membrane specializations indicative of axodendritic synapses remain visible (arrows) without concentration of vesicles at the presynaptic sites. Characteristic tongue-like elongation is presumably caused by gradual shrinkage of the terminal (asterisk). **C.** Advanced stage of osmiophilic shrinkage with synaptic vesicles (sv) that remain distinguishable three months after peripheral axotomy. **D.** Dorsal root terminal in the substantia gelatinosa in advanced stage of osmiophilic shrinkage. Three months after peripheral axotomy, synapatic vesicles are still clearly discernible (sv). The structural changes are similar to those seen in early stages of Wallerian degeneration. **E.** Electron lucent light type of TDA: two dense sinusoid (DSA) axon terminals in the substantia gelatinosa at the L6 level of the spinal cord in an adult monkey sacrificed one month after transection of the sciatic nerve. Note conspicuous decrease of numbers of synaptic vesicles (depletion) and the fine-granular matrix material that fills both DSA. AP: attachment plaque. The two terminals attached with such plaques usually belong to the arborization of the same afferent fiber. **F.** Electron lucent (light) type of TDA. Note increased number of beaded tubular structures (arrow) within the atrophic terminal. One month after transection of the sciatic nerve. **G.** Engulfment by a glial cell of the osmiophilic debris of a primary sensory terminal in the Rolando substance, three months after transection of the sciatic nerve. Gl: glial cytoplasm; sv: synaptic vesicles in the shrunken fragments of the terminal are still discernible. This kind of glial phagocytosis is unfrequent in the course of transganglionic degenerative atrophy.

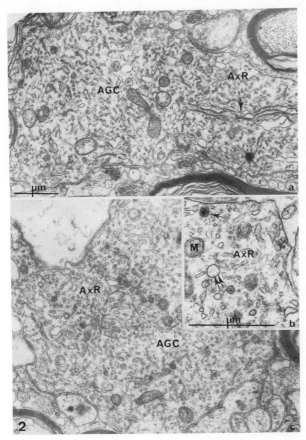

Fig. 2. Axonal growth cones (AGC) of sprouting primary cenral afferent terminals in the upper dorsal horn of the adult rhesus monkey, three months after crushing the segmentally related, ipsilateral peripheral nerve (sciatic). Note large amounts of axoplasmic reticulum (AxR) consisting of densely packed cisterns of sER (arrow in **A**) characterizing axonal growth cones. Arrow in **B** points at a dense-core vesicle; double arrow indicates growth cone vesicle.

dendritic and glial elements (Knyihár and Csillik, 1976c). In the monkey spinal cord, the number of such axonal labyrinths is 300/square microns in the related upper dorsal horn, 3 months after transecting the sciatic nerve (Fig. 4).

(2) The dark type of TDA (Fig. 2) results in shrunken electron-dense osmophilic terminals which are occasionally engulfed by glial elements; glial phagocytosis is, however, far less common than in the course of Wallerian degeneration (Knyihár-Csillik et al., 1987).

(3) The marker enzymes FRAP (Knyihár-Csillik and Csillik, 1981) and TMPase (Knyihár-Csillik et al., 1986) that characterize nociceptive neurons (Hanzely et al., 1983; Szönyi et al., 1979) are depleted from primary afferent central terminals in lamina II (in rat and mouse) on days 4-6 after surgery (Devor and Claman, 1980; Ferencsik, 1986; Fitzgerald and Swett, 1983; Fitzgerald and Vrbova, 1985; Kovacs and Ferencsik, 1986; Mihaly et al., 1980).

(4) Depletion of putative transmitter neuropeptides like substance P (SP) and somatostatin (SOM) from the inflicted primary afferent terminals in Lamina I and II also starts on the 4th-6th postoperative day (Barbut et al., 1981; Csillik et al., 1984a; Jessell et al., 1979; Léránth et al., 1984).

(5) Lectin-binding glycoconjugates are depleted from primary central terminals in Lamina IIi and III in humans and in Lamina I and II in rodents 2-4 weeks postoperatively (Fischer and Csillik, 1985; Fischer et al., 1985; Tajti et al., 1988).

(6) The dorsal root potential (DRP) undergoes conspicuous alterations; its latency is increased, its amplitude is decreased and its duration is reduced. The dorsal root reflex disappears (Horch and Lisney, 1981; Pór, 1985; Woolf and Wall, 1982).

All the above mentioned structural alterations (1-5) characterize only the intraspinal portions of the central branch surrounded by a glial environment. The dorsal roots proper, where axons are surrounded by Schwann cells, do not show any alterations. The changes in DRP reflect alterations in the synaptic circuitry of the upper dorsal horn, rather than those in the dorsal root itself.

TDA can be evoked not only by mechanical injury of a peripheral sensory nerve, but also by blockade of retrograde axoplasmic transport by micortubule inhibitors (Csillik et al., 1977a; Fitzgerald et al., 1984). Extremely small amounts of vinblastin, vincristin, vindesin, leurosin, formyl-leurosin, podophyllotoxin or griseofulvin, that do not block anterograde transport and thus do not cause Wallerian degeneration (e.g., 10^{-8} M vinblastin or 10^{-9} M vincristin) are sufficient to induce TDA. Fine structural, histochemical and electrophysiological consequences of TDA induced by blockers of axoplasmic transport do not differ in any respect from TDA induced by mechanical injury of the peripheral axon.

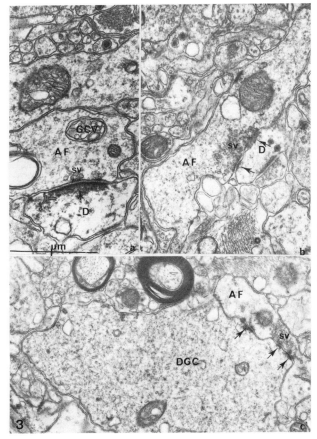

Fig. 3. Filopodia (AF) of axonal growth cones establish synapses with dendritic elements (D). Note accumulation of growth cone vesicles (GCV) and varying numbers of synaptic vesicles (sv) in axonal filopodia. Synapses are marked by arrows. In C, the postsynaptic element is a dendritic growth cone (DGC).

TDA, a reversible process, should not be mistaken for transganglionic degeneration, a Wallerian degeneration of dorsal roots and their terminals, which follows cell death in the dorsal root ganglion (Aldskogius et al., 1985b; Aldskogius et al., 1985a; Arvidsson, 1979; Arvidsson and Grant, 1979; Grant and Ygge, 1981).

Circumstantial evidence suggests that the very causative factor in TDA is the blockade of the retrograde transport of nerve growth factor (NGF; see Discussion).

2. Restoration of Connectivity: Regenerative Proliferation

In striking contrast to the Wallerian degeneration that follows dorsal rhizotomy, TDA in the upper dorsal horn is a reversible process. As soon as retrograde transport of NGF is re-established in the peripheral sensory axon (either by axonal regeneration or by revival of microtubule function), a structural and cytochemical reorganization of full functional value takes place in the dorsal horn, involving regenerative synapto-neogenesis.

Regenerative synapto-neogenesis in the upper dorsal horn starts with the formation of axonal growth cones (Knyihár-Csillik et al., 1985) similar to those observed in the course of embryonic development of the spinal cord (Csillik and Knyihár-Csillik, 1986). Axonal growth cones contain large amounts of smooth axoplasmic reticulum (Fig. 5) which is also present in preterminal myelinated and nonmyelinated portions of the reactivated dorsal root axons (Fig. 6). Filopodia emanating from axonal growth cones establish synapses with dendritic growth cones of intraspinal cells (Fig. 7); this process leads to a complete restoration of the original fine structure and synaptic connectivity of the upper dorsal horn (Fig. 8), which had been derailed by the effect of TDA (Knyihár-Csillik et al., 1986; Majumdar et al., 1983).

Regenerative synapto-neogenesis in the upper dorsal horn is accompanied by cytochemical replenishment of primary central afferent terminals. FRAP and TMPase, which were depleted in the course of TDA, reappear again in the regenerating terminals (Csillik and Knyihár, 1975; Csillik and Knyihár, 1978a; Csillik and Knyihár-Csillik, 1981c; Csillik and Knyihár-Csillik, 1981a; Csillik and Knyihár-Csillik, 1982b; Ferencsik, 1986; Fitzgerald, 1985b). The same applies also for substance P and lectin-binding glycoconjugates. Also functional properties of the formerly incapacitated upper dorsal horn are restored: original parameters of the dorsal root potential (latency, amplitude and duration) gradually return to normal values following regenerative synapto-neogenesis (Horch and Lisney, 1981; Pór, 1985). Circumstantial evidence suggests that regenerative synapto-neogenesis in the upper dorsal horn is subjected to transganglionic regulation by NGF, which is taken up by peripheral receptors at the nerve endings of sensory axons and carried by means of a retrograde axoplasmic transport mechanism to their perikaryon in dorsal root ganglia. Here, NGF promotes synthesis of neuroproteins by means of a second messenger.

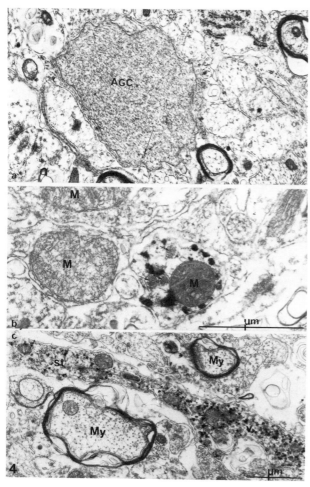

Fig. 4. Electron cytochemical patterns of regenerating TMPase active terminals in the lumbar Rolando substance, 45 days after crushing the ipsilateral sciatic nerve.
A. In axonal growth cones (AGC), only very few, scattered grains of the enzyme reaction end product can be seen (arrow). In preterminal sprouts, both in cross sections (**B**) and in longitudinal sections (**C**) TMPase enzyme reaction is present in high numbers of intraaxonal grains. M: mitochondria. Arrows in (**B**) point at large hatched-core vesicles characterizing regenerating central axons. In (**C**), st stands for the thin stalk of the regenerating sprout, while v designates a varicous swelling which will give rise later to the formation of a sinusoid terminal. My: myelinated axons, devoid of any TMPase activity.

3. Thiamine Monophosphatase Histochemistry

An important development in recent enzyme histochemical investigations of TDA and regenerative proliferation is the use of the specific marker enzyme TMPase. Described originally by Ogawa et al. (Ogawa et al., 1982), TMPase was shown to be a genuine marker of a class of primary sensory neurons, central processes of which terminate in the substantia gelatinosa Rolandi (Knyihár and Csillik, 1976b). Selectivity and specifity of the TMPase reaction is due to two important features of this enzyme, viz. (i) TMPase is contained only in Type C (small) dorsal root ganglion cells and not in any other neurons of the central or the peripheral nervous system, and (ii) TMPase is not present in any other cell organelles except for the rER of small dorsal root ganglion cells (where it is synthesized), their Golgi apparatus (where it is translocated), and at the surface membranes of peripheral and central axons and axon terminals of these cells where it is deposited after axoplasmic transport.

Due to the fact that TMPase is present only and exclusively in perikarya, axons and axon terminals of a well-defined group of dorsal root ganglion cells, this enzyme is superior as a marker for degenerative and regenerative processes of dorsal root terminals as compared to various neuropeptides, like substance

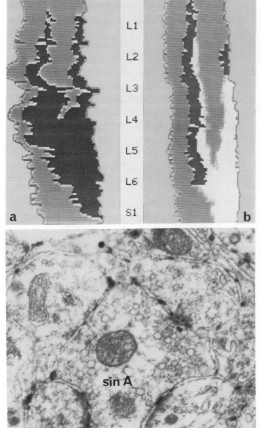

Fig. 5. A. Computer-aided reconstruction of TMPase depletion from the lumbar Rolando substance, 15 days after crushing the ipsilateral sciatic nerve, based on measurements of the rectilinear proection of the TMPase active area in 196 serial sections. Gray: persisting TMPase activity; black: depletion. Note persistence of the marker enzyme in the lateral aspect of the dorsal horn (=projection of the unimpaired postaxial dermatome) and in the tongue-like peninsula intruding into the depleted area (=projection of the unimpaired lumbar plexus).
B. 30 days after sciatic crush. White area represents regeneration which extends according to caudo-rostral and medio-lateral gradients; black and gray: as above.
C. Translocation of TMPase activity in the course of regenerative proliferation, from the initial axoplasmic reaction into a membranous localization, of a recently formed sinusoid axon terminal (sin A). Final result of maturation will be a completely membrane-bound TMPase reaction, similar to that seen under normal conditions, as shown in Fig. 10.

P, known to be present, in addition to dorsal root ganglion cells, also in intraspinal elements and in descending systems (Ljungdahl et al., 1978). On the other hand, since, in contrast to FRAP, TMPase is not exerted by lysosomes at all, there is no need for fluoride inhibition, which, affecting the intensity of the genuine FRAP reaction, may prove detrimental in regeneration studies, where we are dealing with a hardly visible, slight enzyme activity of young axonal sprouts.

The functional importance of TMPase, except for its role in releasing thiamine (vitamine B_1) from its phosphate ester, and, in this context, its contribution to the assumed dual role of thiamine in axonal conduction (Dreyfus, 1985) and the still enigmatic purinergic transmission mechanism (Burnstock, 1978), is essentially unknown.

Plasticity of central primary afferent terminals as revealed by TMPase cytochemistry is evident from the following observations:

(1) Crush injury of the sciatic nerve results, just like transection of the nerve, within 5-6 days in disappearance of TMPase activity from the ipsilateral Rolando substance in the lumbar intumescence. Depletion of TMPase is most distinct in

L5 and L6; it extends, however, as high as L2 in a dichotomized manner (Fig. 5). The tongue-like peninsula of unchanged TMPase reactivity represents the projection of the (unimpaired) lumbar plexus. Also TMPase activity in the lateral edge of the Rolando substance is unchanged since this represents the projection of the unimpaired postaxial dermatome.

(2) Beginning with the fourth postoperative week, a fine granular TMPase reaction starts to appear within the zone of depletion. Replenishment of the TMPase activity proceeds in caudo-rostral and medio-lateral gradients (Fig. 4).

(3) While in the normal, unimpaired Rolando substance, TMPase activity is confined to axolemmal membranes of terminal and preterminal axons, the reaction product is localized within the axoplasms of the regenerating profiles (Fig. 4).

(4) Axonal growth cones in the area of TMPase replenishment contain high amounts of smooth axoplasmic reticulum, tubular elements, several growth cone vesicles but only a few grains of the end product of the TMPase reaction (Fig. 4).

(5) In the course of maturation, intra-axonal TMPase reaction of regenerating axons tends to be translocated to the axolemmal membrane. At the 60th postoperative day, TMPase is present again only at the axolemnal surface membranes. Translocation of TMPase (Knyihár-Csillik et al., 1989) suggests that a hitherto unknown "anchoring factor" might be responsible for maintaining TMPase at the external surface of axolemmal membranes which is absent in early stages of regeneration (Fig. 5).

4. Substance P Immunocytochemistry

In spite of apparent analogies in origin and localization, SP presents remarkable differences to other genuine marker substances of primary sensory neurons with regards to biodynamic reactions. Dorsal rhizotomy, inducing Wallerian degeneration of primary central afferent terminals, results in depletion of the bulk of SP but, due to intraspinal and descending SP-labeled neurons, depletion of SP is never as complete as that of FRAP and especially TMPase, which latter is exerted only and exclusively by primary sensory neurons. The same applies also to peripheral nerve transection; while axotomy results in complete disappearance of FRAP and TMPase from the medial two-thirds of substandia gelatinosa (i.e., from the area of central terminals of neurons innervating the preaxial dermatome), it induces only a partial depletion of SP even in this medial aspect. (The lateral edge of the upper dorsal horn, i.e., the representation of the postaxial dermatome, innervation of which is not impaired by sciatic nerve transection, stays unchanged in both cases.) Similar alterations can be induced by blockade of axoplasmic transport in the peripheral nerve (Léránth et al., 1984).

Reports on the effect of peripheral nerve crush are more controversial. While nerve crush (axocompression) undoubtedly results in Wallerian degeneration in the peripheral stump and, accordingly, induces depletion of the marker enzymes FRAP and TMPase, reportedly it does not affect SP content of the upper dorsal horn (Barbut et al., 1981; McGregor et al., 1984; Wall, 1981b). This

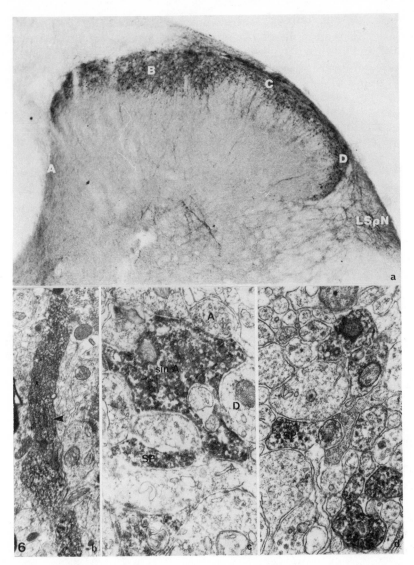

Fig. 6. Substance P in the rat spinal dorsal horn. **A.** Areas of SP reactivity (A, B, C, D and LSpN). **B.** Electron microscopic pattern of SP in axons: end product of the immunocytochemical reaction can be seen to be arranged alongside axonal microtubuli (arrowhead). **C.** SP-positive scallopped axon terminal (sin A), in synaptic contact with dendritic (D) and axonal (A) profiles. Such scallopped terminals are less frequent than simple forms of SP-positive axons (SP). **D.** The great majority of SP-positive structures are simple axons and axonal dilatations.

seemingly minor discrepancy is of major importance with respect to regeneration and the concomittant replenishment of central afferent terminals. Reinvestigation of this problem with a powerful technique of densitometric evaluation of serial immunocytochemical sections by means of an Image Analyser and a computer-aided reconstruction of SP in the lumbar intumescence of the spinal cord, has lead to the following conclusions.

4.1 Normal distribution of SP in the lumbar upper dorsal horn

The pattern of SP-reactivity has been described by Hökfelt et al. (1975, 1976) by means of FITC labeled antiserum, and by Barber et.al. (1979) and Seybold and Elde (1980) with the PAP reaction, and confirmed by numerous studies during the last decade. An accurate topographical analysis in terms of light microscopic anatomy is, however, still missing. In the following, normal distribution of SP in the upper dorsal horn will be described in a medio-lateral sequence (Fig. 6).

Area A consists of a few discretely curved SP axons which mark the border-line between the dorsal column and the head of the dorsal horn.

Area B consists of a homogeneous dense feltwork of SP-positive axons and terminals. This "ledge" occupies nearly half of the upper spinal cord, and comes to an end abruptly.

Area C extends laterally from the "edge of the ledge". The width of the SP-positive area is markedly reduced; it occupies only Lamina I. However, several SP-positive axons running perpendicularly (or rather radially) in the transverse plane of the cord, connect Lamina I and the innermost layer of Lamina II, like stalactites and stalagmites.

Area D, the lateralmost region occupied by SP-positive nerve fibers in the upper dorsal horn, consists of a compact network of axons proceeding mainly transversely in the cross sectional plane of the spinal cord. Bending or rather curling in medially, it tapers off successively; it represents central branches of dorsal root ganglion cells innervating the postaxial dermatome. However, representation of pre- and postaxial dermatomes is less clear-cut than in the case of FRAP or TMPase, since central terminals of SP-positive dorsal root ganglion cells are scattered throughout the entire upper dorsal horn.

Area L (or LSpN, i.e., the "Lateral Spinal Nucleus" or NDLF, i.e., "Nucleus of the Dorso-Lateral Fascicle") is essentially identical with Cajal's "noyeau inter-stitiel". Here, relatively thick SP-positive axons, partly of dorsal root origin, equipped with large bead-like varicosities establish axosomatic and axodendri-tic synapses with large multipolar nerve cells scattered in the dorsolateral aspect of the white matter of the spinal cord. SP-positive axons interconnecting the lateral spinal nucleus with the nucleus proprius of the dorsal horn, represent afferent elements of the nucleus proprius.

4.2 Effect of dorsal rhizotomy

Unilateral rhizotomy of all six lumbar dorsal roots was performed in 4 rats. In serial sections, a uniform and very marked decrease of SP reaction was seen throughout the entire lumbar spinal cord. 6 days after the surgery, depletion of SP is evident in Areas A, B, C and D as well. However, absorbency of the lateral spinal nucleus (Area L) is intriguing: it shows a peculiar alternation with a 200-300 micron periodicity, insomuch as a markedly reduced SP content is apparent in 5-7 sections which is followed by a virtually unchanged reaction in the next 5-7 sections. At the same time, whenever absorption values of Area L are unchanged, reaction of the contralateral LSpN is conspicuously decreased. Absorbency values obtained from densitometric measurements in 96 successi-

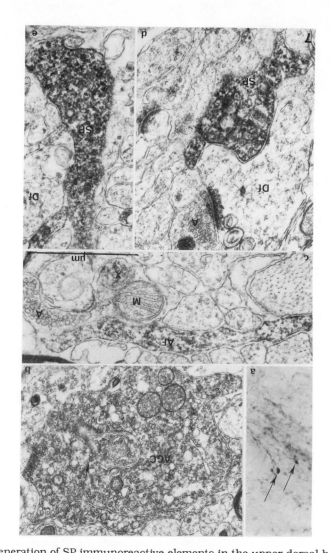

Fig. 7. Regeneration of SP immunoreactive elements in the upper dorsal horn, 25 days after crushing the ipsilateral, segmentally related peripheral nerve.
A. Under light microscopic high power, SP-positive axonal growth cones (arrows) and nerve sprouts can be seen mainly in sagittal sections. **B.** Axonal growth cone (AGC) crowded by vesicles of various sizes, exerts slight SP immunoreactivity (arrow). **C.** Axonal filopodium (AF) emanating from growth cone, contains SP. At X, another SP-positive filopodium can be seen in cross section. M: mitochondrion; A: axon devoid of SP reaction. **D.** SP-positive axonal sprout establishing synapse with a dendritic filopodium (Df) which is engaged also in synaptic contact with an SP-negative axon (A). **E.** SP-positive dichotomizing axonal sprout adjacent to a dendritic filopodium (Df).

ve serial sections are summarized in Table I. Decreased SP content of the dorsal horn does not return to normal even after 2 months postoperatively. Densito-metric measurements of the extinction exerted by SP reacting elements by means of the shade analysis system of the Image Analyser, reveals relative optical density in arbitrary units. In this process, 96 serial sections obtained from normal material were evaluated; in each case, the averages of three vicinal approximately 2000 squaremicrons rectangles of the five main areas (A, B, C, D and L) of the upper dorsal horn, were determined. For the sake of comparison, the highest absorbency value obtained in normal material was regarded as 100%.

4.3 Effect of sciatic nerve transection

Serial sections were obtained in 4 experiments, on postoperative days 11 and 19. The densitometric values obtained from the serial sections are shown in Table I.

4.4 Effect of sciatic nerve crush

Alterations induced by neurocompression, i.e., disruption of the cytological (and physiological) continuity of axons without any major anatomical displacement, are essentially identical to those observed after nerve transection. However, the force exerted by the crushing instruments plays a hitherto unsuspected role in the effect induced in the peripheral stump of the nerve, and, in consequence, also in the central terminals of the inflicted primary sensory neurons.

If applying a standardized crush, controlled by the pinch reflex test (Gutman et al., 1942), the effect of nerve crush is nearly, or, in some cases, completely identical to that of nerve transection. Though the densitometric values are slightly higher than after nerve transection, these differences did not prove statistically significant. Depletion of SP immunoreactivity starts on the 5th postoperative day. Densitometric values obtained on the 11th postoperative day are shown in Table I. These values stay virtually unchanged until the 19th postoperative day. Later on, however, an ostentatious re-establishment of the SP immunoreactivity can be observed which, on the 30th postoperative day, reaches a statistically significant level. The values observed are shown in Table I.

TABLE I

Effects of dorsal rhizotomy, axotomy (transection), axocompression (crush) and peripheral axon regeneration on SP-like immunoreactivity of the upper dorsal horn

	Percentage ± s.e.m. of control average (100%)					
	A	B	C	D	LSpN	x
Rhizotomy (6 D)	18±8	22±9	26±9	34±6	90±9	51±8
Axotomy (11 D)	22±8	35±10	48±9	92±8	78±9	
Crush (11 D)	36±6	44±8	54±9	98±18	62±12	
Regeneration (30 D post crush)	51±9	47±10	72±12	102±20	107±18	
Regeneration	56±7	105±11	91±9	64±8	98±11	

x - sections where LSpN showed markedly decreased reaction

Re-establishment of SP immunoreactivity after crush injury of the segmentally related ipsilateral sciatic nerve never results in totally normal values throughout the entire width of the dorsal horn. This is evident one year after nerve crush; the unimpaired contralateral side serves as self-control (Table I).

4.5 Fine structural correlates of SP replenishment in the upper dorsal horn, following crush-induced depletion

The conspicuous increase in the SP reactivity in the course of the fourth week after sciatic nerve crush, is due to the appearance of young SP-positive nerve sprouts. Such sprouting can most conveniently be studied in horizontal sections of the cord. Fine varicous axons, equipped with axonal growth cones, can be seen in large numbers. Electron microscopy reveals axonal growth cones exerting only slight, if any SP immunoreactivity; filopodia and young axons, however, are SP-positive (Fig. 7).

4.6 Alterations in VIP reactivity of the upper dorsal horn after peripheral nerve transection

Using the methodological parameters outlined above, only slight VIP reactivity can be seen in the upper lumbar spinal cord under normal conditions. In contrast, numerous young axons exerting VIP immunocytochemical staining appear 21 days after transection of the sciatic nerve. These axons are invariably seen in Area A and (mainly) in Area B, alongside the thick myelinated fiber tracts traversing the upper dorsal horn in an arcuate manner. No VIP-positive axons could be seen after crushing the sciatic nerve; neither did VIP-positive axons appear after dorsal rhizotomy.

In order to prove the origin of VIP-positive axons, dorsal roots L2, L3 and L4 were transected in two rats 25 days after transection of the ipsilateral sciatic nerve. 5 and 7 days after rhizotomy, no VIP-positive axons could be seen at all, suggesting their Wallerian degeneration and, *eo ipso*, their dorsal root origin.

5. Immunocytochemical considerations

Though SP is located in dorsal root ganglion cells differing from those exerting FRAP (and/or TMPase) reaction (Nagy and Hunt, 1982), it has been proposed long ago that after peripheral nerve transection, the nociceptive neuropeptide SP (Cuello, 1987) behaves similarly to the above marker enzymes. While a dramatic SP depletion after dorsal rhizotomy was reported by Takahashi and Otsuka (1975) on the basis of bioassay studies, Jessell et al. (1979) were first to show, by means of radioimmunoassay, that transection of the sciatic nerve results in 74% decrease of SP content of the lumbar spinal cord of the rat. Barbut et al. (1981) confirmed this result by the immunocytochemical approach. It is difficult to accept, however, that "the lateral dorsal horn seems to contain less SP immunoreactive material than medial dorsal horn". It is evident from our present studies that regional changes induced by various surgical interventions can be evaluated only by using densitometry of serial sections. Decreased SP content of dorsal horn after peripheral nerve section was reported also by further studies (Léránth et al., 1984; McGregor et al., 1984; Tessler et al., 1985; Wall et al., 1981a) even though, until now, no effort was done to overcome the bias inherent in the visual inspection of randomly selected sections.

Regarding the effect of peripheral nerve crush upon dorsal horn SP reactivity, Barbut et al. (1981) reported "no signs of change", or only "equivocal" or "minor" alterations. This peculiar resistance of spinal SP to peripheral nerve crush offered wide perspectives for speculation. Our results unequivocally prove that crush lesion of the sciatic nerve produces depletion of SP comparable to, though not identical with the effect of peripheral nerve transection. Undoubtedly, several thin peripheral axons (both sympathetic and primary afferent C fibers) may survive the crush injury; such "resistant" non-myelinated axons were observed, described and depleted 60 years ago by Cajal (1918) in the peripheral stump. Obviously, central terminals of such neurons escape TDA and, accordingly, may contribute to the persistence of SP reactivity in the dorsal horn. Therefore, randomly chosen specimens may lead to erroneous conclusions as to the inefficacy of the crush lesion, especially if a less powerful, "mild" crushing procedure was applied.

Since, however, controlled crush lesions *do* induce a marked depletion of SP in the upper dorsal horn, reappearance of SP during the fourth postoperative week has to be due to axonal regeneration. In other words, the behaviour of SP neurons is, essentially, analogous to that of FRAP (and/or TMPase) neurons. In this respect, it is worth noting that axonal growth cones contain only small, if any, amounts of the marker substances TMPase and SP. The process of SP regeneration seems to follow the same laws like FRAP and/or TMPase: it obeys caudo-rostral and medio-lateral gradients.

The fact that SP neurons are subjected to transganglionic regulation is supported by earlier studies proving that NGF, if applied to the proximal stump of a transected nerve, prevents SP depletion from the upper dorsal horn (Csillik, 1984b) and even results in an increased SP content (Fitzgerald et al., 1985a). NGF-dependence of SP-positive dorsal root ganglion cells is supported also by the studies of Kessler and Black (1981) proving that NGF stimulated development of SP in embryonic spinal cord, and by those of Mayer et al. (1982) indicating that antibodies against NGF produce a marked, albeit reversible reduction of SP in dorsal root ganglia of newborn rats. Regenerative propensity of central terminals of SP-positive primary sensory neurons offers new perspectives in the study of factors contributing to CNS regeneration.

Increased VIP reactivity in the upper dorsal horn after peripheral nerve transection had been already noted and described by McGregor et al. (1984) and by Shehab and Atkinson (1986). Our studies prove that these VIP-positive axons are of dorsal root origin. Since there is no indication for any structural sign of ingrowth of axons in the dorsal horn following neurotomy, it stands for reason to assume that some other mechanism might be responsible for this spectacular alteration. The idea seems germane that transmitter plasticity at the molecular level of synthesis might be responsible for the switch in the perikaryal neuropeptide synthesizing machinery, as it has been suggested by Black et al. (1984) in context to other transmitters.

DISCUSSION

Transganglionic regulation is a recently discovered dynamic and plastic gene expression phenomenon of primary sensory neurons. It offers a unique model to study neural plasticity and regeneration in the central nervous system (Csillik et al., 1984c) and obviously calls for reassessment of the law of neuronal trophism. The regulatory role of NGF in the maintenance of central terminals of primary sensory neurons is proved by the following basic features of TDA:

(1) TDA is sequel of blockade of axoplasmic transport mechanism.

This follows from the fact that application of small amounts of microtubule inhibitors around peripheral nerves, induces TDA in the segmentally related ipsilateral Rolando substance (Csillik and Knyihár, 1976; Csillik et al., 1977a; Csillik et al., 1978b; Csillik et al., 1984c). While such small amounts of microtubule inhibitors do not induce Wallerian degeneration in the nerve, higher concentrations of microtubule inhibitors, however may do so.

(2) Blockade of the (fast) retrograde component of axoplasmic transport is responsible for TDA.

Application of 10^{-9} M vincristin or 10^{-8} M vinblastin to the dorsal roots does not induce FRAP depletion in the segmentally related upper dorsal horn. In contrast, application of 10^{-3} M *Vinca* alkaloid around the root, results just like dorsal rhizotomy in rapid disappearance of the FRAP reaction (Csillik et al., 1982a).

(3) TDA involves only terminal and preterminal portions of central sensory axons; the dorsal root proper is not affected.

Neither transection of the sciatic nerve, nor perineural application of microtubule inhibitors, induce degeneration or atrophy in the Schwann-cell surrounded portion of the dorsal root, either at light or at electron microscopic levels (Aldskogius and Risling, 1983; Aldskogius et al., 1985a; Csillik and Knyihár-Csillik, 1981b; Csillik and Knyihár-Csillik, 1986; Csillik et al., 1982; Knyihár-Csillik et al., 1987).

(4) TDA is not related to the conductivity of the peripheral nerve.

Instillation of local anesthetics (Knyihár-Csillik and Csillik, 1981) or application of tetrodotoxin around the nerve (Wall et al., 1982a), does not induce TDA. On the other hand, if TDA was already induced (either by mechanical injury or by a microtubule inhibitor), replenishment of the marker enzyme cannot be obtained by electrical stimulation of the nerve (Knyihár-Csillik and Csillik, 1981).

(5) TDA is the consequence of failure of nerve growth factor to reach dorsal root ganglion cells.

Applying anti-NGF serum in a Gelaspon cuff around the sciatic nerve, results in TDA in the segmentally related, ipsilateral upper dorsal horn (Csillik and Knyihár-Csillik, 1986) evident from the depletion of the marker enzyme FRAP (Fig. 8). In contrast, application of a human anti-D serum does not have any effect (Csillik et al., 1985b).

(6) TDA-inducing effect of *Vinca* alkaloids is due to blockade of retrograde transport of NGF in the peripheral nerve.

The amount of vinblastin, needed to block retrograde transport of experimentally injected, radio-labeled ^{125}I-NGF is identical with the threshold amount sufficient to induce TDA (Csillik et al., 1985a).

(7) TDA can be prevented or at least delayed by local application of NGF.

This follows from application of NGF at the proximal stump of the transected sciatic nerve (Csillik, 1984b; Csillik et al., 1983; Fitzgerald et al., 1985a). NGF is even more effective if applied 40-50 hours after transection, i.e., when pioneer fibers of regenerating axons already grew out; these are known to be equipped with NGF receptors (Csillik et al., 1985a).

(8) Polymyxin and Colimyxin, two cyclic basic peptides induce TDA after perineural application, via competitive inhibition of NGF.

The TDA-inducing action of these drugs (Csillik et al., 1985b) is not related either to histamine release or to interaction with Ca^{2+} in calmodulin-mediated axoplasmic transport mechanisms (Tajti et al., 1989).

(9) The essence of TDA is the derangement of synaptic transmission in the Rolando substance.

Dorsal root potential (DRP) undergoes marked alterations after blockade of retrograde axoplasmic transport in the peripheral nerve: amplitude is decreased, latency is increased (Pór, 1985; Wall, 1982b; Wall and Devor, 1981). TDA is accompanied by the increased latency of the lick reflex in the hot plate test (Szuecs et al., 1983).

(10) TDA can readily be studied in human autopsy or biopsy material.

Due to the fact that a well-defined class of human dorsal root ganglion cells and their central terminals in the upper dorsal horn express a fucose-containing carbohydrate epitope (Fig. 8), subjected to transganglionic regulation, pathological studies of TDA can be extended in the future to include human material as well (Fischer and Csillik, 1985; Fischer et al., 1985).

An important aspect of transganglionic regulation is that small, dark nerve cells in dorsal root ganglia are more dependent on NGF than large, pale nerve cells. Subgroups of small nerve cells are known to be involved in nociception. Therefore, the survival value of TDA to the animal might be that it switches off nociceptive afferentation soon after nerve injury and during the process of peripheral nerve regeneration. Thus, meaningless signals from impaired and incompletely restructured axons are prevented from kindling higher levels of sensory chains.

While the adult mammalian central nervous system is notorious for being refractory in producing new synapses and in replacing degenerated ones, it is possible to induce formation of new synapses experimentally. Such reactive synaptogenesis was observed (i) in central monoaminergic neurons after electrolytic lesions (Katzman et al., 1971); (ii) in the dentate gyrus and hippocampus after transection of the afferents (Lynch et al., 1972; Lynch et al., 1973); (iii) in the olfactory bulb after regeneration of the olfactory nerve (Grazidei and Monti Grazidei, 1978); (iv) in the ventral horn of the spinal cord, rostral from

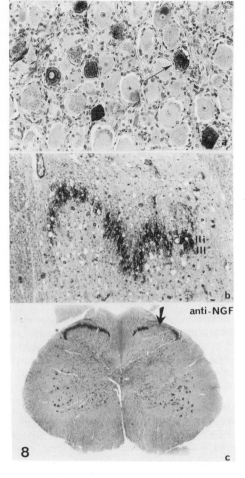

Fig. 8. A. Fucose-containing carbohydrate epitopes in small to medium sized dorsal root ganglion cells (arrow) demonstrated in formalin-fixed, paraffin-embedded human autopsy material by means of Ulex Europaeus lectin histochemistry.
B. Fucose-containing epitopes in axon terminals in Lamina IIi and III in the human dorsal horn. DC: dorsal column (Fischer and Csillik, 1985).
C. Effect of anti-NGF applied around the sciatic nerve. Note depletion of FRAP 14 days after perineural application of the anti-NGF serum (obtained from Prof. H. Thoenen, München-Martinsried). Note depletion of FRAP (arrow). The contralateral sciatic was surrounded by a Gelaspon cuff containing anti-D serum.

the site of transection (Bernstein and Bernstein, 1971; Goldberger and Murray, 1974), and (v) in the upper dorsal horn after TDA of central terminals of primary sensory neurons (this paper). Since this last is the only one that occurs in mammals, including primates, in the course of normal life, i.e., whenever an injured peripheral sensory nerve succeeds in regenerating, it is the most important proof for the regenerative capacity of the mammalian (and, specifically of the primate) central nervous system.

By extrapolating the observations and conclusions outlined in this paper, it can be assumed that essentially similar principles may be operative in the regulation, maintenance and plasticity also of other types of neurons. The concept of such a "transcellular regulation" is supported by Prestige's classical studies (Prestige, 1967) who suggested long ago that the survival of ventral horn motoneurons is due to a maintenance factor carried retrogradely by the motor axon to the perikaryon. Kreutzberg (1985) demonstrated that the receptive field of the neuron (extension of the dendritic tree) is dependent on the structural and functional integrity of the axon. Unsicker et al. (1985) provided evidence for the existence of several trophic factors in the central nervous system, probably each factor being specific for a specific cell type. It seems that, by influencing transcellular regulatory mechanisms, embryonic developmental potencies of otherwise "rigid" adult central neurons, like axonal growth and synaptogenesis can be revived, thus arousing dormant abilities of structural and functional plasticity. Therapeutical application of this principle might help promote regeneration of pathways and synaptic circuitries destroyed by pathological events, injuries and accidents.

Acknowledgements. Investigations were supported by research grant No. 527 from the Hungarian Ministry of Health, grants No. 942, No. 1077 and No. 05-065 from the Hungarian Academy of Sciences and by fellowships from the Max-Planck Gesellschaft. We are indebted to Dr. A. Török and Dr. Cs. Vereczkey for their contributions in computer technology and experimental surgery, as well as to Mrs. Eva Hegyeshalmi and Mrs. Emese Lauly for histochemical assistance. Photographic work of Mr. I. Farago and secretarial assistance of Miss Ibolya Bodi and Miss Katalin Szabo is gratefully acknowledged.

REFERENCES

Aldskogius, H., Cerne, H., and Holmberg, A., 1985a, The effect of sciatic nerve transection on myelinated fibers in the L5 dorsal root and lumbar dorsal column. A Marchi study in the rat, *Anat. Embryol.*, 171:181

Aldskogius, H., Arvidsson, J., and Grant, G., 1985b, The reaction of primary sensory neurons to peripheral nerve injury with particular emphasis on transganglionic changes, *Brain Res. Rev.*, 10:27

Aldskogius, H., and Risling, M., 1983, Preferential loss of unmyelinated axons in the L7 dorsal root of kittens following sciatic neurectomy, *Brain Res.*, 289:358

Arvidsson, J., and Grant, G., 1979, Further observations on transganglionic degeneration in trigeminal primary sensory neurons, *Brain Res.*, 162:1

Arvidsson, J., 1979, An ultrastructural study of transganglionic degeneration in the main sensory trigeminal nucleus of the rat, *J. Neurocytol.*, 8:31

Barber, R. P., Vaughn, J. E., Slemmon, J. R., Salvaterra, P. M., Robert, P. M., and Leeman, S. E., 1979, The origin, distribution and synaptic relationships of substance P axons in rat spinal cord, *J. Comp. Neurol.*, 184:331

Barbut, D., Polak, J. M., and Wall, P. D., 1981, Substance P in spinal cord dorsal horn decreases following peripheral nerve injury, *Brain Res.*, 205:289

Bernstein, J. J., and Bernstein, M. E., 1971, Axonal regeneration and formation of synapses proximal to the side of lesion following hemisection of the rat spinal cord, *Exp. Neurol.*, 30:336

Bezzegh, A., Knyihár-Csillik, E., Boti, Zs., Tajti, J., Zaborski, Z., and Csillik, B., 1986, A computer-aided analysis of the effect of peripheral nerve transection on TMPase activity of substantia gelatinosa Rolandi, *Z. mikrosk.-anat. Forsch.*, 100:428

Black, I. B., Adler, J. E., Dreyfus, C. F., et al, 1984, Neurotransmitter plasticity at the molecular level, *Science*, 225:1266

Burnstock, G., 1978, A basis for distinguishing two types of purinergic receptor, *in:* "Cell membrane receptors for drugs and hormones: a multidisciplinary approach," L. Bolis and R. W. Straub, eds., Raven Press, New York: 107

Cajal, S. R. y, 1918, Degeneration and regeneration of the nervous system, University Press, Oxford: 750

Csillik, B., and Knyihár, E., 1975, Degenerative atrophy and regenerative proliferation in the rat spinal cord, *Z. mikrosk.-anat. Forsch.*, 89:1099-1103

Csillik, B., and Knyihár, E., 1976, "New" features of the trophical entity, *in:* "Neuron concept today," J. Szentágothai and J. Hamori, eds., Tihany: 27

Csillik, B., Knyihár, E., and Elshiekh, A. A., 1977a, Degenerative atrophy of central terminals of primary sensory neurons induced by blockade of axoplasmic transport in peripheral nerves, *Experientia (Basel)*, 33:656

Csillik, B., and Knyihár, E., 1977b, Histochemistry of synapses, *Cell Mol. Biol.*, 22:285-292

Csillik, B., 1980, Infrastructure of the neuron, *in:* "Neurotransmitters. Comparative aspects," J. Salanki and T. M. Turpaev, eds., Akadémiai Kiadó: 149

Csillik, B., and Knyihár, E., 1978a, Biodynamic plasticity in the Rolando substance, *Progr. Neurobiol.*, 10:202

Csillik, B., Knyihár, E., Jojart, I., Elshiek, A. A., and Pór, I., 1978b, Perineural microtubule inhibitors induce degenerative atrophy of central nociceptive terminals in the Rolando substance, *Res. Comm. Chem. Path. Pharm.*, 21:467

Csillik, B., Knyihár, E., and Rakic, P., 1982, Transganglionic degenerative atrophy and regenerative proliferation in the Rolando substancee of the primate spinal cord: discoupling and restoration of synaptic connectivity in the central nervous system after peripheral lesions, *Folia Morph.*, 30:189

Csillik, B., and Knyihár-Csillik, E., 1981a, The spinal projection area of primary nociceptive afferents: Regenerative synaptoneogenesis in the Rolando substance, *in:* "Adv. Physiol. Sci., Vol. 36," O. Feher and F. Joo, eds., Akadémiai Kiadó: 47

Csillik, B., and Knyihár-Csillik, E., 1981b, Reactions of the substantia gelatinosa Rolandi to injury of peripheral sensory axons, *in:* "Spinal cord sensation," A. G. Brown and M. Réthelyi, eds., Scottish Academic Press, Edinburgh: 309

Csillik, B., and Knyihár-Csillik, E., 1981c, Regenerative synaptoneogenesis in the mammalian spinal cord: Dynamics of synaptochemical restoration in the Rolando substance after transganglionic degenerative atrophy, *J. Neural Transmiss.*, 52:303

Csillik, B., Knyihár-Csillik, E., and Tajti, J., 1982a, Blockade of retrograde axoplasmic transport induces tranganglionic degenerative atrophy of central terminals of primary nociceptive neurons, *Acta Biol. Acad. Sci. Hung.*, 33:149

Csillik, B., and Knyihár-Csillik, E., 1982b, Reversibility of microtubule inhibitor-induced transganglionic degenerative atrophy of central terminals of primary nociceptive neurons, *Neuroscience*, 7:1149

Csillik, B., Schwab, M. E., and Thoenen, H., 1983, Transganglionic regulation by nerve growth factor of the primary nociceptive analyzer, *Neurosci. Lett. Suppl.*, 14:79

Csillik, B., Knyihár-Csillik, E., Bezzegh, A., Kiss, J., Léránth, Cs., Pór, I., and Záborszky, Z., 1984a, Transganglionic degenerative atrophy of central terminals of primary sensory neurons after perineural application of vinblastin: decay of dorsal root potential and depletion of neuropeptides, *in:* "Antidromic vasodilatation and neurogenic inflammation", L. A. Chahl, J. Szolcsanyi, and F. Lembeck, eds., Akadémiai Kiadó, Budapest: 141

Csillik, B., 1984b, Nerve growth factor regulates central terminals of primary sensory neurons, *Z. mikr.-anat. Forsch.*, 98:11

Csillik, B., Knyihár-Csillik, E., Pór, I., et al., 1984c, Role of nerve growth factor (NGF) in structural and functional regulation of the primary nociceptive analyzer, *Acta Physiol. Hung.*, 63:311

Csillik, B., and Knyihár-Csillik, E., 1986, "The protean gate. Structure and plasticity of the primary nociceptive analyzer," Akadémiai Kiadó

Csillik, B., Schwab, M. E., and Thoenen, H., 1985a, Transganglionic regulation of central terminals of dorsal root ganglion cells by nerve growth factor (NGF), *Brain Res.*, 331:11

Csillik, B., Kovacs, K., Penke, B., Tajti, J., and Szilard, J., 1985b, Transganglionic effect of basic peptides on the primary nociceptive analyzer, *in:* "Receptors and centrally acting drugs," E. S. Vizi, S. Fürst, and G. Zsilla, eds., Proc. 4th Congr. Hung. Pharmacol. Soc.,Vol. 2, Budapest

Csillik, B., Knyihár-Csillik, E., and Bezzegh, A., 1986, Comparative electron histochemistry of thiamine monophosphatase and substance P in the upper dorsal horn, *Acta histochem.*, 80:125

Cuello, A. C., 1987, Peptides as neurormodulators in primary sensory neurons, *Neuropharmacol.*, 26:971

Devor, M., and Claman, D., 1980, Mapping and plasticity of acid phosphatase afferents in the rat dorsal horn, *Brain Res.*, 190:17

Dreyfus, P. M., 1985, The neurochemistry of vitamin deficiencies. - Thiamin, *in:* "Handbook of neurochemistry, Vol. 10," A. Lajtha, ed., Plenum Press, New York: 759

Ferencsik, M., 1986, Dynamism of tranganglionic degenerative atrophy following crush injury to the peripheral nerve, *Z. mikrosk.-anat. Forsch.*, 100:490

Fischer, J., and Csillik, B., 1985, Lectin binding: a genuine marker for transganglionic regulation of human primary sensory neurons, *Neurosci. Lett.*, 54:263

Fischer, J., Klein, P. J., and Csillik, B., 1985, *Ulex Europaeus I* lectin-binding glycoprotein in primary sensory terminals of human spinal cord, *in:* "Lectins, Vol. 4," T. C. Bog-Hansen and J. Breborowicz, eds., Walter de Gruyter and Co., Berlin-New York: 117

Fitzgerald, M., and Swett, J., 1983, The termination pattern of sciatic nerve afferents in the substantia gelatinosa of neonatal rats, *Neurosci. Lett.*, 43:149

Fitzgerald, M., Woolf, C. J., Gibson, S. J., and Mallaburn, P. S., 1984, Alterations in the structure, function and chemistry of C fibres following local application of vinblastine to the sciatic nerve of the rat, *J. Neurosci.*, 4:430

Fitzgerald, M., and Vrbova, G., 1985, Plasticity of acid phosphatase (FRAP) afferent terminal fields and of dorsal horn cell growth in the neonatal rat, *J. Comp. Neurol.*, 240:414

Fitzgerald, M., Wall, P. D., Goedert, M., and Emson, P. C., 1985a, Nerve growth factor counteracts the neurophysiological and neurochemical effects of chronic sciatic nerve section, *Brain. Res.*, 332:131

Fitzgerald, M., 1985b, The sprouting of saphenous nerve terminals in the spinal cord following early postnatal sciatic nerve section in the rat, *J. Comp. Neurol.*, 240:407

Gobel, S., and Binck, J. M., 1977, Degenerative changes in primary trigeminal axons and in neurons in nucleus caudalis following tooth pulp extirpations in the cat, *Brain Res.*, 132:347

Gobel, S., 1984, An electron microscopic analysis of the transsynaptic effects of peripheral nerve injury subsequent to tooth pulp extirpations on neurons in Lamina I and II of the medullary dorsal horn, *J. Neuroscience*, 4:2281

Goldberger, M. E., and Murray, M., 1974, Restitution of function and collateral sprouting in the rat spinal cord: the deafferentated animal, *J. Comp. Neurol.*, 158:37

Grant, G., and Ygge, J., 1981, Somatotopical organization of the thoracic spinal nerve in the dorsal horn demonstrated with transganglionic degeneration, *J. Comp. Neurol.*, 202:357

Grazidei, P. P. C., and Monti Grazidei, G. A., 1978, Continuous nerve cell renewal in the olfactory system, in: "Handbook of sensory Physiology, Vol. IX," M. Jacobson, ed., Springer-Verlag, New York: 55

Gutman, E., Gutman, C., Medawar, P. B., and Young, J. Z., 1942, The rate of regeneration of nerve, *J. Exp. Biol.*, 19:14

Hanzely, B., Knyihár-Csillik, E., and Csillik, B., 1983, Fluoride-resistant acid phosphatase (FRAP) activity of nociceptive nerve terminals in the dental pulp, *Z. mikrosk.-anat. Forsch.*, 97:43

Horch, K. W., and Lisney, S. J. W., 1981, Changes in primary afferent depolarization of sensory neurons during peripheral nerve regeneration in the cat, *J. Physiol. (Lond.)*, 313:287

Hökfelt, T., Kellerth, J. O., Nilsson, G., and Pernow, B., 1975, Experimental immunohistochemical studies on the localization and distribution of substance P in cat primary sensory neurons, *Brain Res.*, 100:235

Hökfelt, T., Elde, R., Johansson, O., Luft, R., and Nilsson, G., 1976, Immunohistochemical evidence for separate populations of somatostatin-containing and substance P-containing primary afferent neurons in the rat, *Neuroscience*, 1:131

Jessell, T., Tsunoo, A., Kanazawa, I., and Otsuka, M., 1979, Substance P: depletion in the dorsal horn of rat spinal cord after section of the peripheral processes of primary sensory neurons, *Brain Res.*, 168:247

Johnson, L. R., and Westrum, L. E., 1980, Brainstem degeneration patterns following tooth extractions: visualization of dental and periodontal afferents, *Brain Res.*, 194:489

Johnson, L. R., Westrum, L. E., and Canfield, R. C., 1983, Ultrastructural study of transganglionic degeneration following dental lesions, *Exp. Brain Res.*, 52:226

Katzman, R., Björklund, A., Owman, Ch., Stenevi, U., and West, K. A., 1971, Evidence for regenerative axon sprouting of central catecholamine neurons in rat mesencephalon following electrolytic lesions, *Brain Res.*, 25:579

Kessler, J. A., and Black, I. B., 1981, Nerve growth factor stimulates development of substance P in the embryonic spinal cord, *Brain Res.*, 208:135

Knyihár, E., and Csillik, B., 1976a, Representation of cutaneous afferents by fluoride-resistant acid phosphatase (FRAP)-active terminals in the rat substantia gelatinosa Rolandi, *Acta neurol. scand.*, 53:217

Knyihár, E., and Csillik, B., 1976b, Effect of peripheral axotomy on the fine structure and histochenistry of the Rolando substance: degenerative atrophyj of central processes of pseudounipolar cells, *Exp. Brain Res.*, 26:73

Knyihár, E., and Csillik, B., 1976c, Axonal labyrinths in the rat spinal cord. A consequence of degenerative atrophy, *Acta Biol. Acad. Sci. Hung.*, 27:299

Knyihár, E., and Csillik, B., 1977, Regional distribution of acid phosphatase-positive axonal systems in the rat spinal cord and medulla, representing central terminals of cutaneous and visceral nociceptive neurons, *J. Neural Transmiss.*, 40:227

Knyihár-Csillik, E., Bezzegh, A., Böti, Zs., and Csillik, B., 1986, Thiamine monophosphatase: A genuine marker for tansganglionic regulation of primary sensory neurons, *J. Histochem. Cytochem.*, 34:363

Knyihár-Csillik, E., and Csillik, B., 1981, FRAP: Histochemistry of the primary nociceptive neuron, *Progr. Histochem. Cytochem.*, 14:1

Knyihár-Csillik, E., Rakic, P., and Csillik, B., 1985, Fine structure of growth cones in the upper dorsal horn of the adult spinal cord in the course of reactive synaptoneogenesis, *Cell Tiss. Res.*, 239:633

Knyihár-Csillik, E., Rakic, P., and Csillik, B., 1986, Reactive synaptoneogenesis in the upper dorsal horn of the adult primate: regenerative or collateral sprouting?, in: "Development and Plasticity of the mammalian spinal cord," M. Goldberger, A. Gorio, and M. Murray, eds., Fidia Res. Series, Vol. III, Liviana Press, Padova

Knyihár-Csillik, E., Rakic, P., and Csillik, B., 1987, Transganglionic degenerative atrophy in the substantia gelatinosa of the spinal cord after peripheral nerve transection in rhesus monkeys, *Cell Tiss. Res.*, 247:599

Knyihár-Csillik, E., Kreutzberg, G. W., and Csillik, B., 1989, Enzyme translocation in the course of regeneration of central primary afferent terminals in the substantia gelatinosa of the adult rodent spinal cord, *J. Neurosci. Res. (in press)*,

Kovacs, A., and Ferencsik, M., 1986, Mapping of spinal projection of primary nociceptive neurones in the rat, *Acta Morph. Hung.*, 34:187

Kreutzberg, G. W., 1985, The motoneuron and its microenvironment responding to axotomy, *in:* "Neural transplantation and regeneration," G. D. Das and R. B. Wallace, eds., Springer-Verlag, New York: 271

Léránth, Cs., Csillik, B., and Knyihár-Csillik, E., 1984, Depletion of substance P and somatostatin in the upper dorsal horn after blockade of axoplasmic transport, *Histochem.*, 81:391

Ljungdahl, A., Hökfelt, T., and Nilsson, G., 1978, Distribution of substance P-like immunoreactivity in the central nervous system of the rat, *Neuroscience*, 3:861

Lynch, G., Matthews, D. A., Mosko, S., Parks, T., and Cotman, C. W., 1972, Induced acetylcholinesterase-rich layer in rat dentate gyrus following entorhinal lesions, *Brain Res.*, 42:311

Lynch, G., Stanfield, B., and Cotman, C. W., 1973, Development differences in postlesion axonal growth in the hippocampus, *Brain Res.*, 59:155

Majumdar, S., Mills, E., and Smith, P. G., 1983, Degenerative and regenerative changes in central projections of glossopharyngeal and vagal sensory neurons after peripheral axotomy in cats: a structural basis for central reorganization of arterial chemoreflex pathways, *Neuroscience*, 10:841

Mayer, N., Lembeck, F., Goedert, M., and Otten, U., 1982, Effects of antibodies against nerve growth factor on the postnatal development of substance P - containing sensory neurons, *Neurosci. Lett.*, 29:47

McGregor, G. P., Gibson, S. J., Sabate, I. M., Blank, M. A., and Christofides, N. D., 1984, Effect of peripheral nerve section and nerve crush on spinal cord neuropeptides in the rat; increased VIP and PHI in the dorsal horn, *Neuroscience*, 13:207

Mihaly, A., Pór, I., Bencze, Gy., and Csillik, B., 1980, Effects of perineurally applied cytostatic, cytotoxic and chelating agents upon peripheral and central processes of primary nociceptive neurons, *Z. mikrosk.-anat. Forsch.*, 94:531

Nadelhaft, I., 1984, The sessile drop method immunohistochemical processing of unmounted sections of nervous tissue, *J. Histochem. Cytochem.*, 32:1344

Nagy, J. I., and Hunt, S. P., 1982, Fluoride-resistant acid phosphatase-containing neurones in dorsal root ganlgia are separate from those containing substance P or somatostatin, *Neurosci.*, 378:188

Ogawa, K., Sakai, M., and Inomata, K., 1982, Recent findings on ultracytochemistry of thiamin phosphatases, *Ann. N. Y. Acad. Sci.*, 378:188

Pór, I., 1985, Alterations of dorsal root potential in the course of tranganglionic degenerative atrophy, *Acta Physiol. Hung.*, 65:255

Prestige, M. C., 1967, The control of cell number in the lumbar spinal ganglia during the development of *Xenopus laevis* tadpoles, *J. Embryol. Morph.*, 17:453

Risling, M., Aldskogius, H., Hildebrand, C., and Remahl, S., 1983, Effects of sciatic nerve resection on L7 spinal root and dorsal root ganglia in adult cats, *Exp. Neurol.*, 82:568

Seybold, V., and Elde, R., 1980, Immunohistochemical studies of peptidergic neurons in the dorsal horn of the spinal cord, *J. Histochem. Cytochem.*, 28:367

Shehab, S. A. S., and Atkinson, M. E., 1986, Vasoactive intestinal polypeptide (VIP) increases in the spinal cord after peripheral axotomy of the sciatic nerve originate from primary afferent neurons, *Brain Res.*, 372:37

Szuecs, A., Csillik, B., and Knyihár-Csillik, E., 1983, Functional impairment of the primary nociceptive analyzer in the course of transganglionic degenerative atrophy, *Acta Biol. Acad. Sci. Hung.*, 34:267

Szönyi, G., Knyihár, E., and Csillik, B., 1979, Extra-lysosomal fluoride-resistant acid phosphatase active neuronal system subserving nociception in the rat cornea, *Z. mikrosk.-anat. Forsch.*, 93:974

Tajti, J., Fischer, J., Knyihár-Csillik, E., and Csillik, B., 1988, Tranganglionic regulation and fine structural localization of lectin-reactive carbohydrate epitopes in primary sensory neurons of the rat, *Histochemistry*, 88:213

Tajti, J., Penke, B., Kovacs, K., and Csillik, B., 1989, Competitive mechanisms of tranganglionic degenerative atrophy, *Acta Morph. Hung. (in press)*,

Takahashi, T., and Otsuka, M., 1975, Regional distribution of substance P in the spinal cord and nerve roots of the cat and the effect of dorsal root section, *Brain Res.*, 87:1

Tessler, A., Himes, B. T., Krieger, N. R., Murray, M., and Goldberger, M. E., 1985, Sciatic nerve transection produces death of dorsal root ganglion cells and reversible loss of substance P in spinal cord, *Brain Res.*, 332:209

Unsicker, K., Skaper, S. D., and Varon, S., 1985, Developmental changes in the responses of rat chromaffine cells to neurotrophic and neurite-promoting factors, *Develop. Biol.*, 111:425

Wall, P. D., Fitzgerald, M., and Gibson, S. J., 1981a, The response of rat spinal cord cells to unmyelinated afferents after peripheral nerve section and after changes in substance P levels, *Neuroscience*, 6:2205

Wall, P. D., 1981b, The nature and origins of plasticity in adult spinal cord, *in:* "Spinal cord sensation," A. G. Brown and M. Réthelyi, eds., Scottish Academic Press, Edinburgh: 297

Wall, P. D., Mills, R., Fitzgerald, M., and Gibson, S. J., 1982a, Chronic blockade of sciatic nerve transmission by tetrodotoxin does not produce central changes in the dorsal horn of the spinal cord of the rat, *Neurosci. Lett.*, 30:315

Wall, P. D., 1982b, The effect of peripheral nerve lesions and of neonatal capsaicin in the rat on primary afferent depolarization, *J. Physiol. (Lond.)*, 329:21

Wall, P. D., and Devor, M., 1981, The effect of peripheral nerve injury on dorsal root potentials and on transmission of afferent signals into the spinal cord, *Brain Res.*, 209:95

Westrum, L. E., and Canfield, R. C., 1977, Light and electron microscopy of degeneration in the brain stem spinal trigeminal nucleus following tooth pulp removal in adult cats, *in:* "Pain in trigeminal region," D. J. Anderson and B. Matthews, eds., Elsevier, Amsterdam-New York: 171

Westrum, L. E., Johnson, L. R., and Canfield, R. C., 1984, Ultrastructure of transganglionic degeneration in brain stem trigeminal nuclei during normal primary tooth exfoliation and permanent tooth eruption in the cat, *J. Comp. Neurol.*, 230:198

Woolf, C. J., and Wall, P. D., 1982, Chronic peripheral nerve section diminished the primary afferent A-fibre mediated inhibition of rat dorsal horn neurones, *Brain Res.*, 242:77

CONTRIBUTORS

Andres, K. H., Ruhr-Universität Bochum, Institut für Anatomie, Abteilung Neuroanatomie, Universitätsstrasse 150, D-4630 Bochum, F.R.G.

Bankoul, S., Anatomisches Institut der Universität Zürich-Irchel, Winterthurerstrasse 190, CH-8057 Zürich, Switzerland

Barakat, I., Institut d'Histologie et d'Embryologie, Université de Lausanne, rue du Bugnon 9, CH-1005 Lausanne, Switzerlan

Clerc, N., C.N.R.S., Laboratoire de Neurobiologie, 31, chemin J. Aiguier, F-13402 Marseille Cedex 9, France

Csillik, B., Department of Anatomy, Albert Szent-Györgyi Medical University, Szeged, Hungary

Davies, A. M., Department of Anatomy, St. George's Hospital Medical School, Cranmer Terrace, Tooting, London SW17 ORE, England

Droz, B., Institut d'Histologie et d'Embryologie, Université de Lausanne, rue du Bugnon 9, CH-1005 Lausanne, Switzerland

Drukker, J., Capaciteitsgroep Anatomy/Embryology, Rijksuniversiteit Limburg, Biomedische Centrum, Beeldsnijdersdreef 101, NL-6200 MD Maastricht, The Netherlands

Düring, M. v., Ruhr-Universität Bochum, Institut für Anatomie, Abteilung Neuroanatomie, Universitätsstrasse 150, D-4630 Bochum, F.R.G.

Grant, G., Department of Anatomy, Karolinska Institutet, P. O. Box 60400, S-104 01 Stockholm 60, Sweden

Halata, Z., Anatomisches Institut der Universität Hamburg, Abteilung für funktionelle Anatomie, Martinistrasse 52, D-2000 Hamburg 20, F.R.G.

Heppelmann, B., Physiologisches Institut der Universität Würzburg, Röntgenring 9, D-8700 Würzburg, F.R.G.

Irminger, D., Anatomisches Institut der Universität Zürich-Irchel, Winterthurerstrasse 190, CH-8057 Zürich, Switzerland

Kazimierczak, J., Institut d'Histologie et d'Embryologie, Université de Lausanne, rue du Bugnon 9, CH-1005 Lausanne, Switzerland

Knyihár-Csillik, E., Department of Anatomy, Albert Szent-Györgyi Medical University, Szeged, Hungary

Kopp, M., Institut für Anatomie der Universität Wien, 3. Lehrkanzel, Währingerstrasse 13, A-1090 Wien, Austria

Mense, S., Anatomisches Institut III der Universität Heidelberg, Im Neuenheimer Feld 307, D-6900 Heidelberg, F.R.G.

Meßlinger, K., Physiologisches Institut der Universität Würzburg, Röntgenring 9, D-8700 Würzburg, F.R.G.

Molander, C., Department of Anatomy, Karolinska Institutet, P. O. Box 60400, S-104 01 Stockholm 60, Sweden

Neiss, W. F., Anatomisches Institut der Universität zu Köln, Joseph-Stelzmann-Strasse 9, D-5000 Köln 41, F.R.G.

Neuhuber, W. L., Anatomisches Institut der Universität Zürich-Irchel, Winterthurerstrasse 190, CH-8057 Zürich, Switzerland

Philippe, E., Institut d'Histologie et d'Embryologie, Université de Lausanne, rue du Bugnon 9, CH-1005 Lausanne, Switzerland

Réthelyi, M., 2nd Department of Anatomy, Semmelweis University Medical School, Tuzolto utca 58, H-1094 Budapest IX, Hungary

Rochat, A., Institut d'Histologie et d'Embryologie, Université de Lausanne, rue du Bugnon 9, CH-1005 Lausanne, Switzerland

Schaden, G., Institut für Anatomie der Universität Wien, 3. Lehrkanzel, Währingerstrasse 13, A-1090 Wien, Austria

Schmidt, R. F., Physiologisches Institut der Universität Würzburg, Röntgenring 9, D-8700 Würzburg, F.R.G.

Strasmann, T., Anatomisches Institut der Universität Hamburg, Abteilung für funktionelle Anatomie, Martinistrasse 52, D-2000 Hamburg 20, F.R.G.

Szabolcs, M., Institut für Anatomie der Universität Wien, 3. Lehrkanzel, Währingerstrasse 13, A-1090 Wien, Austria

van der Wal, J. C., Capaciteitsgroep Anatomy/Embryology, Rijksuniversiteit Limburg, Biomedische Centrum, Beeldsnijersdreef 101, NL-6200 MD Maastricht, The Netherlands

Weihe, E., Anatomisches Institut der Johannes Gutenberg-Universität Mainz, Saarstrasse 19-21, D-6500 Mainz, F.R.G.

Zenker, W., Anatomisches Institut der Universität Zürich-Irchel, Winterthurerstrasse 190, CH-8057 Zürich, Switzerland

INDEX